Learn to Use Microsoft Project 2016

Michelle N. Halsey, PMP, CSM

ISBN-10: 1-64004-263-6

ISBN-13: 978-1-64004-263-6

Silver City Publications & Training, L.L.C.
P.O. Box 1914
Nampa, ID 83653
https://www.silvercitypublications.com/shop/

Table of Contents

Chapter 1 – Opening and Closing Project

In this chapter, you will learn to open and close Project and Project files. You will also explore the Project interface. Finally, you will learn to create a blank project and a project using a template.

Opening Project

Use the following procedure to open Project.

Step 1: Press the Windows key.

Step 2: Select Microsoft Project 2016.

Understanding the Interface

Project 2016 has a new interface. Project 2016 uses the Ribbon interface that was introduced in Microsoft Office 2007 applications. Each Tab in the Ribbon contains many tools for working with your document. To display a different set of commands, click the tab name. Buttons are organized into groups according to their function.

In addition to the Tabs, Project 2016 also makes use of the Quick Access Toolbar from the MS Office 2007 applications.

Shown here is the Project interface, including the Ribbon, the Timeline area, the Task List, the Gantt chart, the Quick Access toolbar, and the Status bar.

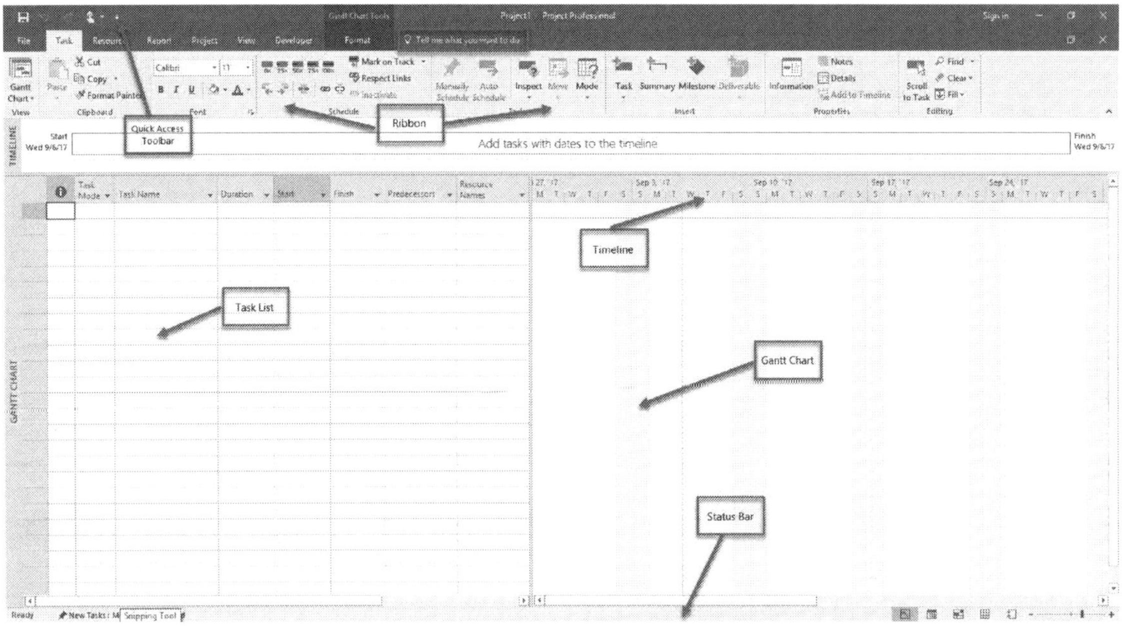

The Quick Access Toolbar appears at the top of the Project window and provides you with one-click shortcuts to commonly used functions. You may customize the contents of the toolbar by clicking the arrow icon immediately to the right of the toolbar.

By default, the Quick Access Toolbar contains buttons for Save, Undo and Redo.

To customize the toolbar, select the arrow next to the Quick Access Toolbar.

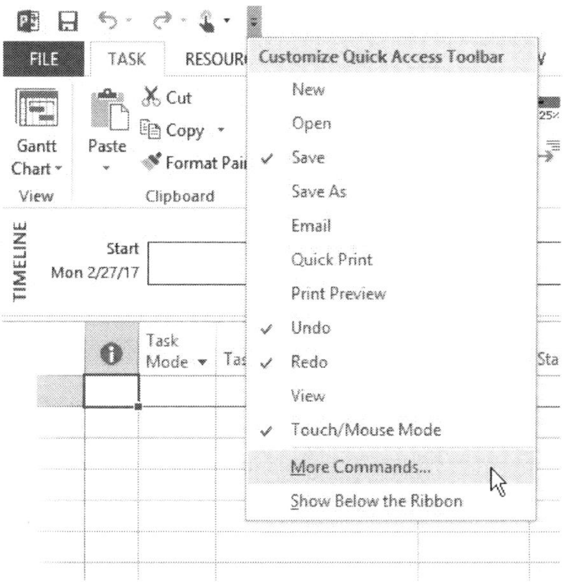

Add an item to the Quick Access Toolbar by selecting it from the list. You can remove an item by reopening the list and selecting the item again.

If you select More Commands, Project opens the Project Options window.

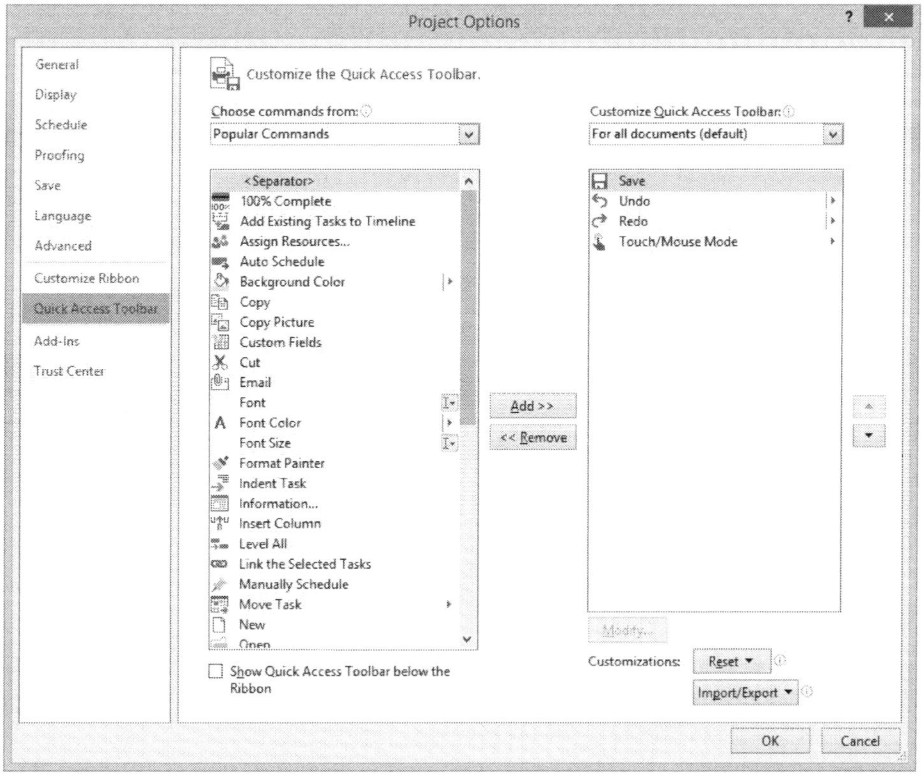

To add a command, select the item from the list on the left and select Add. Select Ok when you have finished.

Creating a Blank Project

Use the following procedure to create a blank project.

Step 1: Select the File tab.

Step 2: Select the New tab on the Backstage View.

Step 3: Select Blank Project.

Step 4: Select Create.

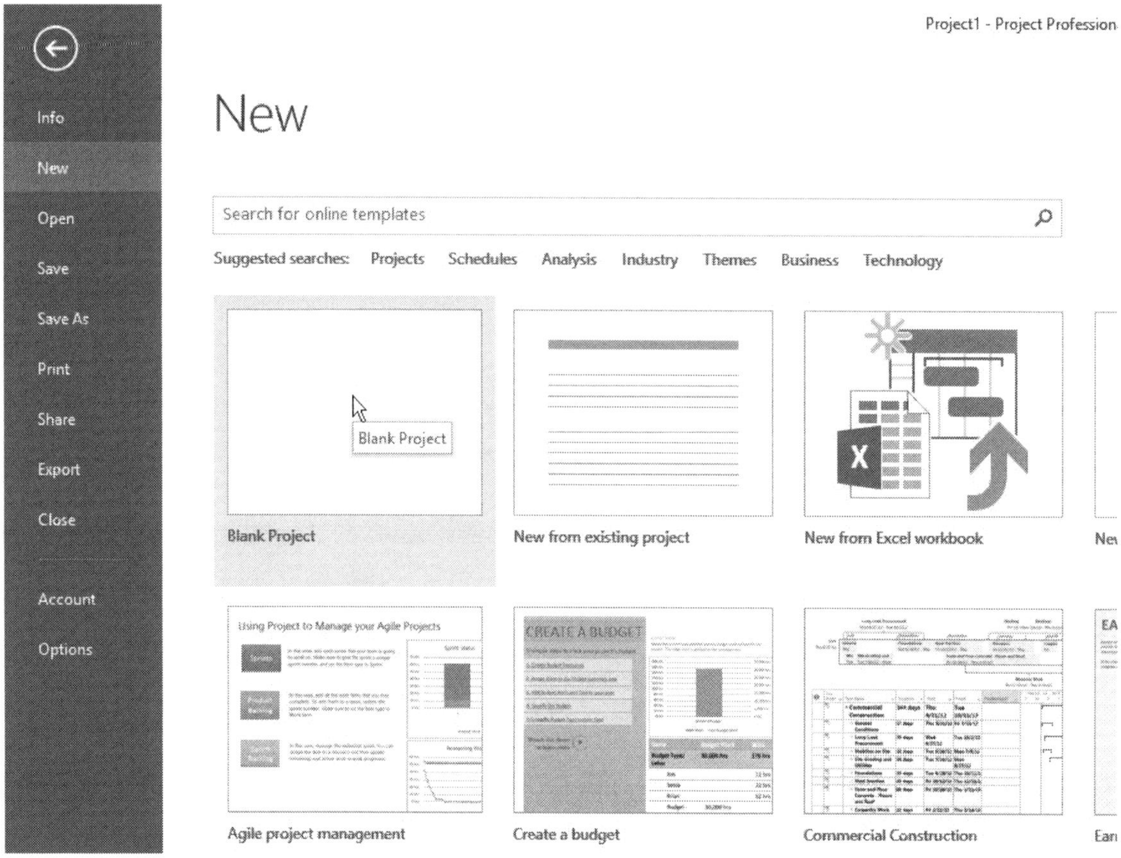

Creating a Project from a Template

Templates help to save time when creating a project. They can also provide consistency across several related projects. The New tab of the Backstage View provides links to several different templates. Some templates are created by Microsoft, while others are submitted by the community.

You can use or download these templates at any time. The new project can be modified to suit your needs.

Use the following procedure to create a blank project from an Office.com template.

Step 1: Select the File tab on the Ribbon.

Step 2: Select the New tab in the Backstage View.

Step 3: Select a template category.

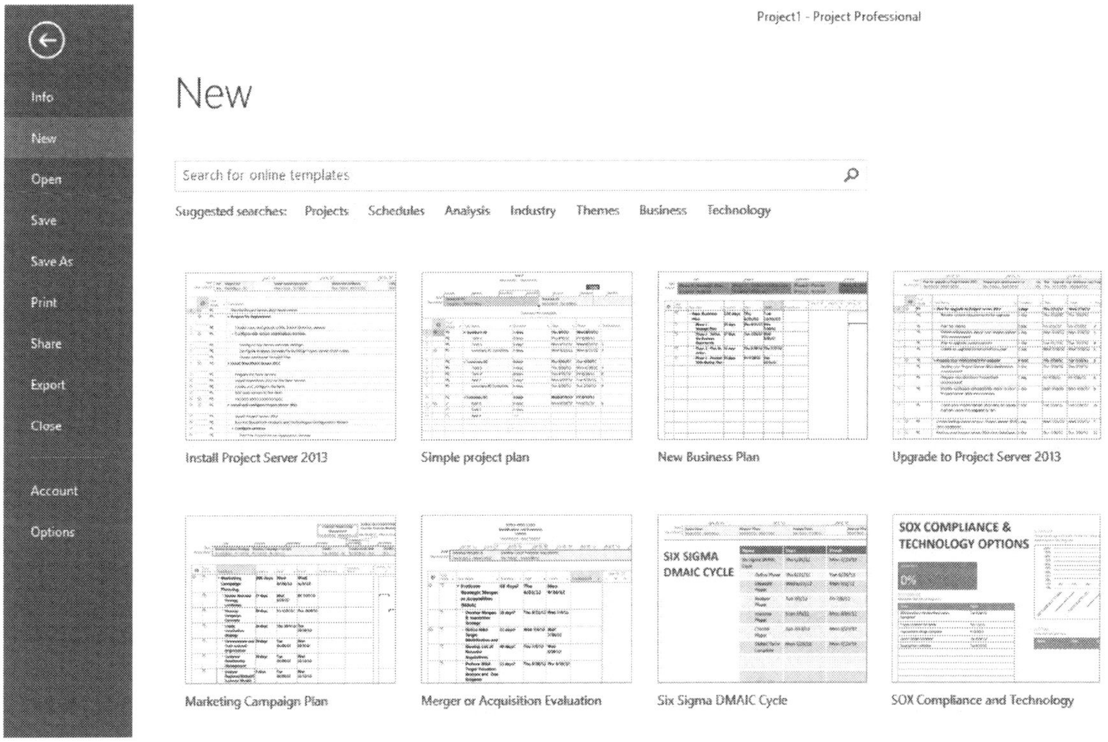

Step 4: Select a template.

Step 5: Select Create.

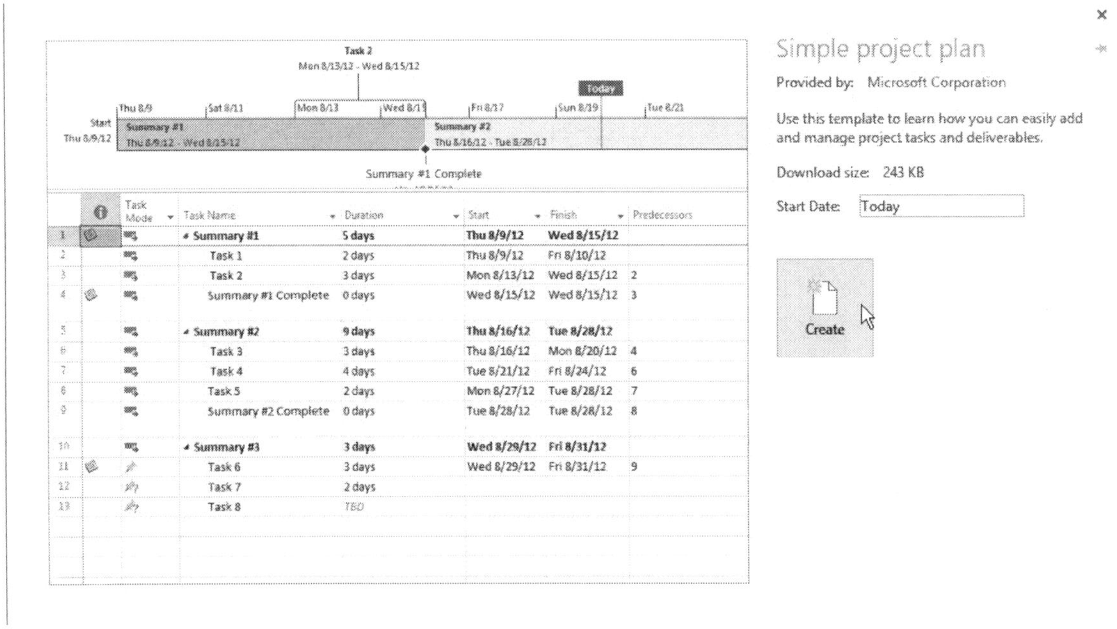

The Backstage view returns to the background after the new operation is complete.

Opening and Closing Files

To open a file in Project, use one of the following commands:

- The Open command on the Backstage View (on the File tab)
- Ctrl + O shortcut

To close a file in Project, use the Close command on the Backstage View or select the X in the top right-hand corner.

Use the following procedure to open a file in Project.

Step 1: Select the File tab.

Step 2: Select Browse to browse to a file on OneDrive or on the PC.

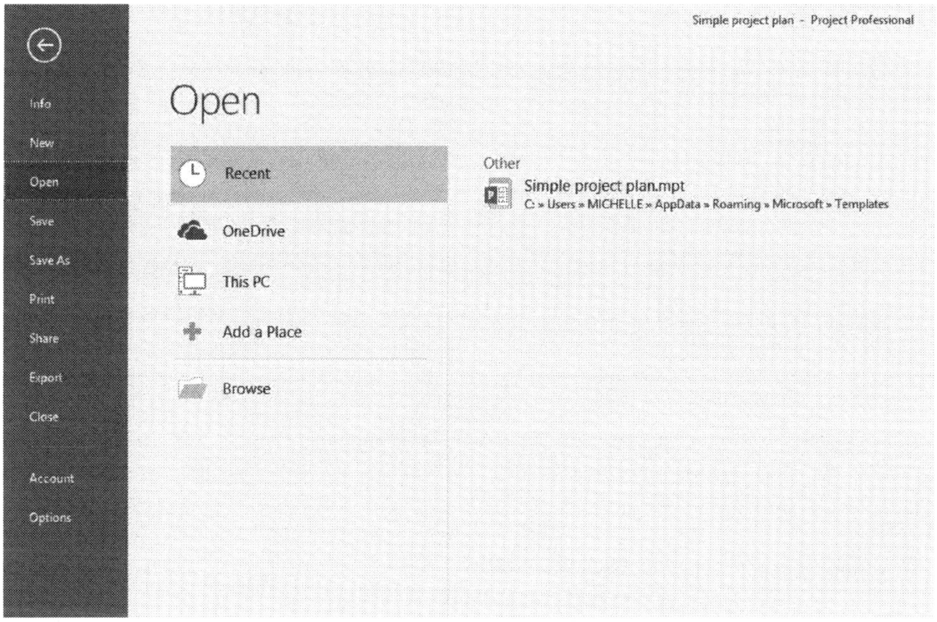

There are several options for opening files available:

- Recent shows the most recent files you have opened.
- OneDrive, a cloud drive. You will need to login to your Microsoft account.
- This PC, the computer hard drive.
- Add a Place, allows you to add locations such as SharePoint sites.
- Browse, will let you browse to files on the computer or on OneDrive.

Step 3: The system opens the Open dialog box to allow you to locate the file on your computer.

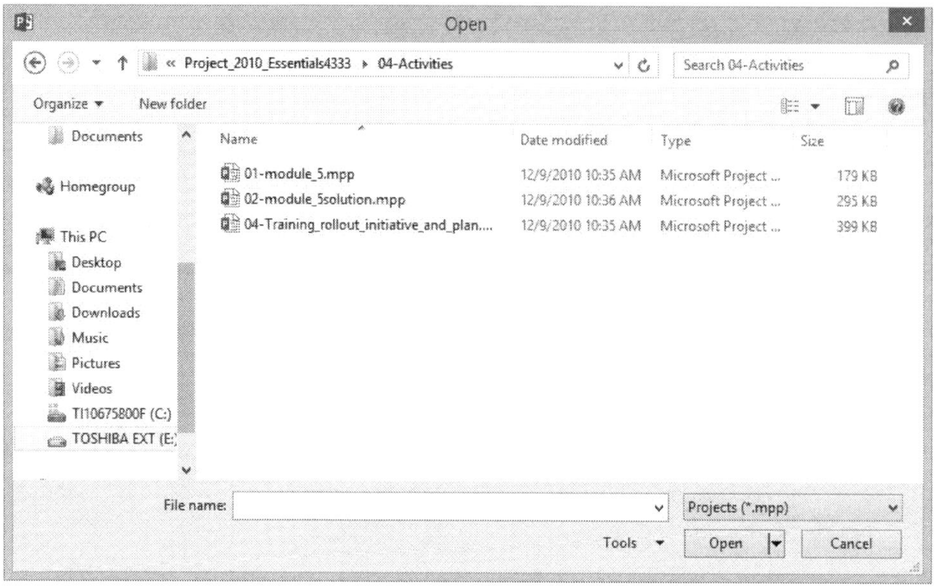

Step 4: Highlight the file you want to open and select Open.

Use the following procedure to close a file.

Step 1: There are two X icons at the top right-hand corner. The top one is for closing Project. Click on the other X to close the current Project file without closing Project.

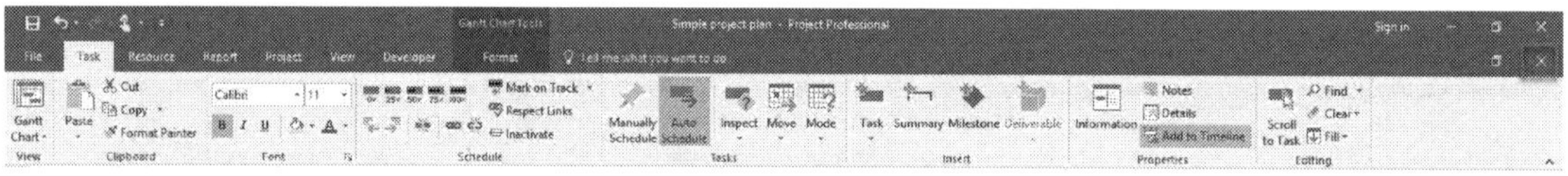

Closing Project

Use the following procedure to close Project from the Backstage View.

Step 1: Select the File tab on the Ribbon.

Step 2: Select the Close command in the Backstage View.

Chapter 2 – Your First Project

In this chapter, you will get started using Microsoft Project by creating a basic project and setting up the project schedule. You will add tasks to your project and set constraints on those tasks to further customize your project.

Creating a Basic Project

The Project Information dialog box allows you to determine the start (or finish) date for your project to determine accurate scheduling.

Use the following procedure to schedule a project start date.

Step 1: Select the Project tab.

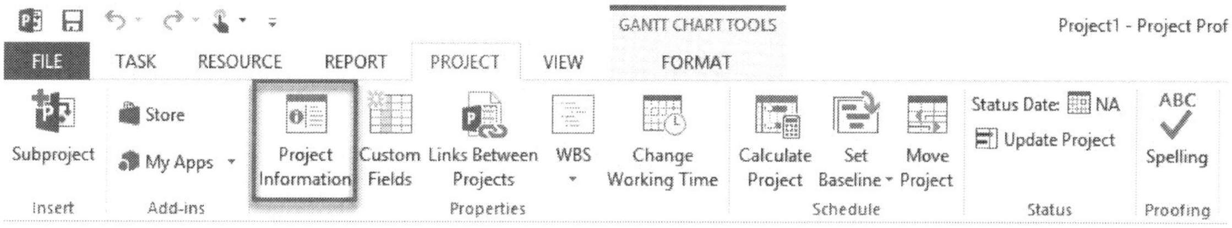

Step 2: Select Project Information.

The system opens the Project Information dialog box.

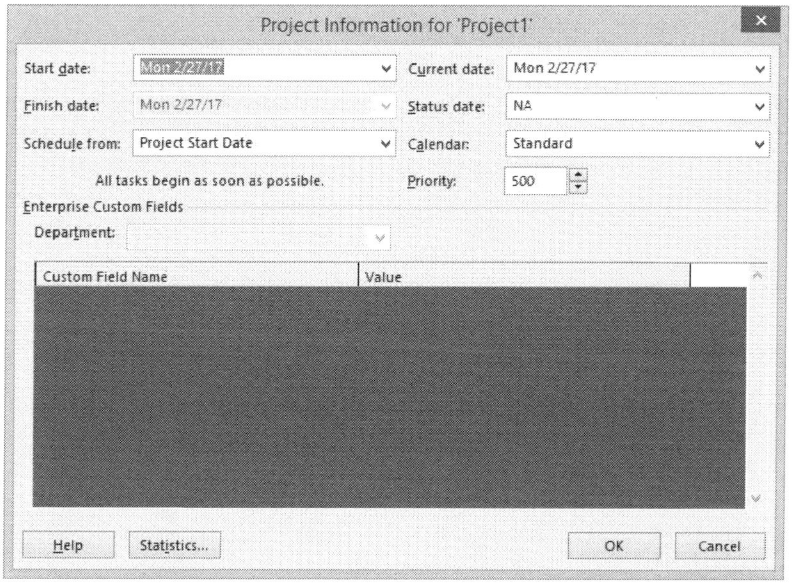

Step 3: Select the arrow next to Start date to choose a date from the calendar.

While you have the Project Information dialog box open, discuss some of the other options. You can select Project Finish Date for Schedule from and choose the date you would like the project finished instead. The Current Date and Status Date allow you to change the viewpoint when reviewing or reporting on details about your project. The Calendar sets the default calendar for determining the schedule.

Step 4: Select Ok to set your schedule start date.

Adding Tasks to Your Project

The left side of the Gantt chart in the default view works like a spreadsheet. You enter each task on a new row. You can add as much or as little information about the task as you like to get started. You can enter the tasks manually or paste the information from another program (such as Word or an email message).

Use the following procedure to add tasks to a project.

Step 1: In the Task Name column on the left side of the Gantt chart, enter the name of the task and press Enter.

Step 2: Continue entering task names for as many tasks as you want to enter.

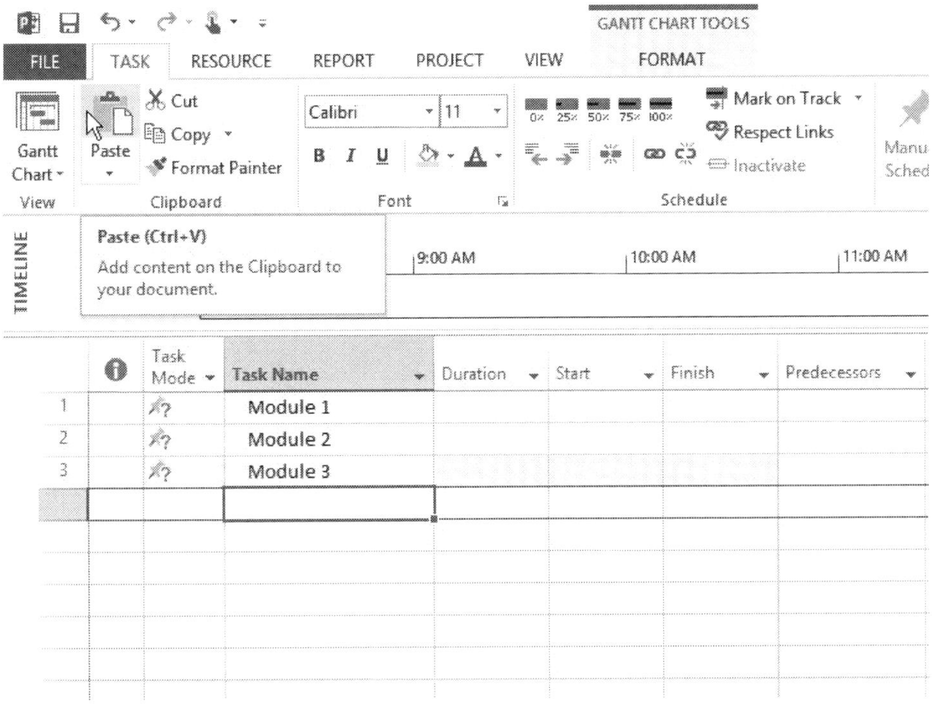

Use the following procedure to add tasks to a project by pasting.

Step 1: Copy the information from another program.

Step 2: Place your cursor in an empty row in the Task Name column.

Step 3: Select the Task tab from the Ribbon.

Step 4: Select Paste.

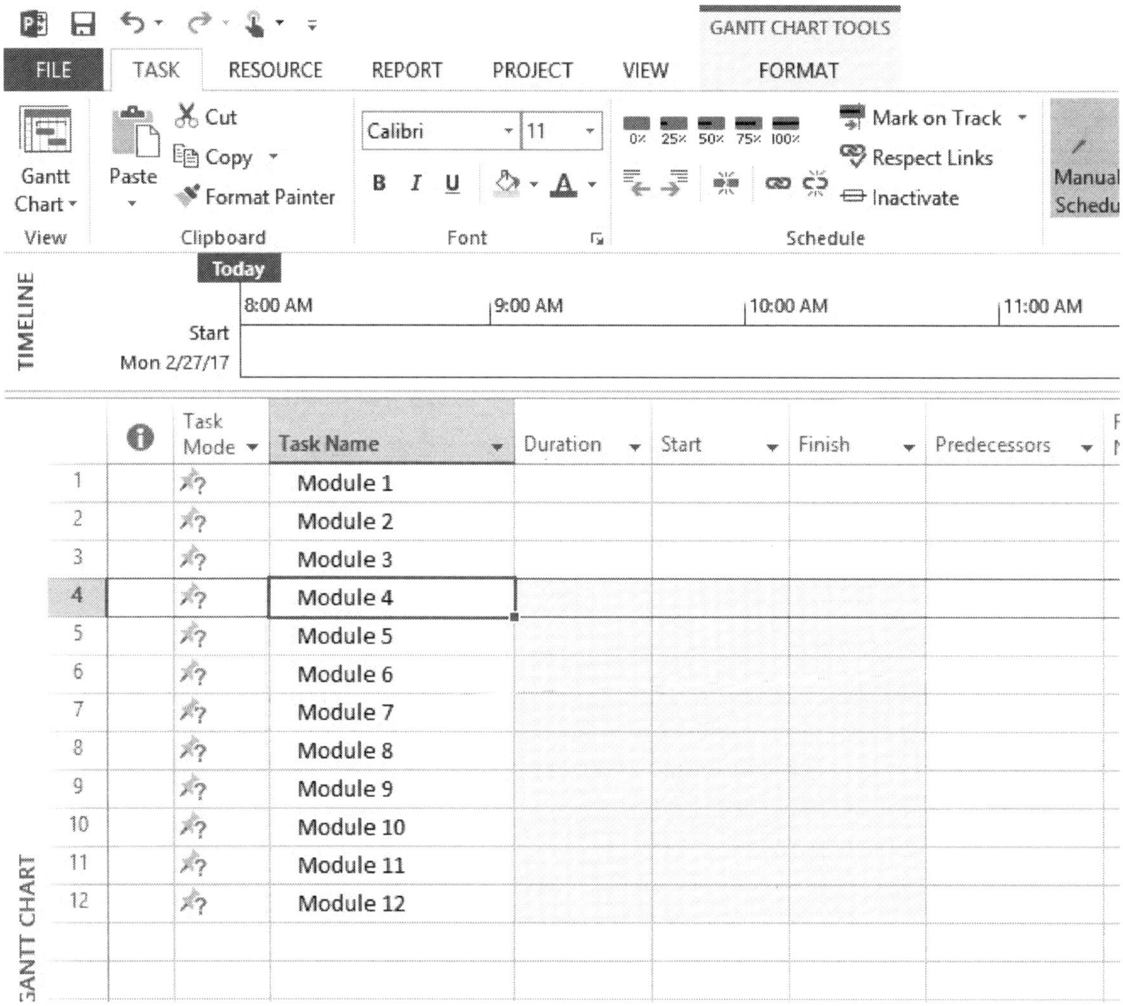
GANTT CHART TOOLS
FILE
TASK
RESOURCE
REPORT
PROJECT
VIEW
FORMAT
Gantt Chart
Paste
Cut
Copy
Format Painter
Calibri
11
Mark on Track
Respect Links
Inactivate
View
Clipboard
Font
Schedule
TIMELINE
Today
8:00 AM
9:00 AM
10:00 AM
11:00 AM
Start
Mon 2/27/17
Task Mode
Task Name
Duration
Start
Finish
Predecessors
Module 1
Module 2
Module 3
Module 4
Module 5
Module 6
Module 7
Module 8
Module 9
Module 10
Module 11
Module 12
GANTT CHART

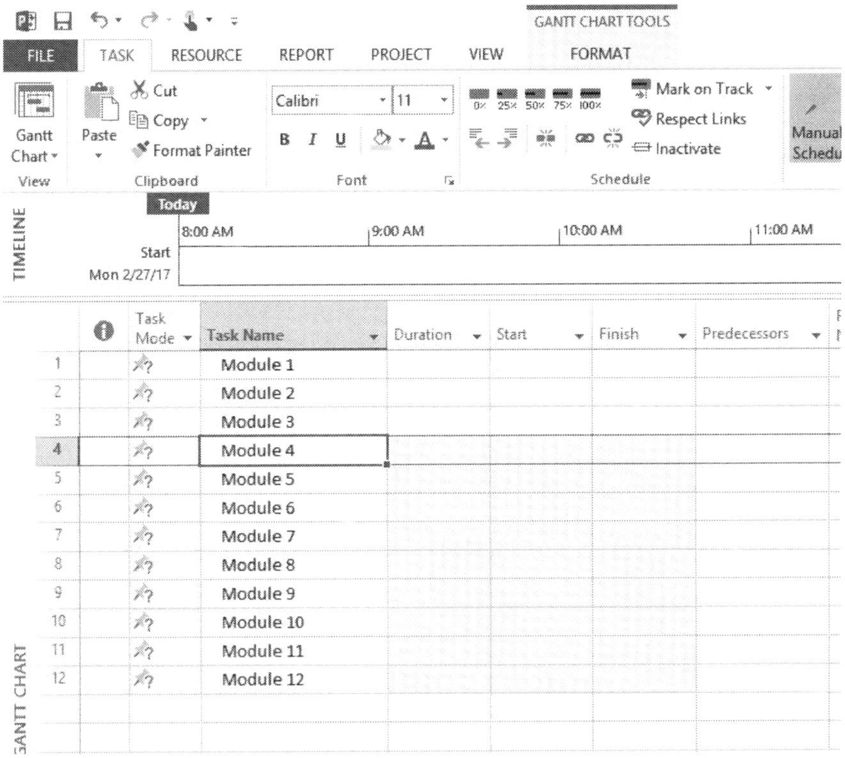

Setting Constraints on Tasks

The Task Details form is another way to enter task information. The form includes information on constraints. Constraints are restrictions set on the start or finish date of a task.

There are several types of constraints:

- As soon as possible
- As late as possible
- Finish no earlier than
- Finish no later than
- Must Finish On
- Must Start On

- Start No Earlier Than
- Start No Later Than

Use the following procedure to open the Task Details form and set a constraint type and date.

Step 1: Select the Display Task Details icon from the Task tab on the Ribbon. It is in the Properties area.

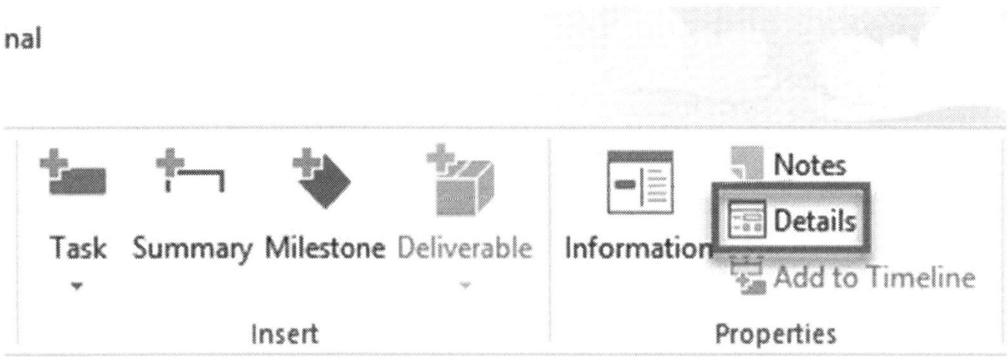

The system displays the Task Details Form in the bottom area of the screen.

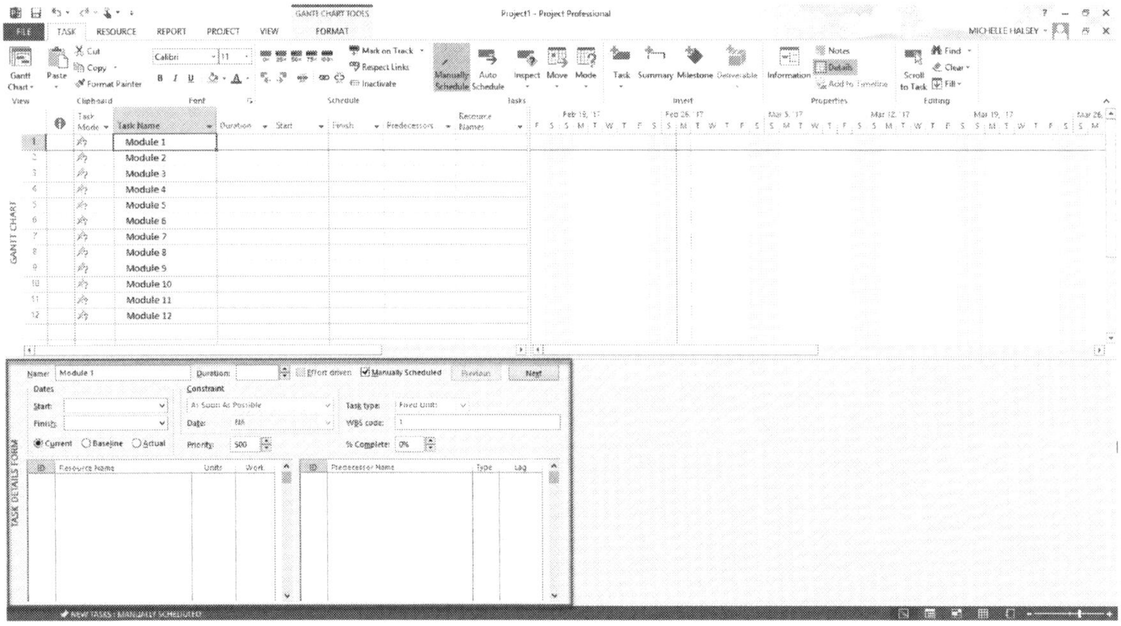

Step 1: You can use the Next or Previous buttons to select the task for which you want to set a constraint.

Step 2: Uncheck the Manually Scheduled box. Constraints are ignored for manually scheduled tasks.

Step 3: Select the Constraint type from the drop-down list.

Step 4: Select the arrow next to the Constraint date and select the constraint date from the calendar.

Step 5: Select Ok.

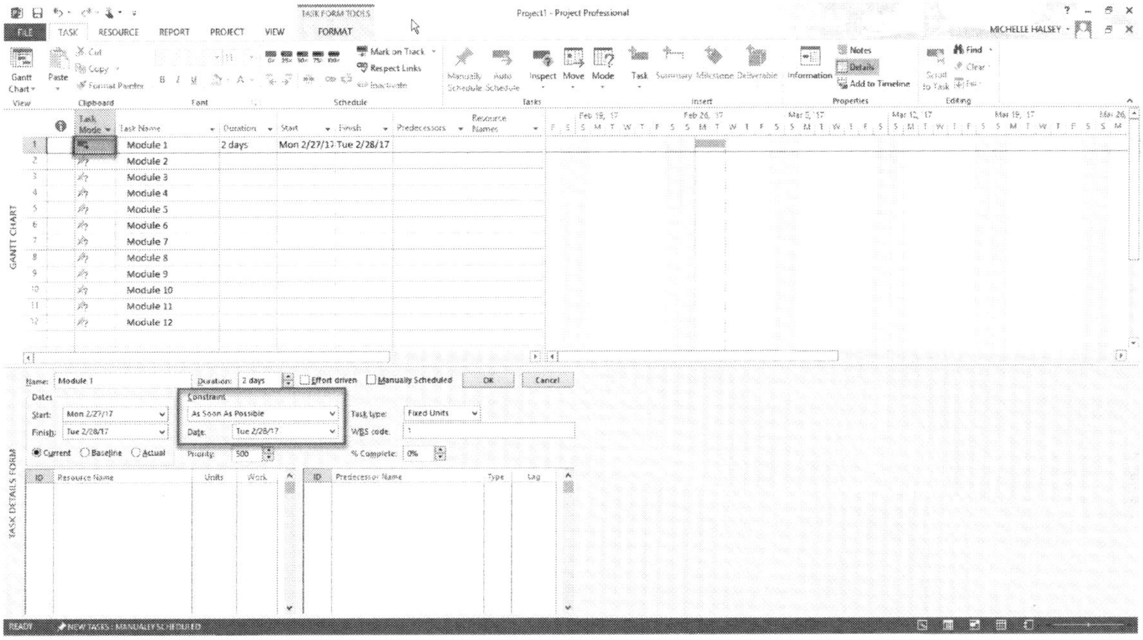

Chapter 3 – Adding Tasks

This chapter will delve a little deeper into understanding tasks. Project 2016 introduces manually scheduled tasks. You can also schedule tasks automatically using the Project scheduling engine. We will discuss the key terms for understanding tasks in this chapter. We will also learn how to view task information and sort and filter tasks. Finally, we will look at task indicators.

Understanding Key Terms

The key terms for understanding tasks in Project are:

Duration – is the amount of time a task will take to complete. When you are using manually scheduled tasks, you can enter anything that will help you schedule the tasks, such as "a couple of days" or "need to talk to Bob." For automatically scheduled tasks, you must put a number with a unit indicator, such as d for days, h for hours, or w for weeks.

Start Date – is when the project starts. For manually scheduled projects, this can be any text that will help you schedule tasks. For automatically scheduled tasks, it must be a date.

Finish Date – is when the project will be completed. For manually scheduled projects, this can be any text that will help you schedule tasks. For automatically scheduled tasks, it must be a date.

Resources – are the materials, people, and other costs associated with completing the project. Resources do not affect manually scheduled tasks, but do help Project to determine a schedule for automatically scheduled tasks.

Gantt Chart – the default view is the Gantt Chart. The right side of this view includes bars that represent the task duration.

Viewing Task Information

Use the following procedure to view task information on the Task Information dialog box.

Step 1: Select the task you want to view. You can select the number to the left to highlight the task.

Step 2: Select Information from the Task tab on the Ribbon.

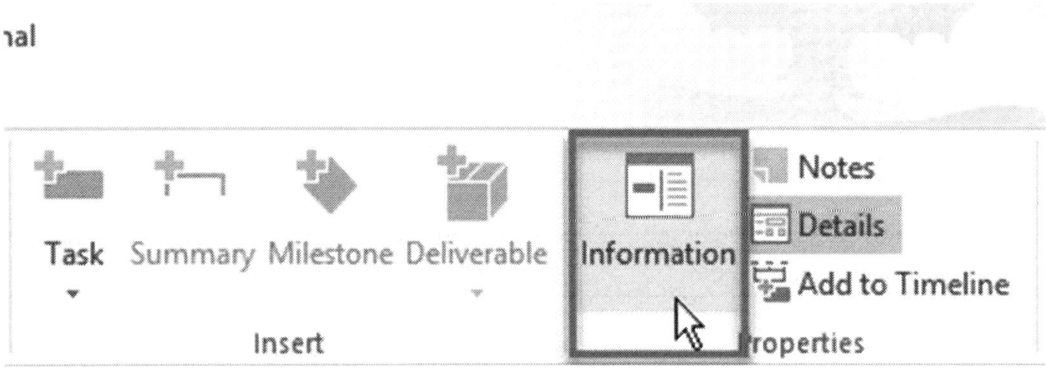

Review the Task Information dialog box.

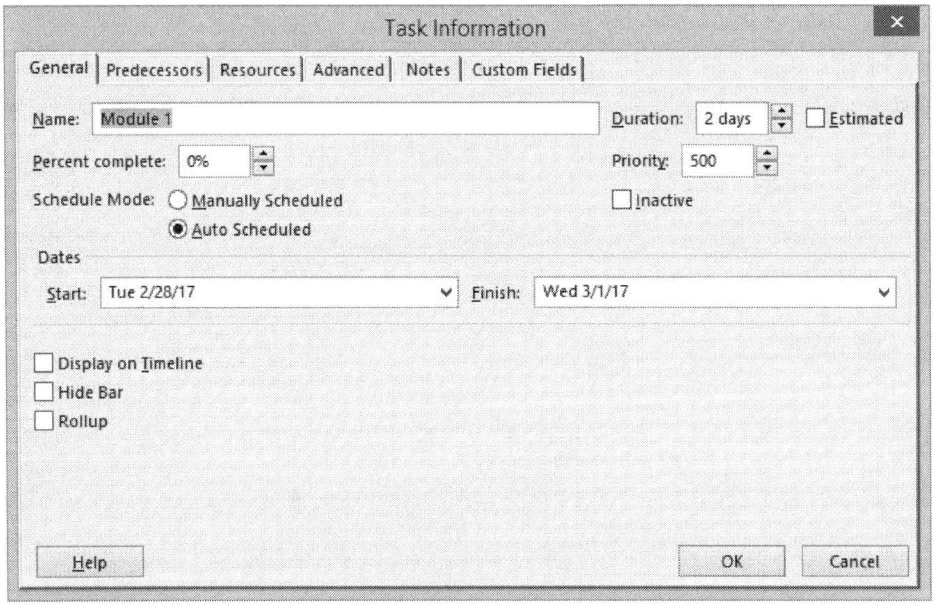

The General tab includes the following information:

- Task Name – the name of the task
- Percent complete – represents a percentage of how much of the task is complete
- Schedule Mode – select Manually or Auto
- Duration – indicates how long the task will take to complete
- Estimated – indicates that the selected duration is an estimate
- Priority – indicates a priority level in comparison with other tasks
- Inactive – makes the task inactive
- Start and Finish dates – indicates the start and finish date of the task, according to the indicated duration
- Display on Timeline – indicates that Project should display the task on the timeline
- Hide Bar – indicates that Project should hide the Gantt chart bar
- Rollup – indicates that Project should rollup the task with its summary task

Review the other tabs briefly.

The Predecessors tab allows you to work with linked tasks and task relationships.

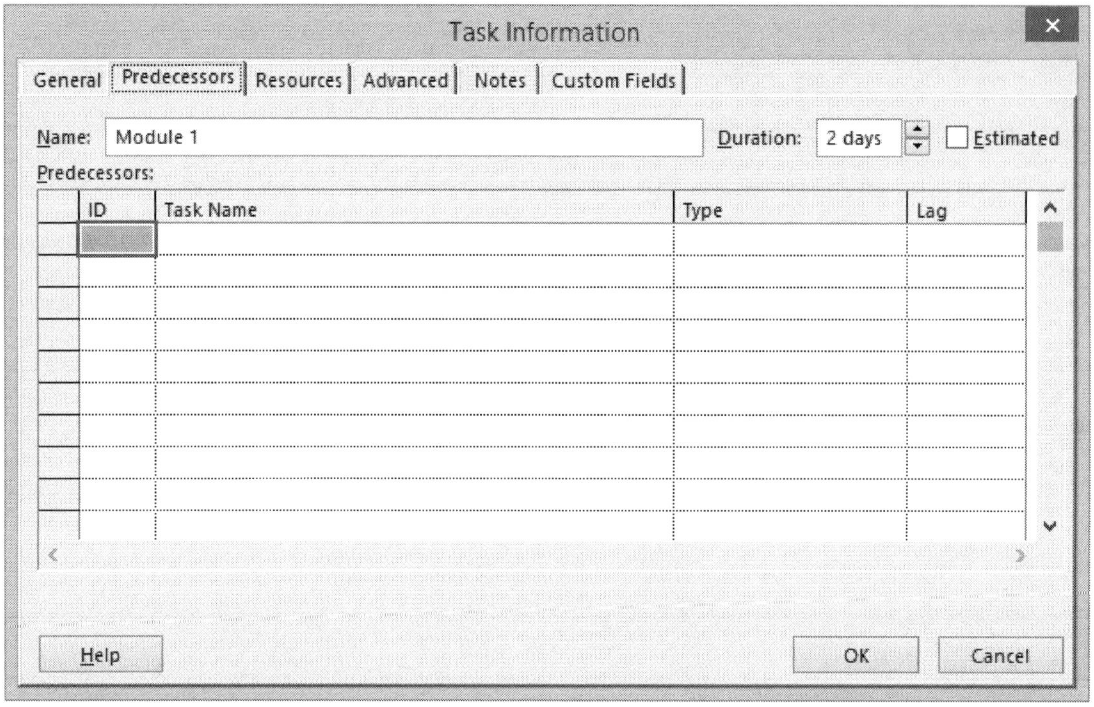

The Resources tab lists resources assigned to the task. We will talk about resources in Chapter six.

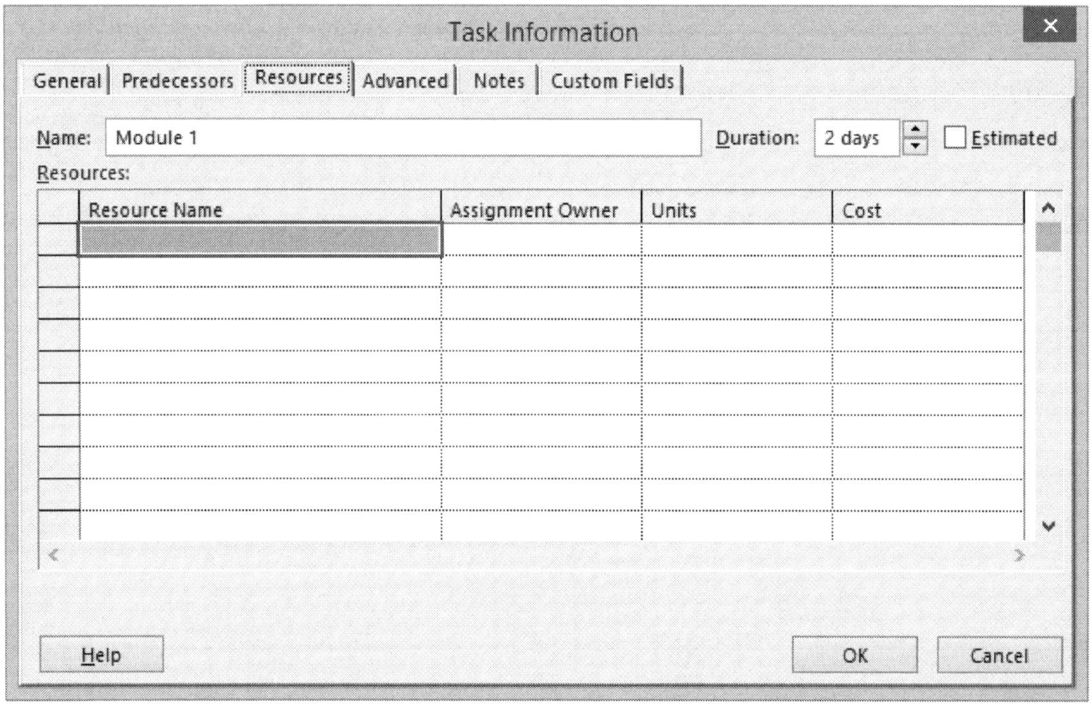

The Advanced tab allows you to set deadlines and constraints. It also includes the task type, WBS code (work breakdown structure) and the earned value method.

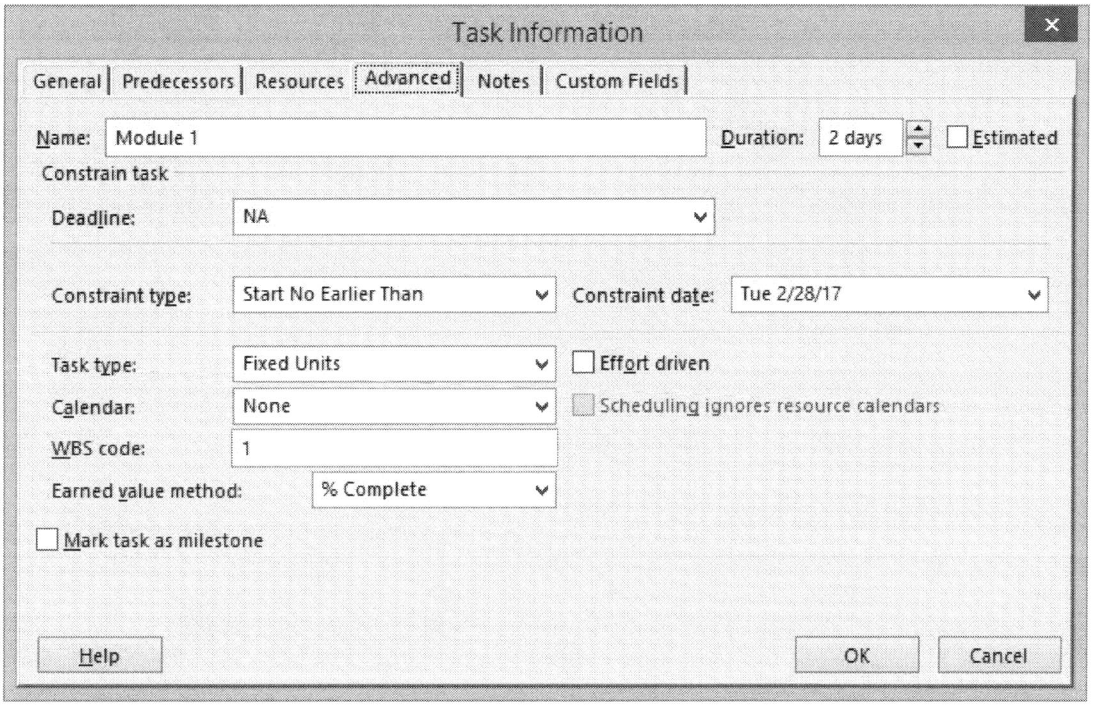

The Notes tab allows you to keep textual notes about the task.

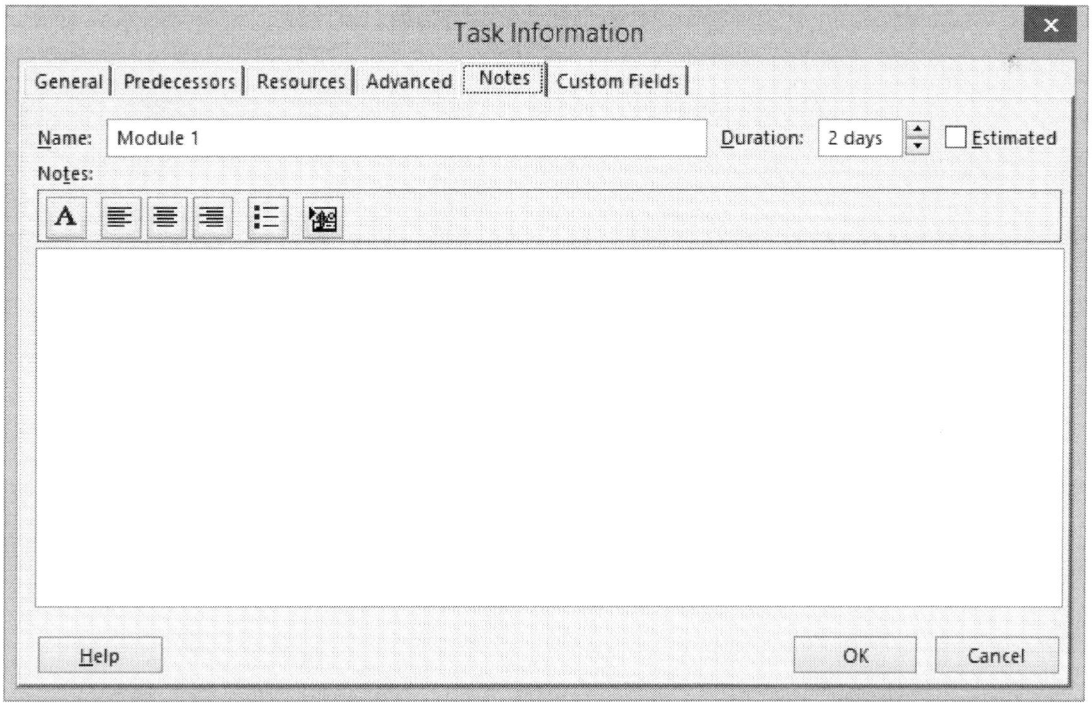

The Custom Fields tab allows you to include information in fields you have customized for the project.

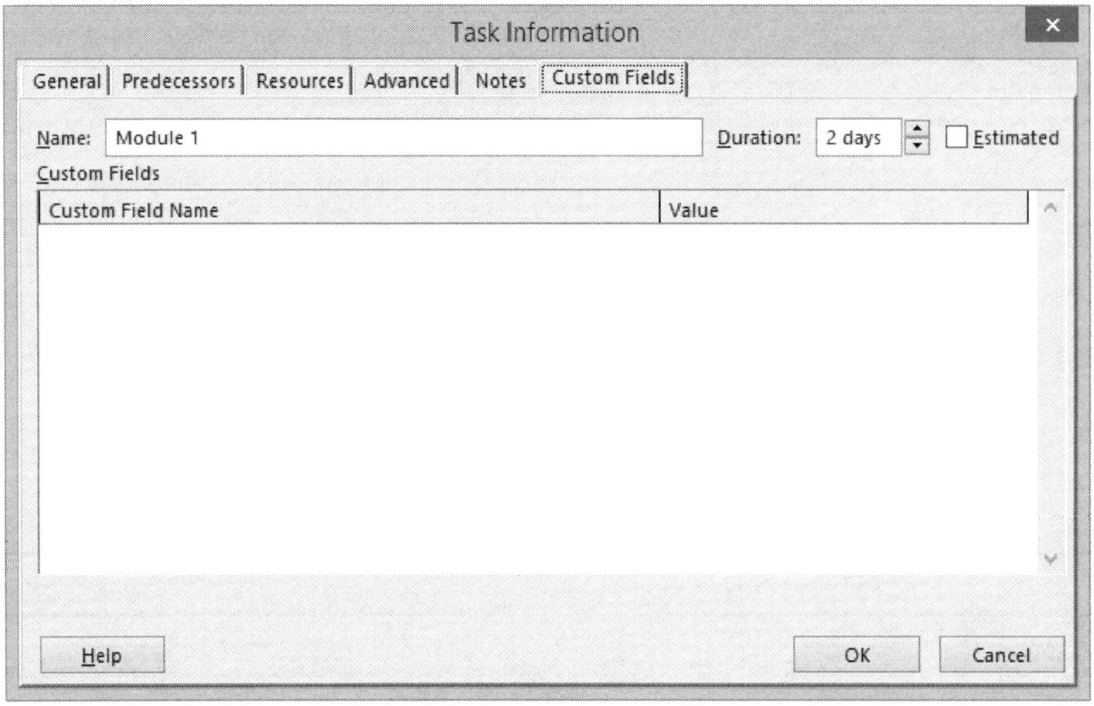

Sorting and Filtering Tasks

The View tab includes tools to sort and filter your task list. This can help you find a task in a long list of items, or filter the view to only include certain parts of the project.

Use the following procedure to see how to sort tasks.

Step 1: Select the View tab from the Ribbon.

Step 2: Select Sort.

Step 3: Select a sorting option.

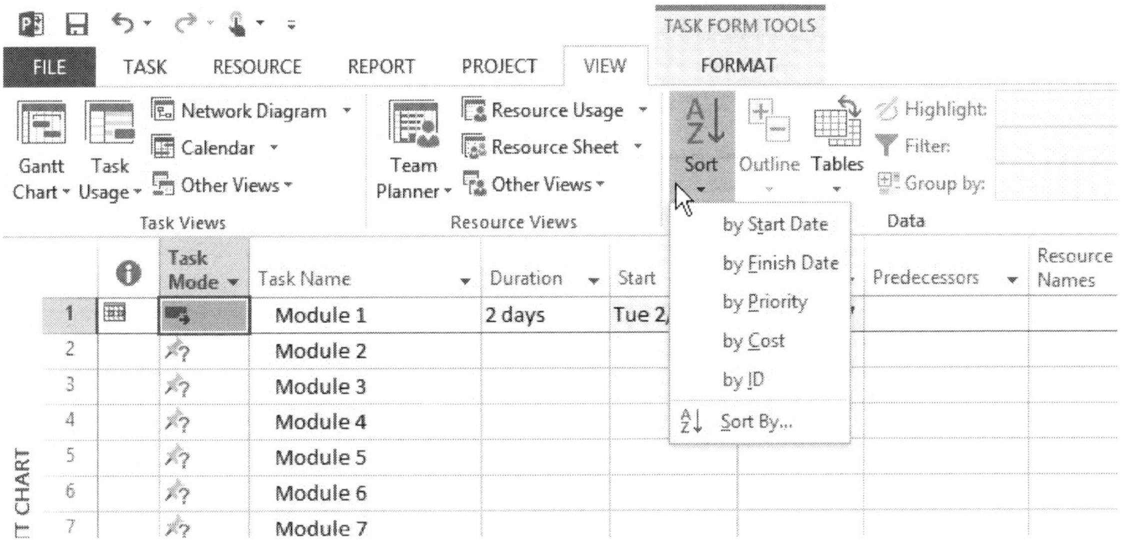

Use the following procedure to view the Sort By dialog box.

Step 1: Select the View tab from the Ribbon.

Step 2: Select Sort.

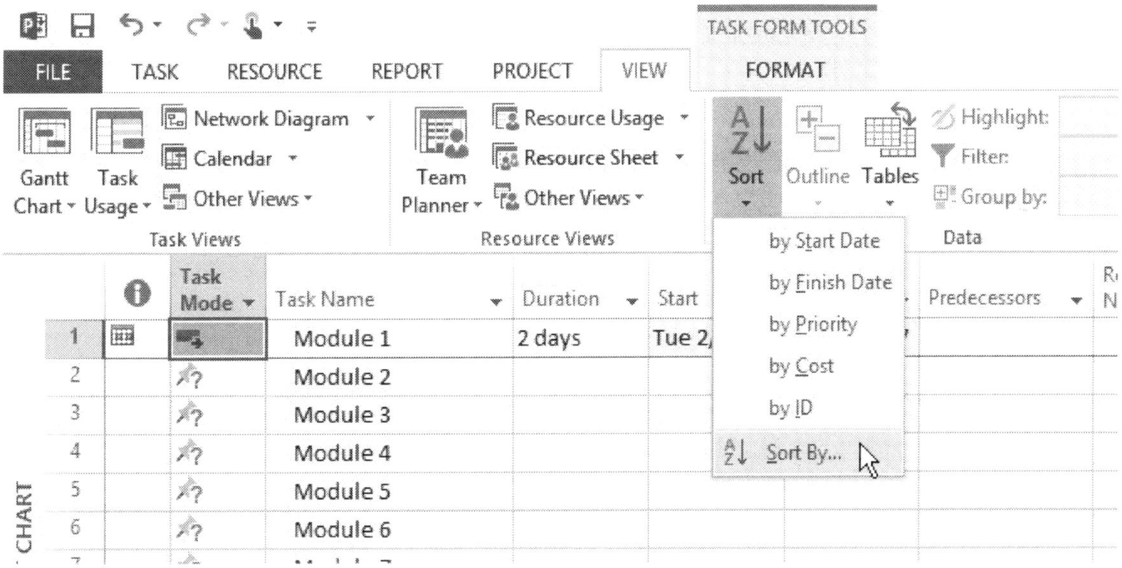

Step 3: Select Sort By.

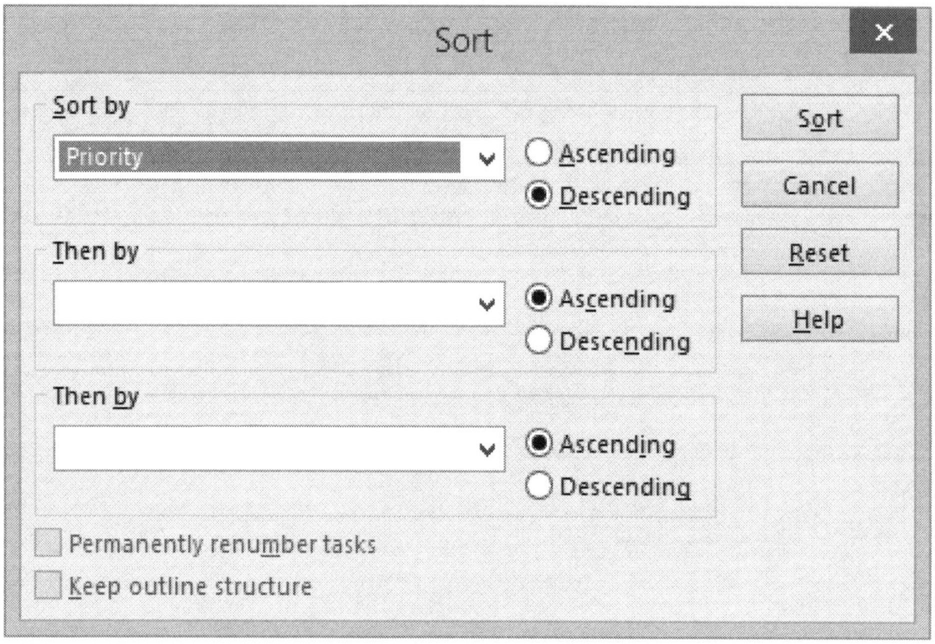

Step 4: You can select up to three options for sorting, and determine whether they should be sorted ascending or descending. Select Sort when you have finished.

Understanding Task Indicators

The following list explains some of the indicators associated with tasks:

- Note Task
- Hyperlink Task
- Deadline Task
- Inflexible Constraint
- Flexible Constraint
- Recurring Task
- Complete Task

- Task Calendar

Indicators provide valuable information. Remember to check the indicator column to find out details about a task. Point out that other views include indicators as well, such as the Resource view. For some views, you may need to add the Indicator field to show this information.

Think of the indicators in the sample project file. When you hover over the indicator, Project displays a hint for the meaning of the indicator.

Chapter 4 – Advanced Task Operations

In this chapter, you will learn some advanced techniques for working with tasks. You will learn how to split tasks to account for time away from a task. You will learn to link and unlink tasks, which can be used to create dependencies between tasks (such as when one task cannot begin until another is completed). This chapter also explains summary and sub tasks and how they can help you organize your project. Finally, you will learn how to create recurring tasks.

Splitting Tasks

You can split tasks to allow for time away from a task, such as when the resource needed for that task is temporarily reassigned to another task or must take an unplanned leave. You can split tasks multiple times.

Use the following procedure to split tasks.

Step 1: In the Gantt chart view, select the bar for the task you want to split.

Step 2: Select the Split Tasks tool from the Task tab on the Ribbon.

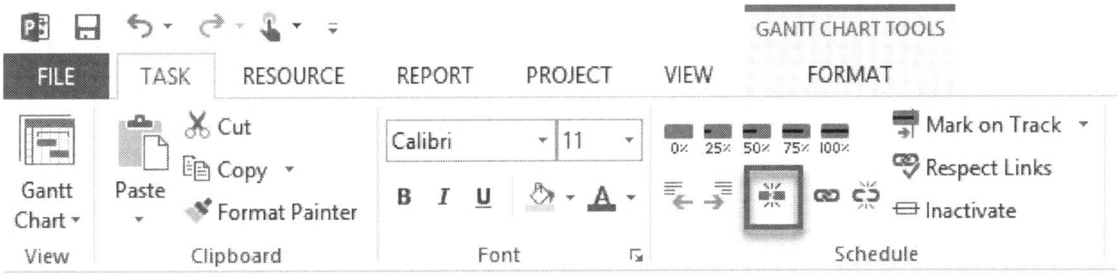

Project displays the Split Tasks dialog.

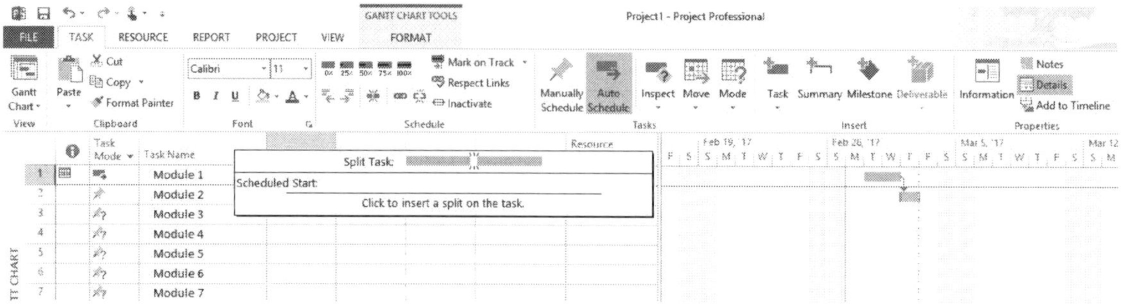

Step 3: Hover your mouse over the task. Project displays the corresponding date in the Split Tasks dialog.

Step 4: When the date corresponds to the date where you need to split the task, click the mouse.

Project inserts a break in the task, as illustrated below.

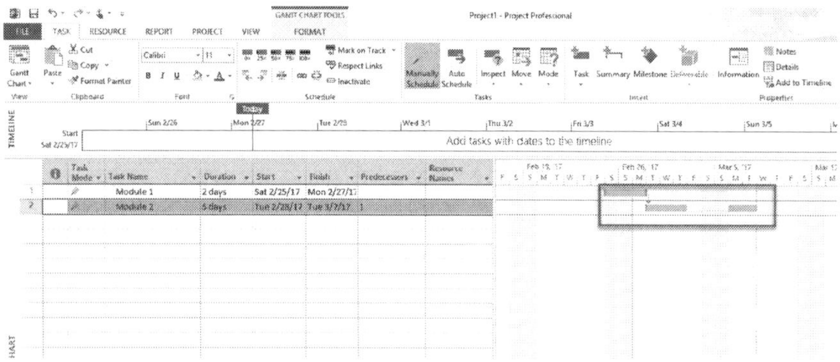

Step 5: You can drag the second part of the task to any start date.

The following illustration shows a task that has been split, and then resumed on the following Tuesday.

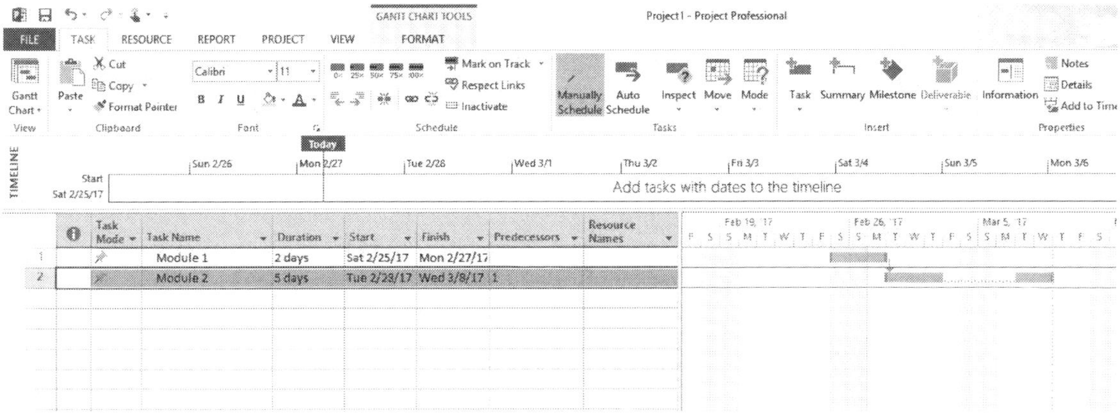

Linking and Unlinking Tasks

Linking tasks allows you to create dependencies between two or more tasks. To link tasks, select the tasks you want to link and use the Link tasks tool. To unlink the tasks, select the tasks and use the Unlink tasks tool.

Use the following procedure to link tasks.

Step 1: Select the tasks you want to link by highlighting them in the task list.

Step 2: Select the Link Tasks tool.

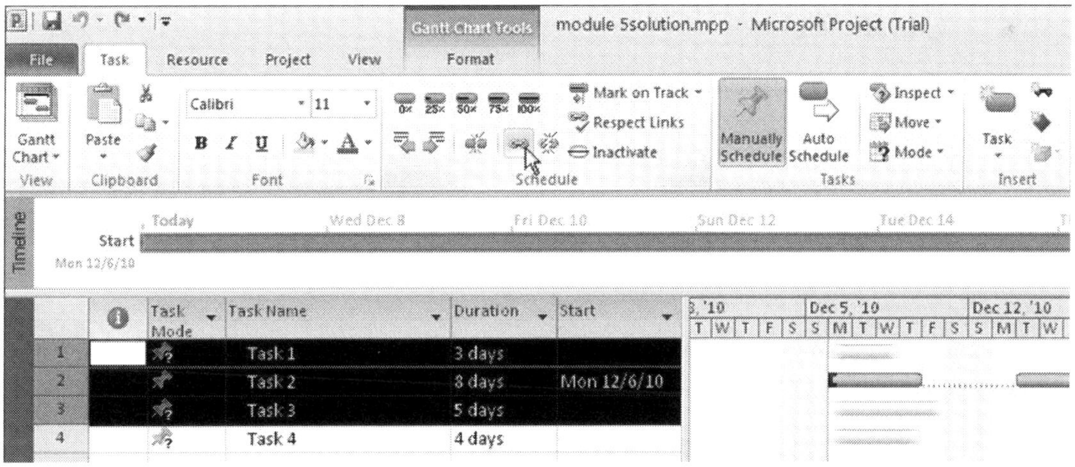

Notice the new start dates and the way the task duration bars on the Gantt chart have changed.

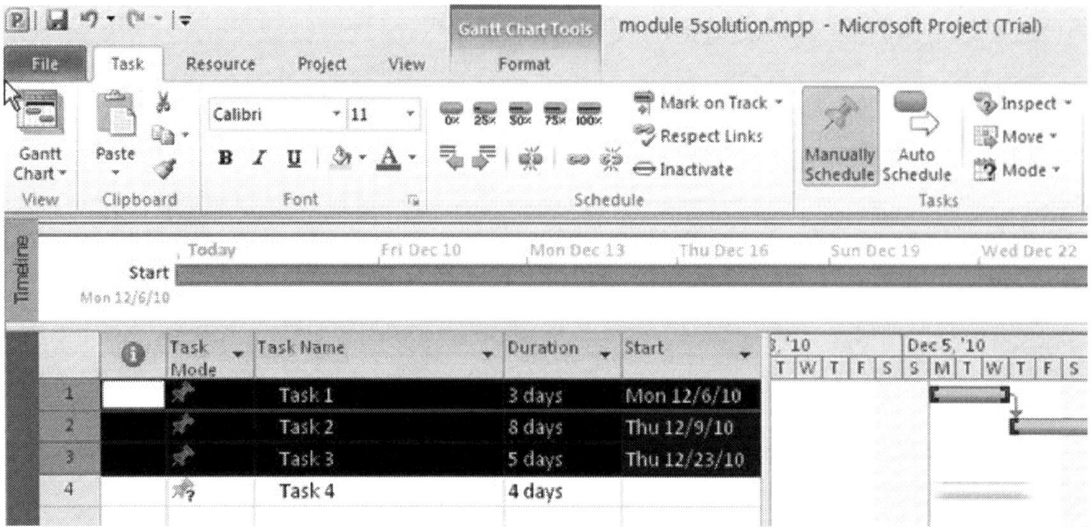

Use the following procedure to unlink tasks. Select the tasks you want to unlink by highlighting them in the task list.

Step 1: Select the Unlink Tasks tool.

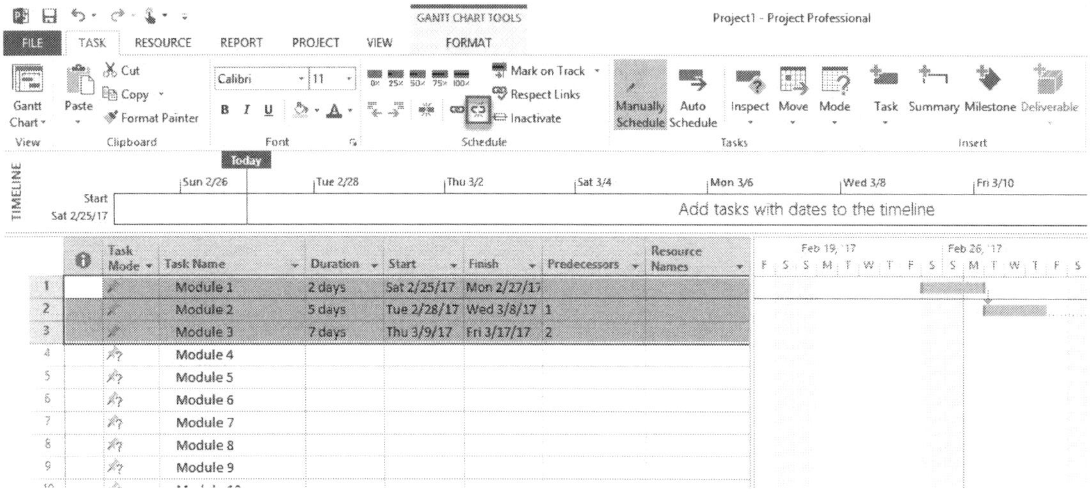

Creating Summary and Sub Tasks

Summary tasks allow you to see certain information about your project "rolled up" into one task that summarizes several related tasks. You can insert summary tasks or make existing tasks summary tasks. You can use the indent and outdent tools to make existing tasks summary tasks or subtasks.

Use the following procedure to insert a summary task.

Step 1: Place your cursor in the task that will be a subtask to the new summary task, or in a blank row on the task list.

Step 2: Select the Insert Summary Task tool from the Task tab on the Ribbon.

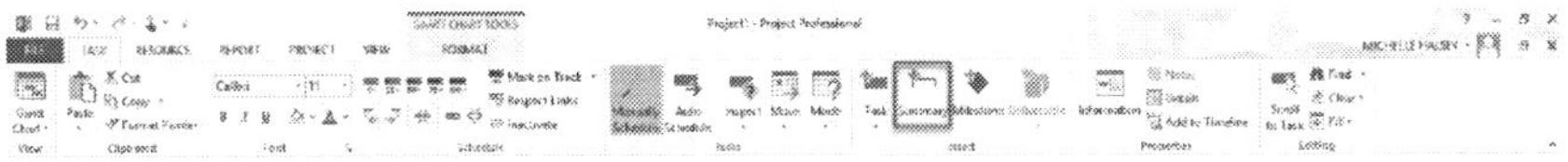

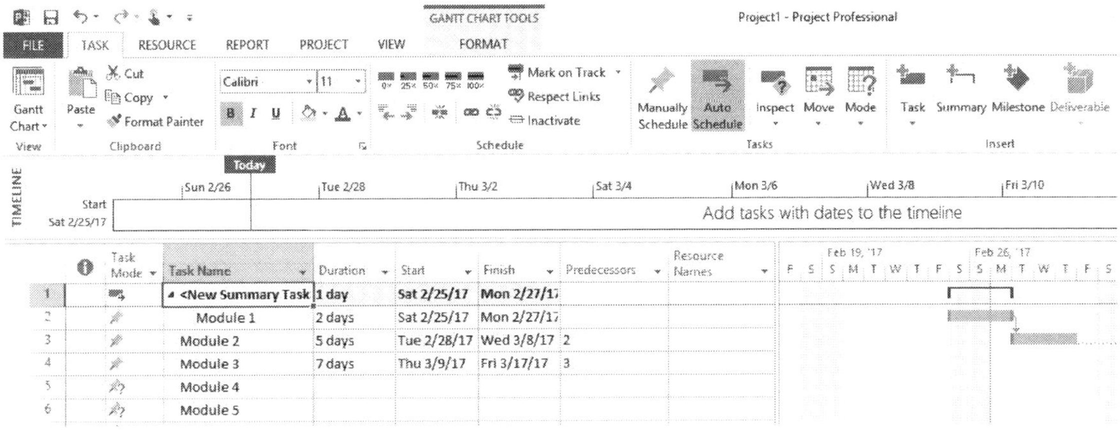

Step 3: Enter the name of the summary task.

Use the following procedure to make subtasks using the Indent tool.

Step 1: Highlight one or more tasks that will be subtasks to the summary task preceding them.

Step 2: Select the Indent tool.

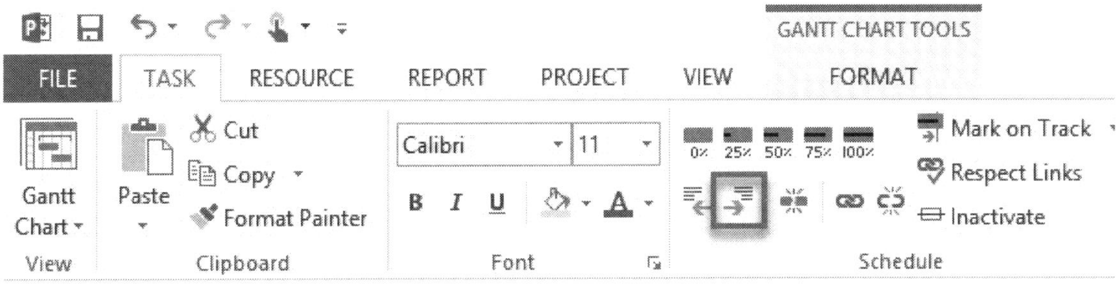

Creating Recurring Tasks

If you have a task that is repeated at regular intervals, it will save time to enter it as a recurring task. You enter the recurrence information when you insert the task.

Use the following procedure to create a recurring task.

Step 1: Place your cursor on the row below where you want the recurring task to appear in the task list.

Step 2: Select Task from the Task tab on the Ribbon.

Step 3: Select Recurring Task.

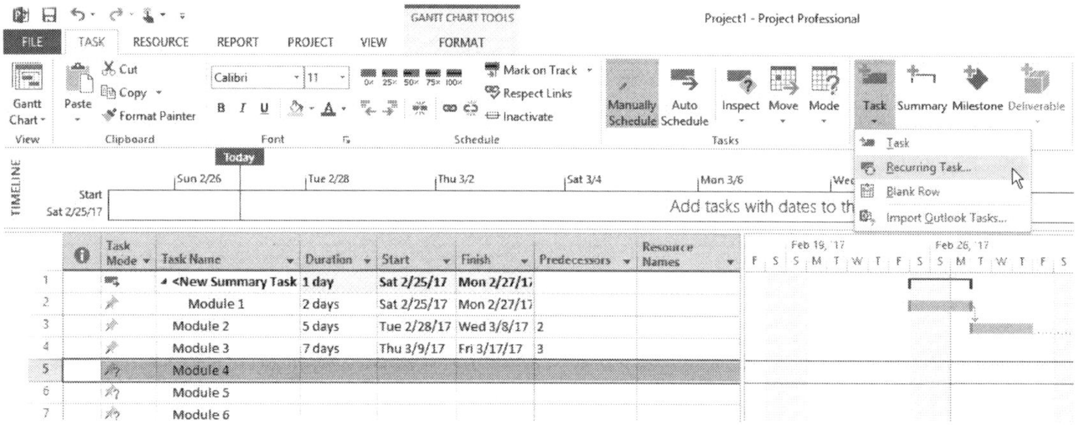

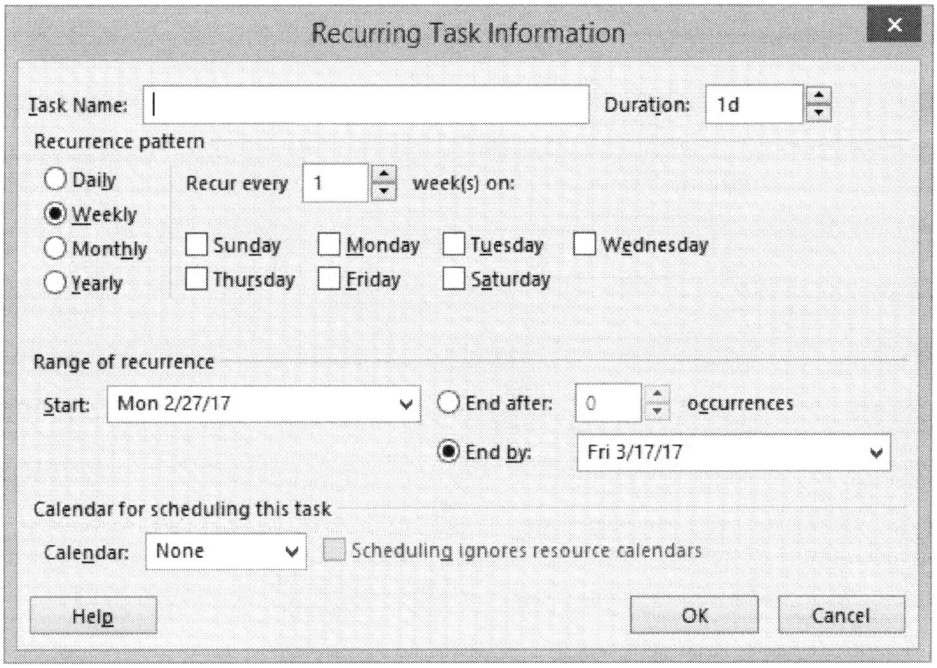

Step 4: Enter the Task Name.

Step 5: Select the Recurrence Pattern and enter the pattern details according to your selection.

Step 6: Enter the Start and End information for the Range of Recurrence.

Step 7: Select the Calendar for the task.

Step 8: Select OK.

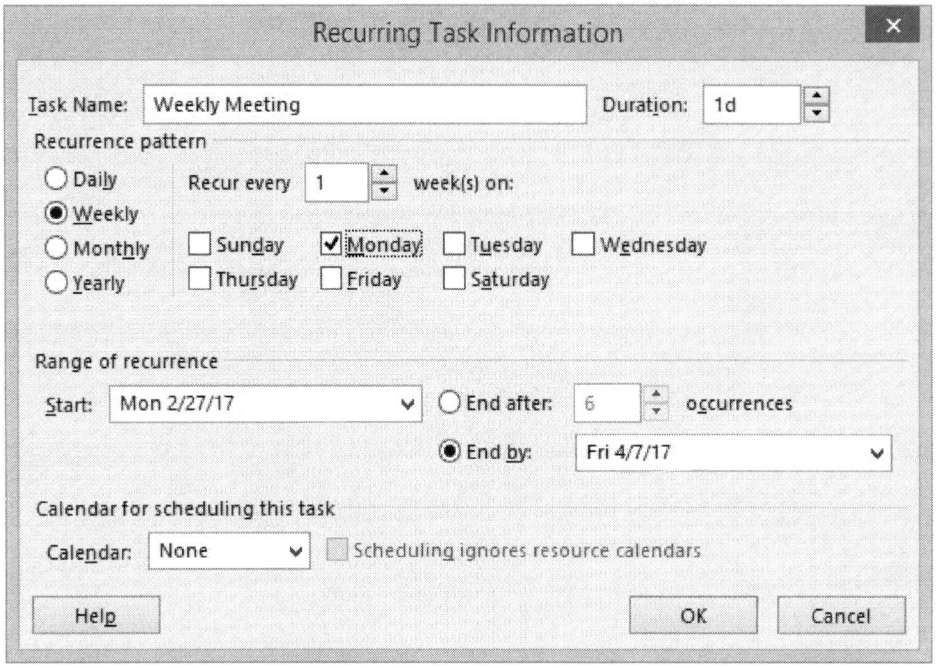
Recurring Task Information
Task Name: Weekly Meeting
Duration: 1d
Recurrence pattern
Daily
Weekly
Monthly
Yearly
Recur every 1 week(s) on:
Sunday
Monday
Tuesday
Wednesday
Thursday
Friday
Saturday
Range of recurrence
Start: Mon 2/27/17
End after: 6 occurrences
End by: Fri 4/7/17
Calendar for scheduling this task
Calendar: None
Scheduling ignores resource calendars
Help
OK
Cancel

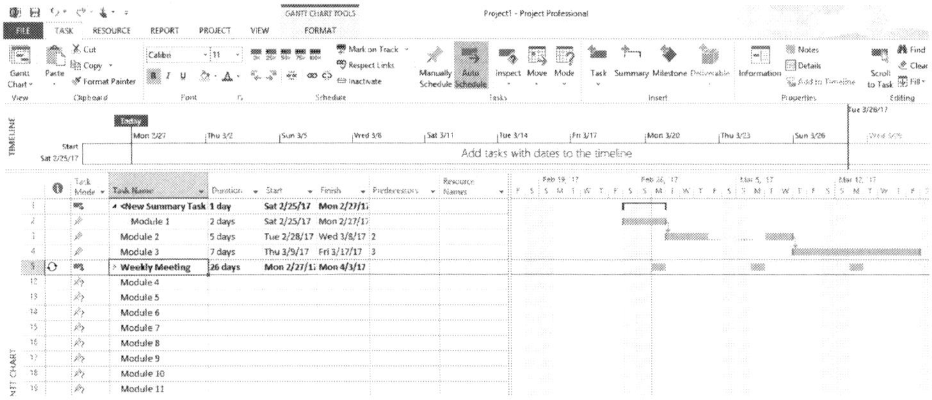

Chapter 5 – Adding Resources

This chapter introduces resources, the people, equipment, and materials needed to complete your project. This chapter will give an overview of how resources are used in Project 2016. You will learn how to add resources and view resource information. You will also learn how to assign resources to tasks. Finally, this chapter presents an introduction to leveling resources, which is a way to reschedule tasks so that your resources are not over scheduled.

Understanding Resources

The resources must be added to the project before they are available to assign to tasks. There are three different types of resources in Project 2016:

Work resources are the people included in your project plan. These are the people who will do the work of completing tasks related to the project.

Material resources are the items that are required to complete your project that are measured in units rather than work hours, such as cases of roofing shingles or gallons of paint.

Cost resources are fixed costs associated with your project, such as airfare or equipment rental.

Adding Resources

You will need to create your data base of resources before you can do anything else with your resource information. The Resource Sheet allows you to enter your resource information.

Use the following procedure to enter resources.

Step 1: Select the View tab from the Ribbon.

Step 2: Select Resource Sheet.

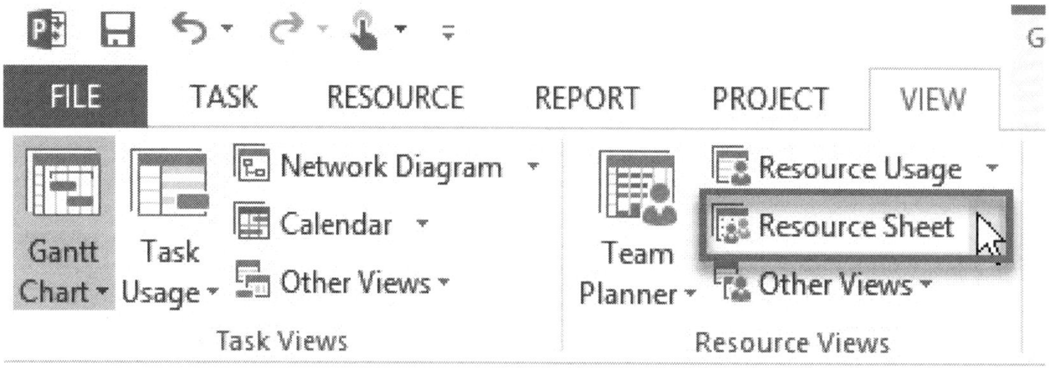

Step 3: Select the Resource tab from the Ribbon.

Step 4: Select Add Resources.

Step 5: You can add resources from your Active Directory or your Outlook Address Book. Or select Work Resource, Material Resource, or Cost Resource to add a resource of that type. You can also add a resource by entering information directly into the Resource sheet. Enter the Resource Name, the resource Type, and the Standard Rate for the resource in the Resource sheet. The standard rate can be indicated in different ways, including an hourly or annual rate for work resources and a per unit price for material resources.

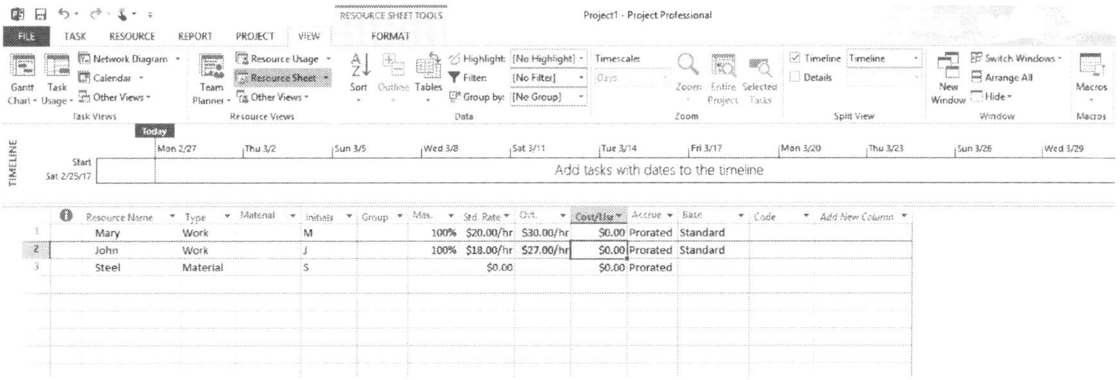

Step 6: Make sure to save your work.

Viewing Resource Information

The Resource Information dialog box includes information about the resource, including availability and costs. You can also store notes about the resource and set up custom information.

Use the following procedure to view the Resource Information dialog box.

Step 1: Highlight the resource you want to view in the Resource Sheet.

Step 2: Select Information from the Resource tab on the Ribbon.

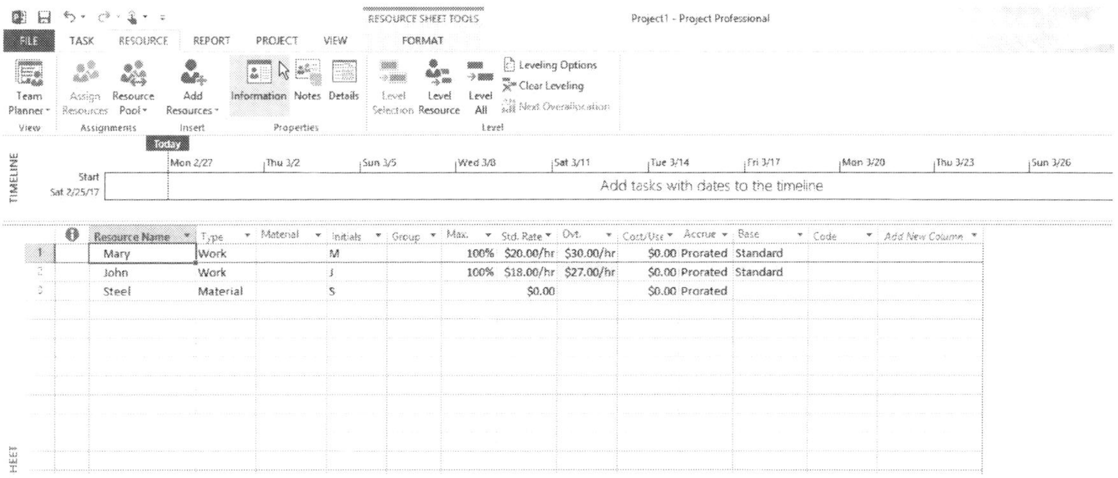

The General tab of the Resource Information dialog box includes basic information about the resource, including the name, email address, and resource type. Some more advanced options including the calendar and availability are also available on this tab.

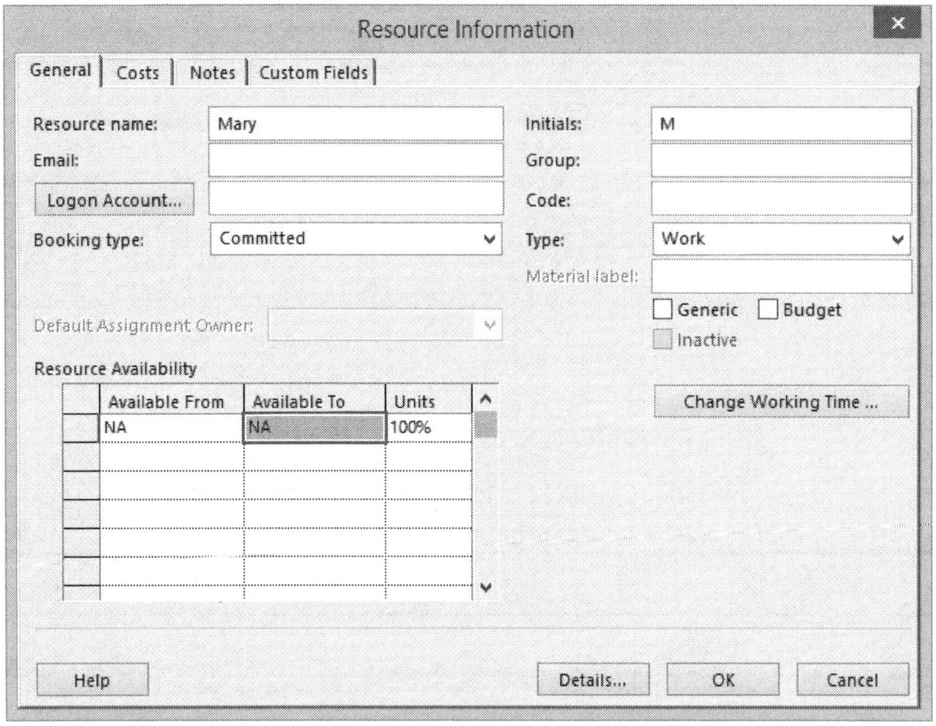

The Costs tab of the Resource Information dialog box includes information about the resource pay rates or costs.

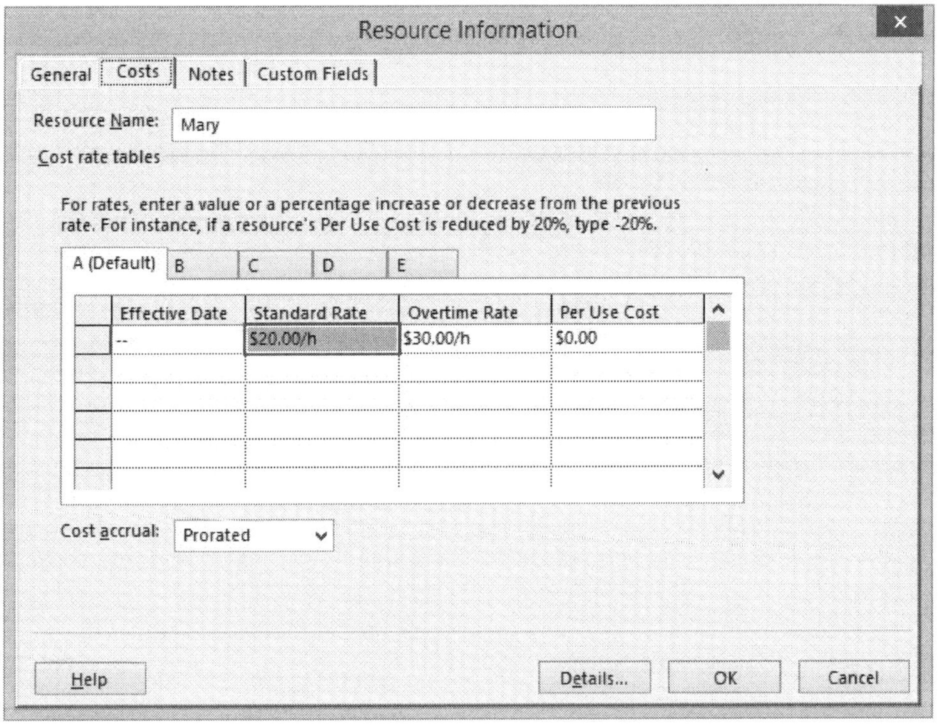

The Notes tab of the Resource Information dialog box allows you to include notes about the resource.

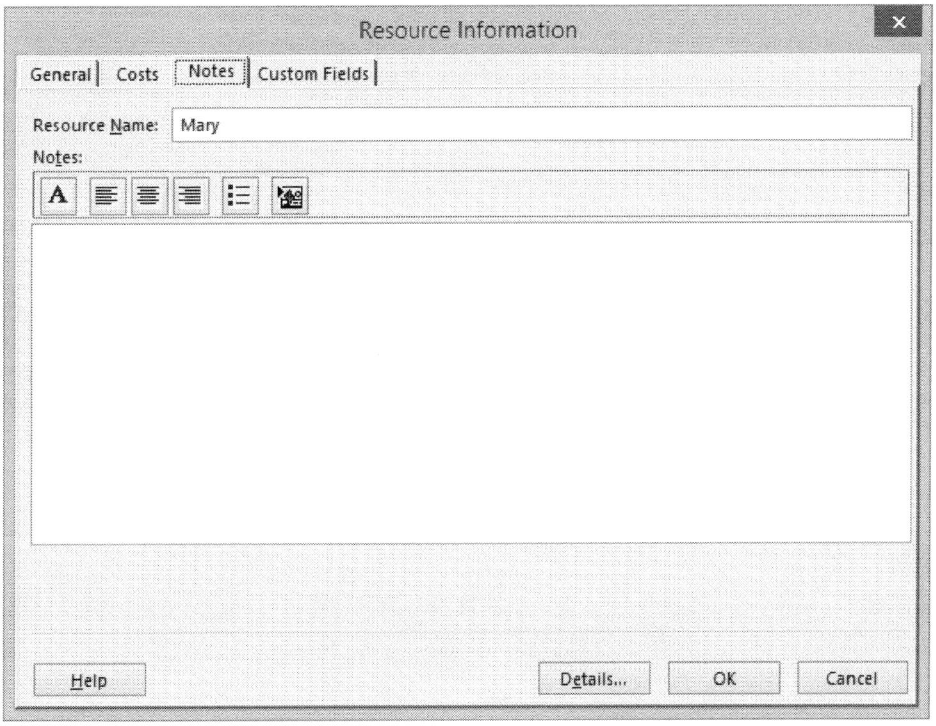

The Custom Field tab of the Resource Information dialog box allows you to add custom fields to the resource.

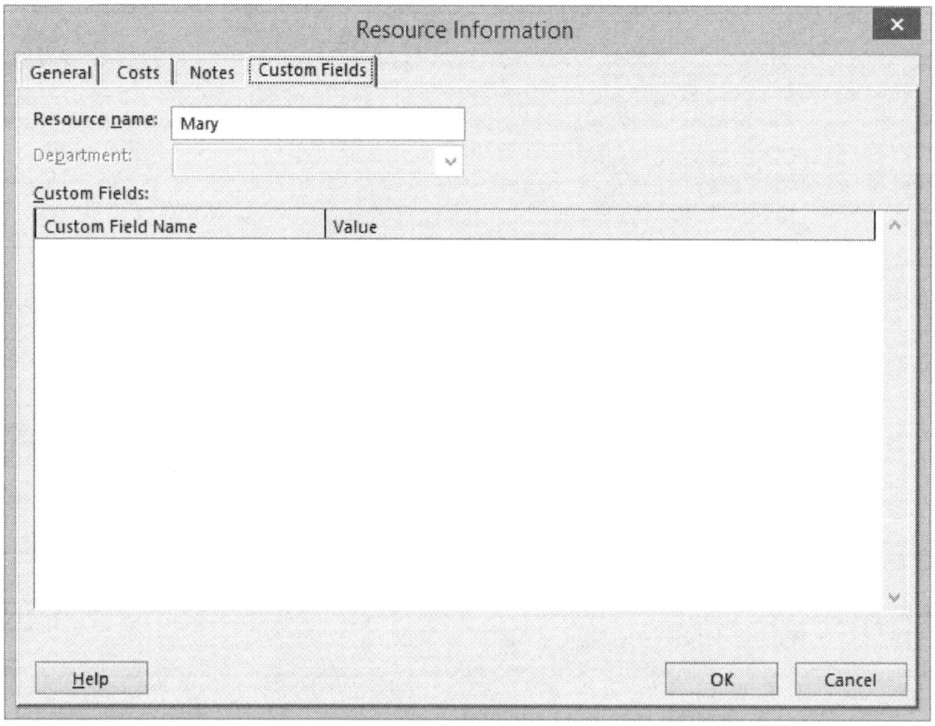

Assigning Resources to Tasks

Once you have added resources to your project, you can assign resources to your tasks. You return to the task view to assign resources to individual tasks. You can assign more than one resource to each task.

Use the following procedure to assign a resource to a task.

Step 1: Select Gantt Chart to return to the Gantt Chart view.

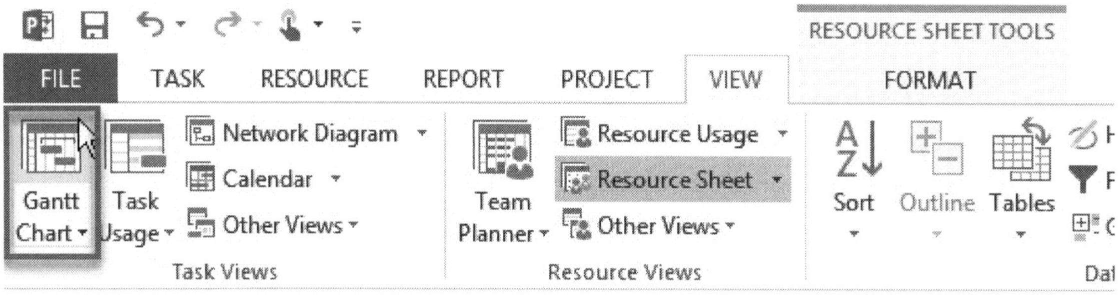

Step 2: Select the Resource tab from the Ribbon.

Step 3: Select a task from the task list.

Step 4: Select Assign Resources.

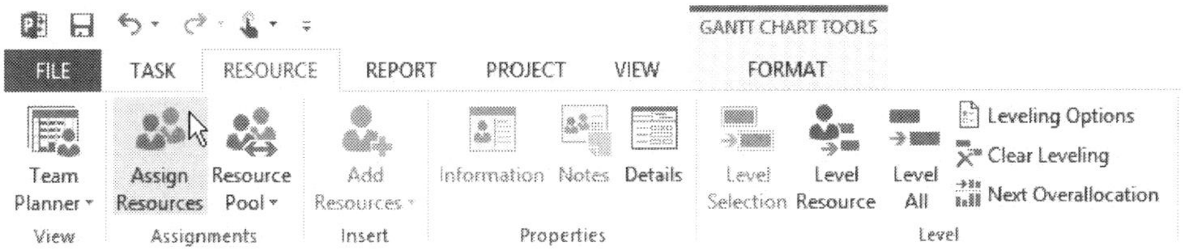

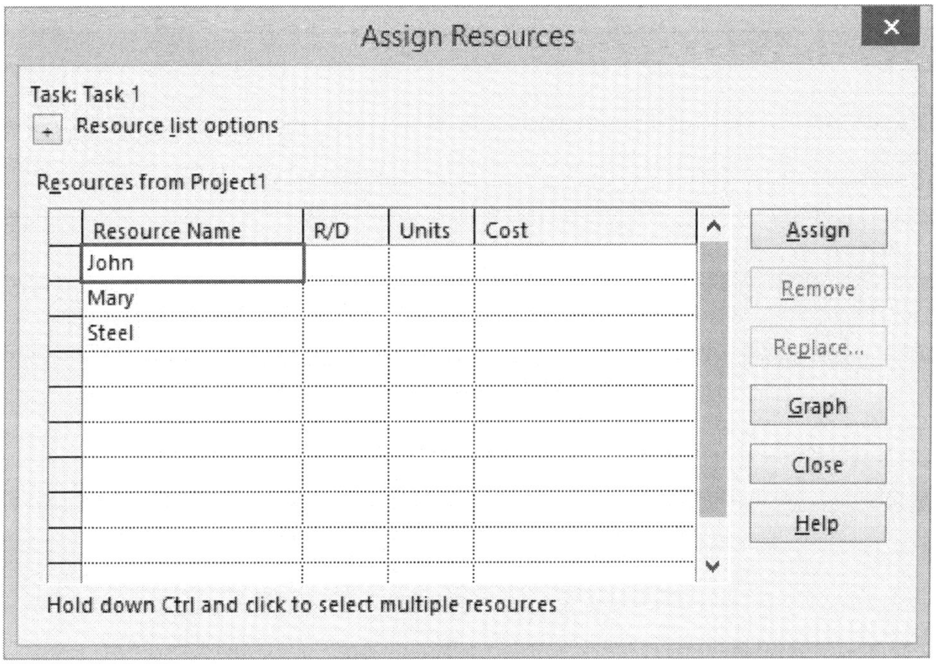

Step 5: Highlight one or more resources from the list. To select multiple resource, press the Shift or Ctrl key while selecting the resource name.

Step 6: Select Assign. You can also assign multiple resource individually by highlighting each resource and then clicking Assign.

Step 7: When you have finished assigning resources, select Close.

Notice the resource information has been added to the Gantt chart next to the duration bars.

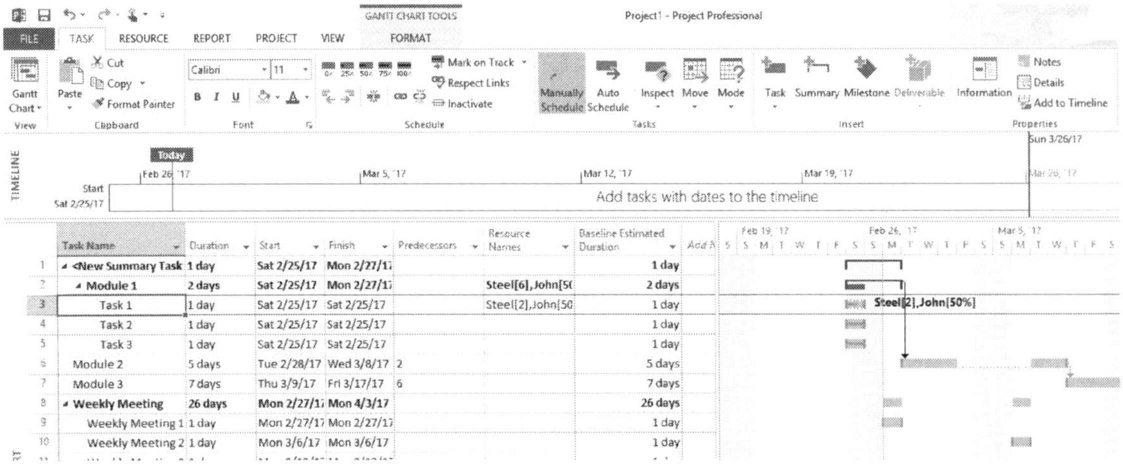

Leveling Resources

The Level Resources tool in Project 2016 looks at task assignments for your resources and makes sure that a resource is not over-scheduled. If task assignments for a given day cause a resource to be over-scheduled, the affected tasks are rescheduled (sometimes by splitting or delaying tasks) so that the resource is not over allocated.

Use the following procedure to level resources.

Step 1: Select Level Resource from the Resource tab on the Ribbon.

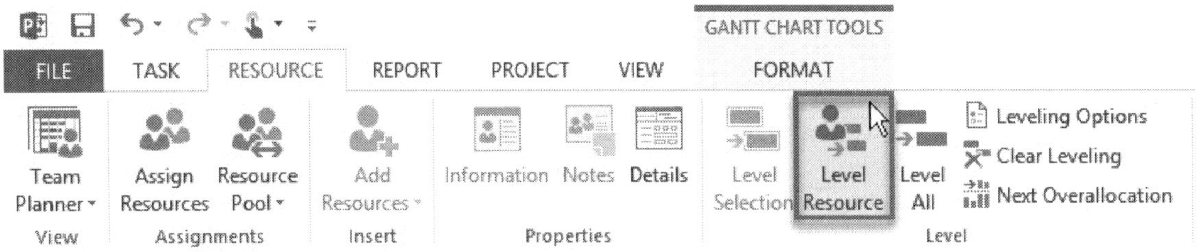

Step 2: Select the Resource you want to level.

Step 3: Select Level Now.

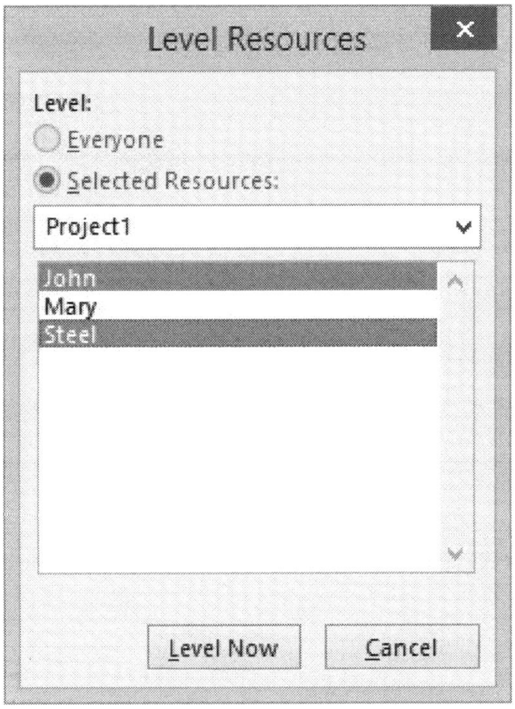

Notice how Task 2 has been split to accommodate Joe's time at the Weekly meeting.

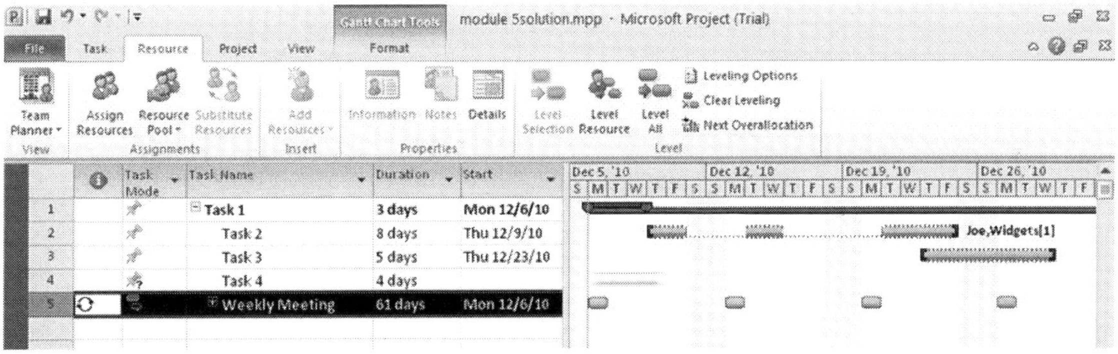

Chapter 6 – Other Ways to View Project Information

In this chapter, we will investigate some of the other ways to view information about your project in Project 2016. First, we will look at a new feature in Project 2016, the Team Planner. We will look at the important task views and the important resource views. You will learn how about formatting your view to get it to look just like you want. Finally, we will look at how to format the Timescale.

The Team Planner

The Team Planner is a new feature in Project 2016 that allows you to quickly see your resource task assignments, as well as unassigned tasks and resources that are free. You can drag and drop tasks to quickly reassign them.

The Team Planner shows the resources in your project, with bars to represent the tasks to which they have been assigned. The bottom of the view allows you to easily spot tasks that do not yet have resources assigned. You can also see which resources are free.

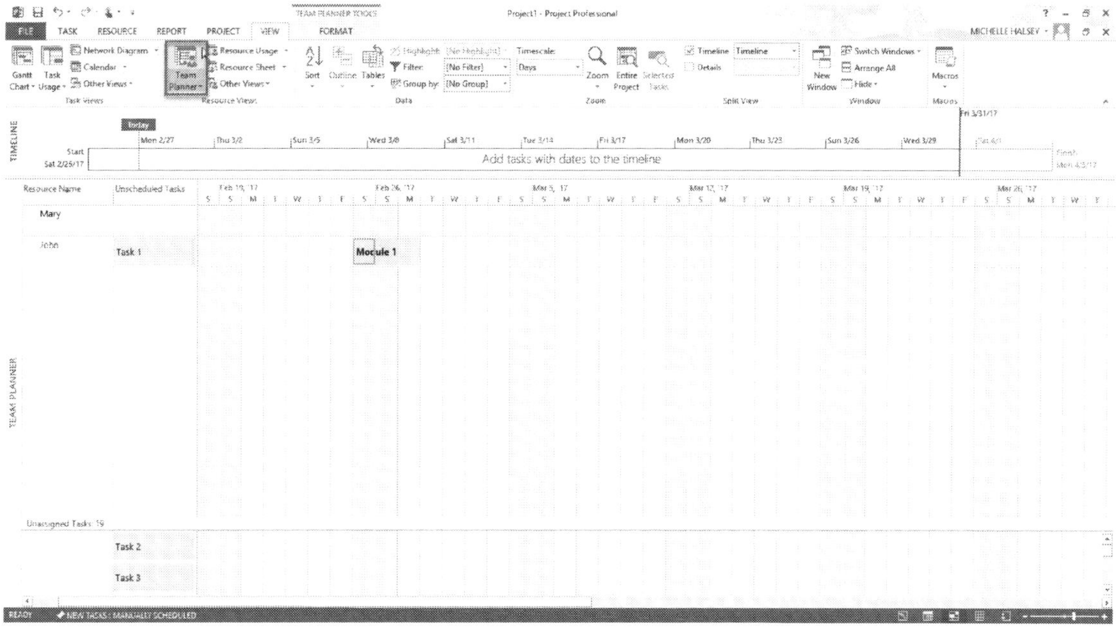

Practice dragging task assignments to new dates or resources.

Important Task Views

The Task Views area of the View tab on the Ribbon includes tools for the most common task views. You can also access other task views.

Shown here is the different task views available from the View tab.

Step 1: The Gantt Chart is the default task view. There is also a Tracking Gantt available, which shows the percentage completion for the tasks, instead of the resources assigned.

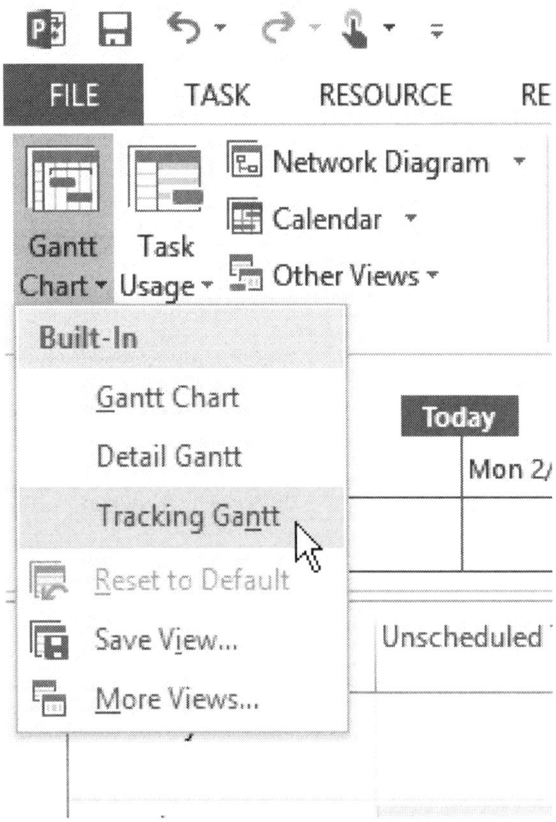

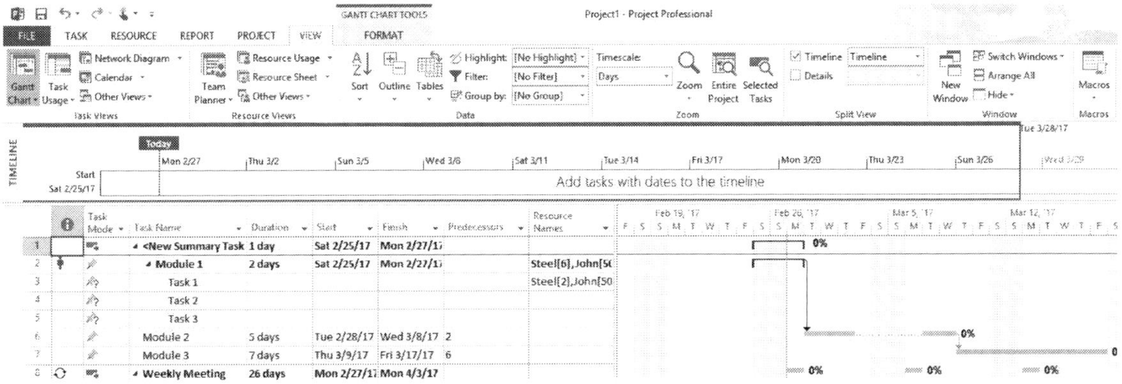

The Task Usage view shows a chart of the work assigned for tasks.

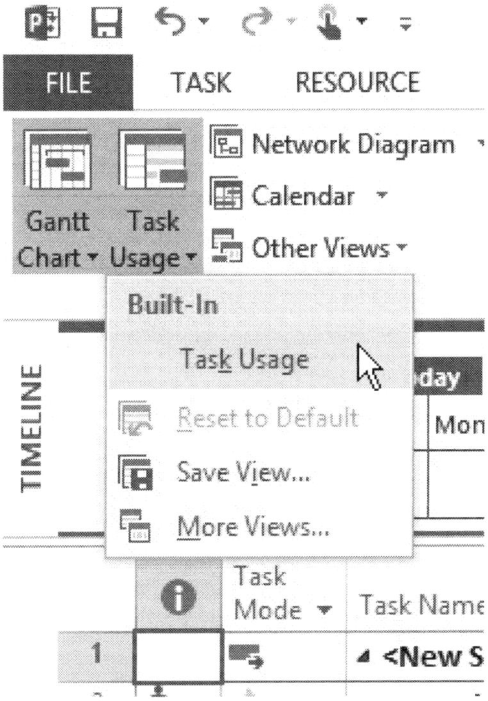

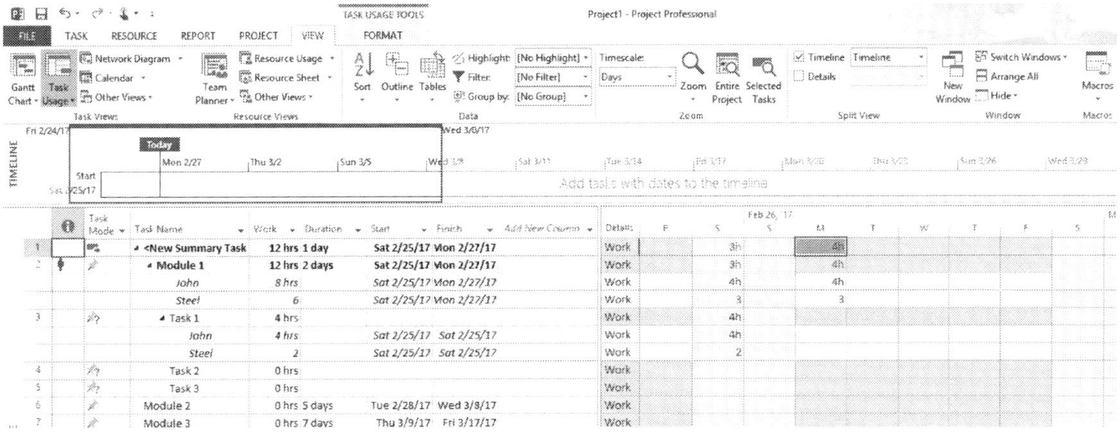

Here is an example of the Network Diagram.

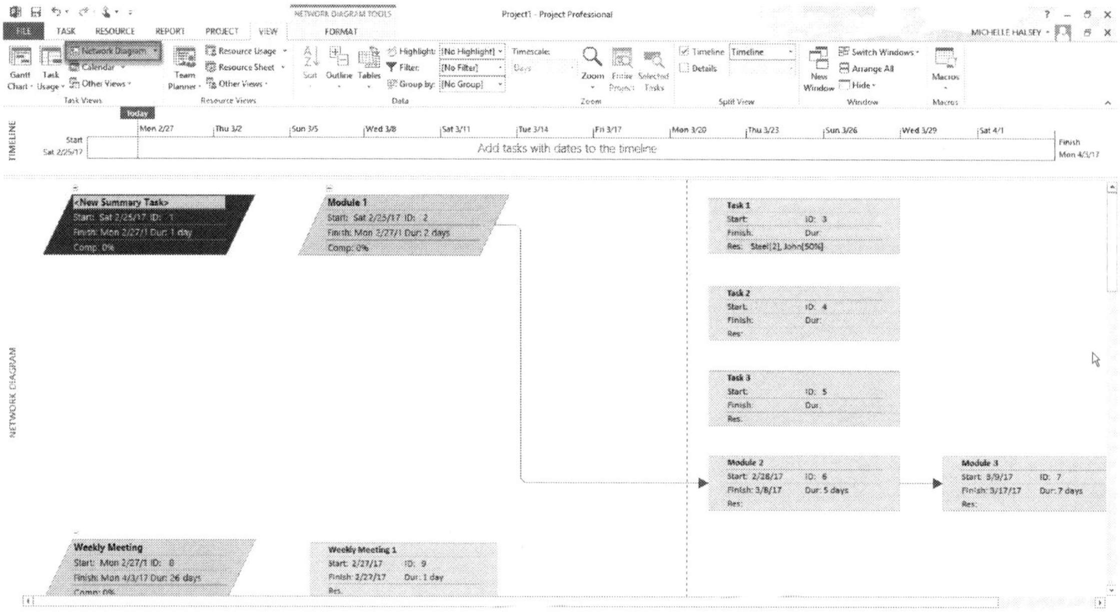

Here is an example of the Calendar View.

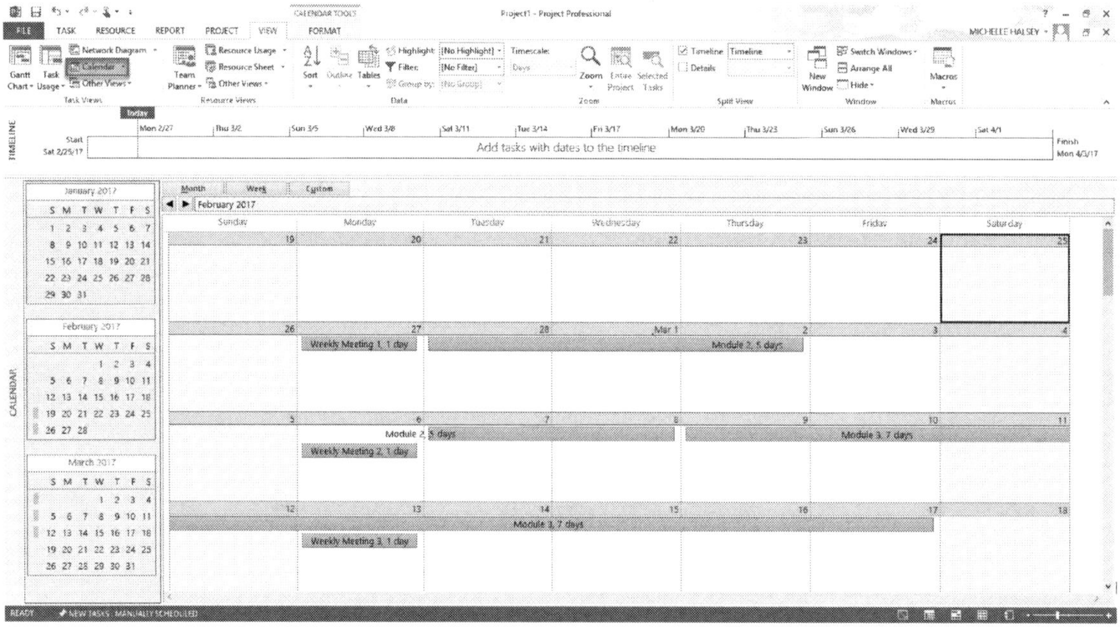

Here is an example of the Task Form view.

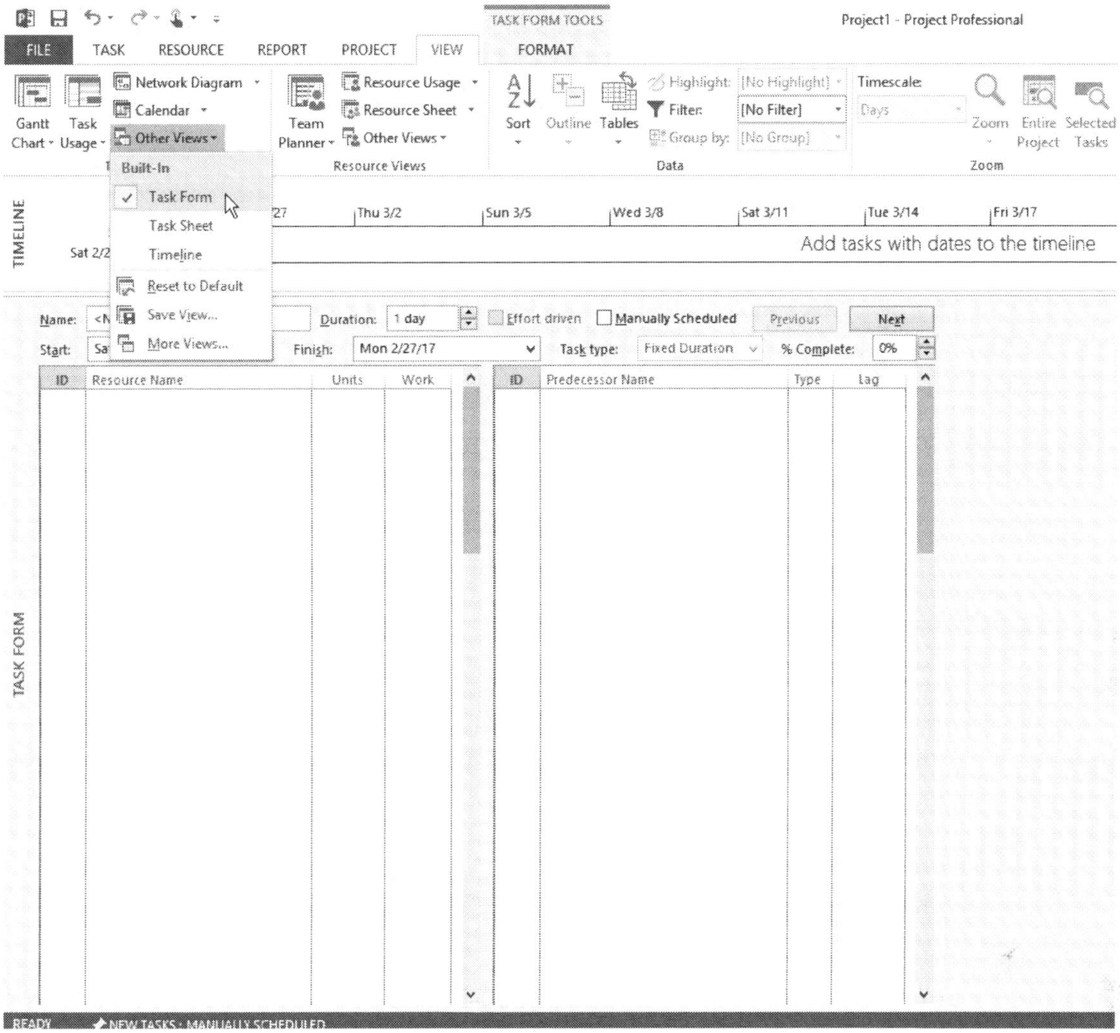

Use the following procedure to see the other task views.

Step 1: Select one of the Task Views tools.

Step 2: Select More Views.

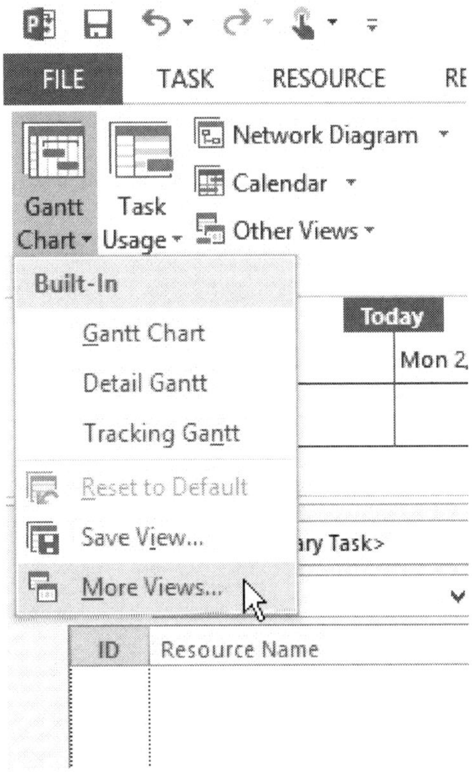

Step 3: Select the view you want to use from the list and select Apply.

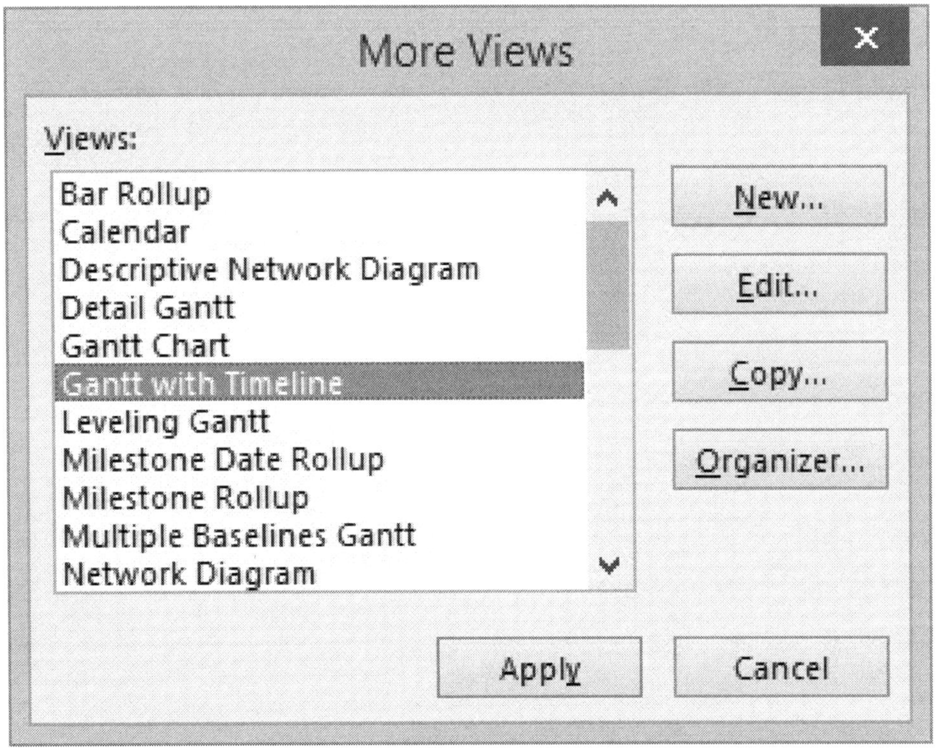

Important Resource Views

The Resource Views area of the View tab on the Ribbon includes tools for the most common resource views. The most important Resource Views are:

- The Team Planner
- The Resource Usage worksheet
- The Resource Sheet
- The Resource Form
- The Resource Graph

Shown here is the different resource views available from the Resource tab.

You have already seen the Team Planner and the Resource Sheet, where you add resources to your project.

The Resource Usage view shows a chart of the work assigned for resources.

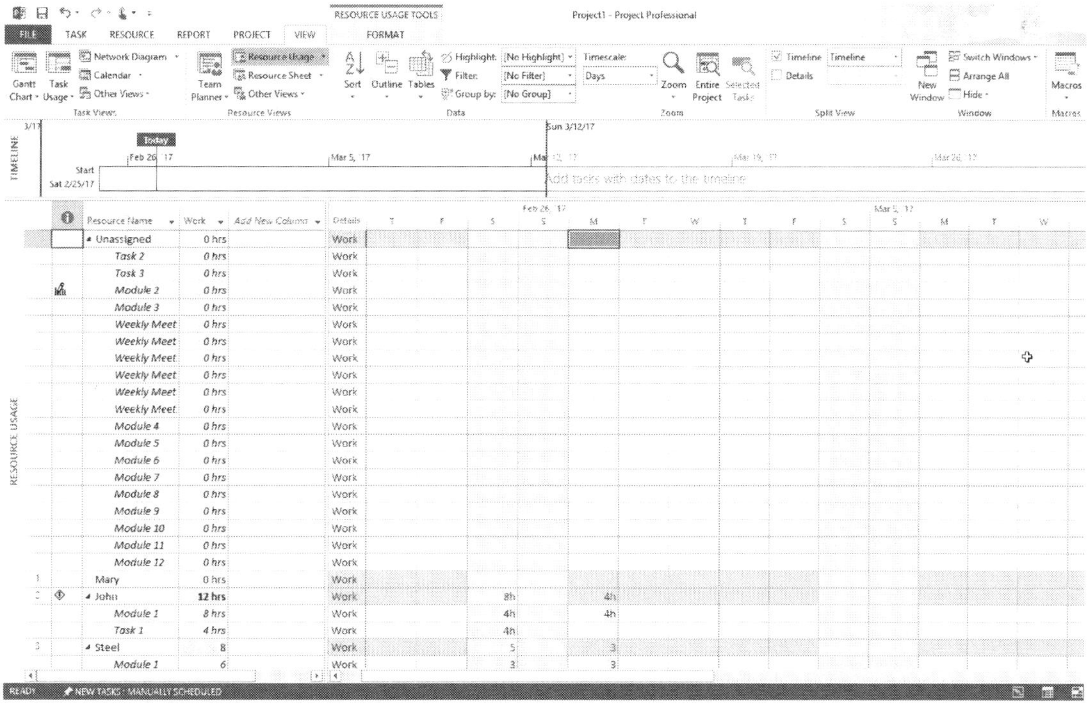

To access the Resource Form, select the Resource Usage drop-down and then select Other Views. Then select Resource Form.

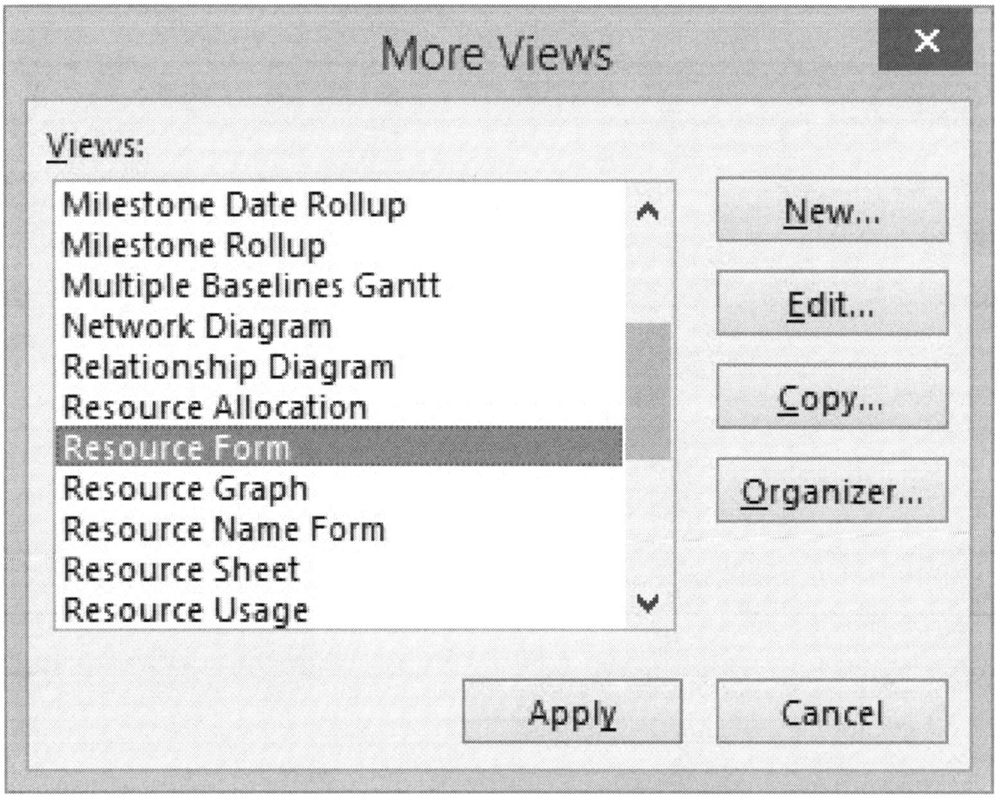

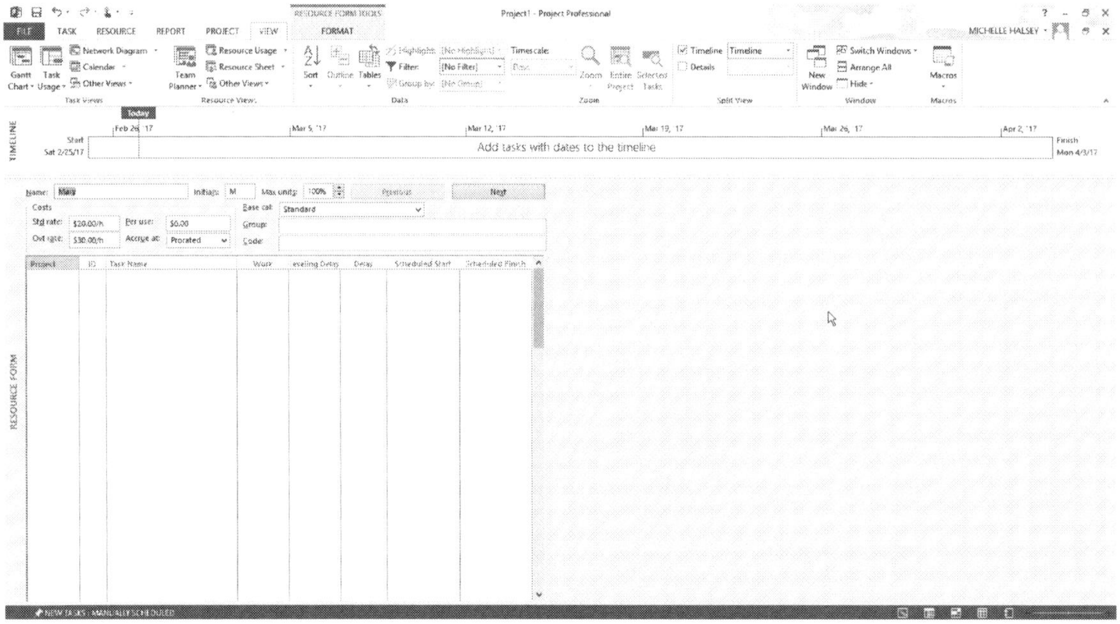

To access the Resource Graph, select Other Views. Then select Resource Graph.

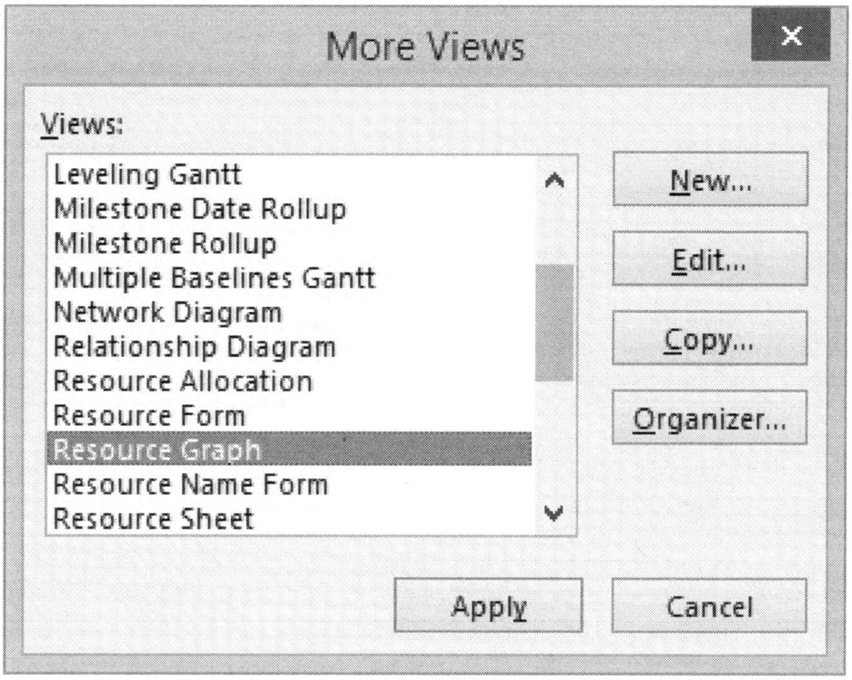

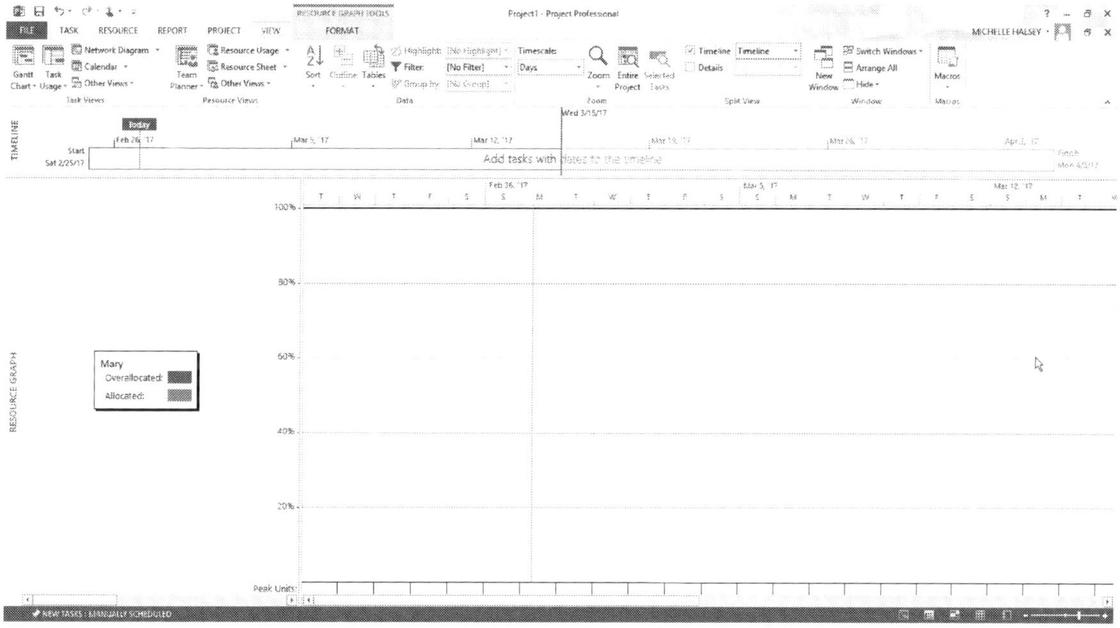

Using the Tools Tabs

Each task and resource view in Project 2016 includes a context-sensitive Tools tab that displays when you use that view. The Tools tabs include formatting tools to help you customize that view.

Use the following procedure to use the Gantt Chart Tools tab.

Step 1: Select the Gantt Chart view.

Step 2: Select the Format tab.

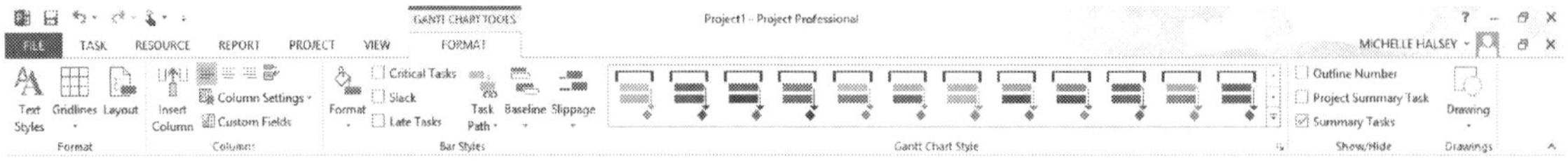

Shown here is the Team Planner Tools tab.

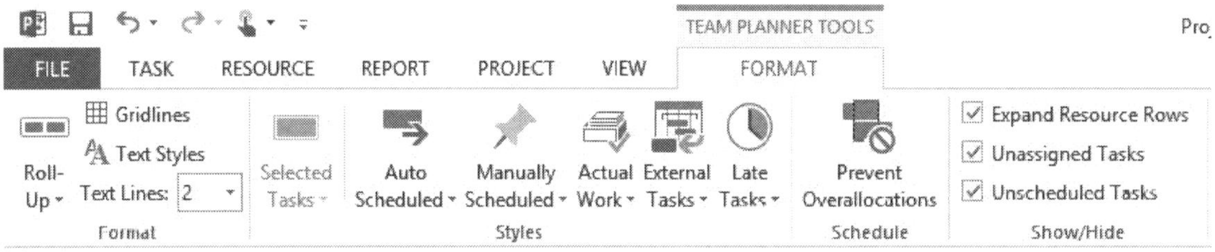

Shown here is the Task Usage Tools and the Resource Usage Tools tabs.

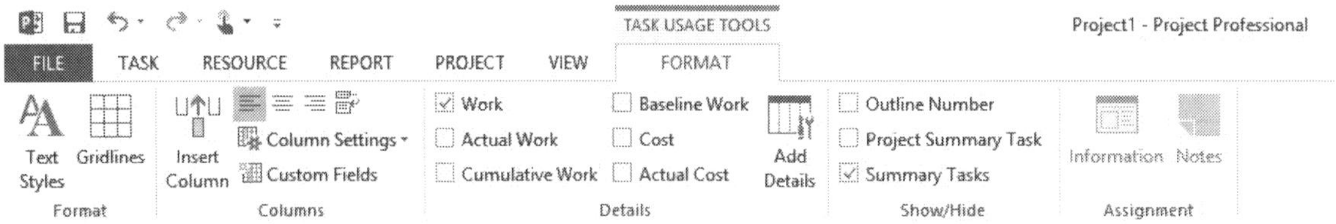

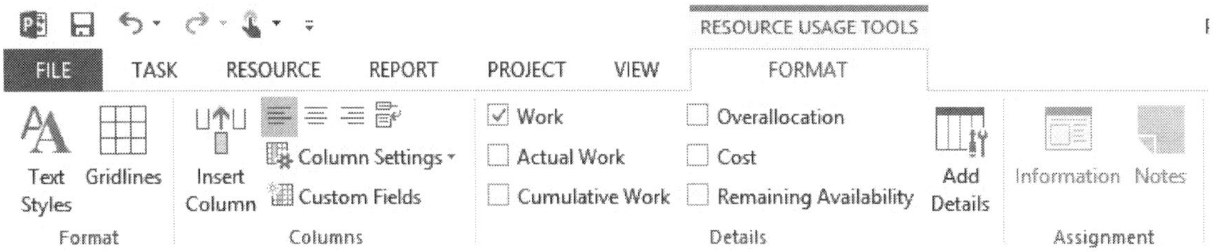

Shown here is the Calendar View Tools tab.

Shown here is the Task Form Tools tab.

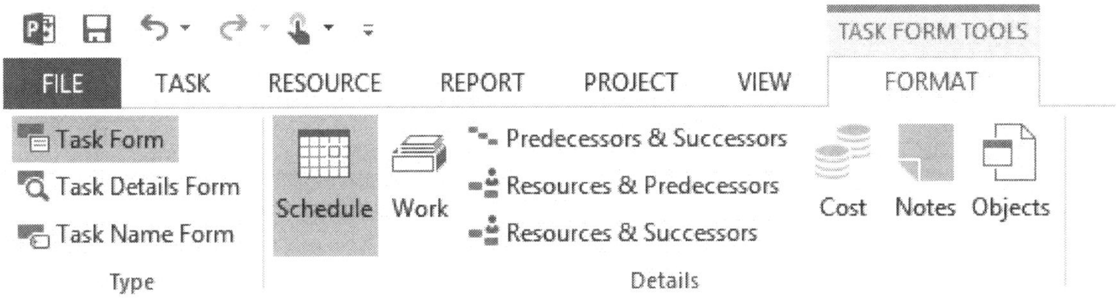

Formatting the Timescale

The timescale shows the tasks in your project on a timeline. The timescale depends on how long your project and individual tasks take. You can view the timescale by different units of time to get a different view of your project. You can scale down to see the details for one task. Or you can zoom out to get a bigger picture view of your entire project.

Use the following procedure to change the timescale zoom.

Step 1: Select the Timescale dropdown list from the View tab on the Ribbon.

Step 2: Select a timescale option.

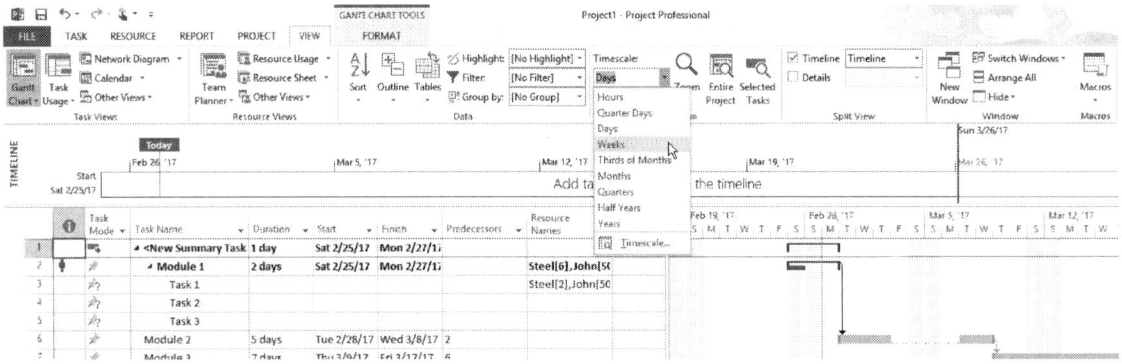

Chapter 7 – Managing Your Project Status

This chapter will help you learn how to manage your project status. First, you will learn how to set a baseline. Then you will learn two different ways for updating tasks. You will also learn how to update the entire project.

Creating a Baseline

A baseline plan allows you to evaluate your project's progress against that baseline later. You can set several different baselines, giving each a different name so that you can manage the baselines.

Use the following procedure to set a baseline for selected tasks.

Step 1: Enter the tasks, durations, and other details of your base project before setting a baseline.

Step 2: To set a baseline for selected tasks, select the tasks you want to track from the Gantt chart view.

Step 3: Select the Project tab from the Ribbon.

Step 4: Select Set Baseline. Select Set Baseline.

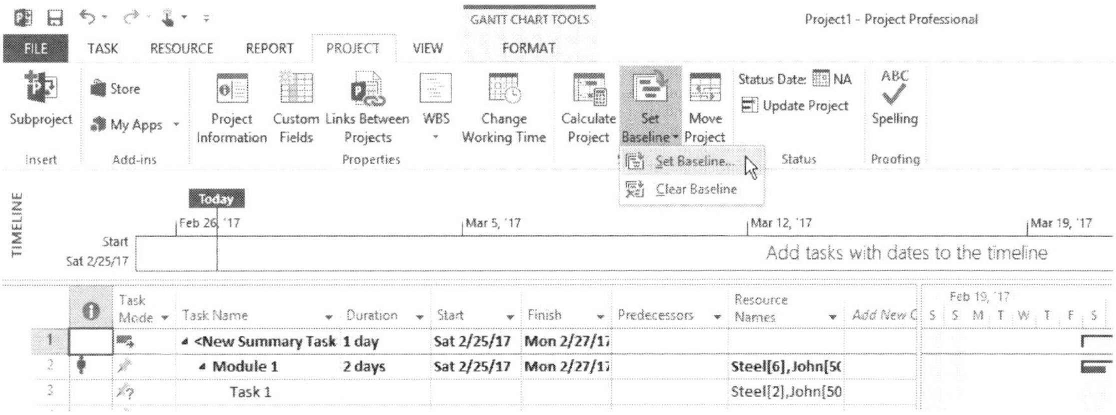

Project displays the Set Baseline dialog box.

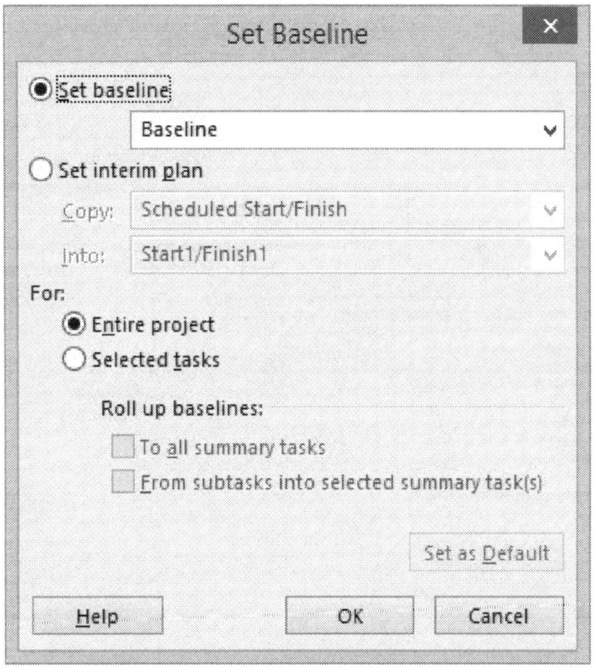

Step 5: From the Set baseline drop-down list, select which baseline you want to set. You have the choice of the default baseline, or up to 10 other baselines, which are numbered 1 – 10.

Step 6: Select Selected tasks.

Step 7: Indicate how to roll up the baselines by checking the to all summary tasks box and/or the from subtasks into selected summary task(s) box.

Step 8: Select Ok.

The Baseline column stores the baseline information. Use the following procedure to add the Baseline column to see the baseline information.

Step 1: Right-click a column in the Gantt Chart view and select Insert Column from the context menu.

Step 2: Select a field to enter from the Field name drop-down list. For each baseline you can set, you can enter the following columns to your table:

- Baseline Budget Costs
- Baseline Budget Work
- Baseline Cost
- Baseline Duration
- Baseline Finish
- Baseline Fixed Cost
- Baseline Fixed Cost Accrual
- Baseline Start
- Baseline Work

Step 3: Select Ok. Repeat to add more baseline information to your current view.

The selected baseline column(s) display the baseline values for the selected task(s). In the following example, the Baseline Estimated Duration column was added.

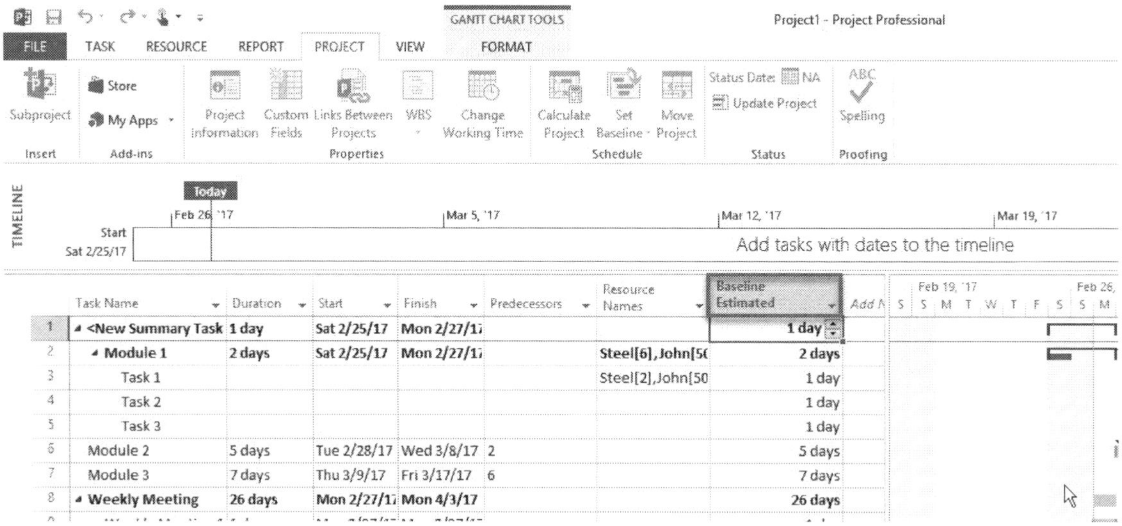

Updating Tasks

To report the status of a project, you must tell Project how much of the tasks have been completed. There are two ways to update tasks. You can use the tools in the Schedule area of the Task tab on the Ribbon. Or you can use the Update Task dialog box. When you have updated a task, the bars on the Gantt Chart include progress bars.

Use the following procedure to update a task using the Schedule tools on the Task tab of the Ribbon.

Step 1: Select the task(s) you want to update. You can hold the Shift key while selecting to select multiple consecutive tasks. You can hold the Ctrl key while selecting to select multiple non-consecutive tasks.

Step 2: Select the desired Percentage Complete tool from the Task tab on the Ribbon.

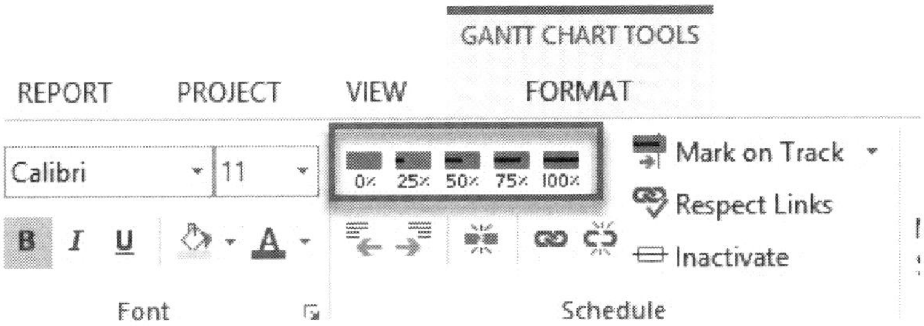

Use the following procedure to update a task using the Update Tasks dialog box.

Step 1: Select the task(s) you want to update. You can hold the Shift key while selecting to select multiple consecutive tasks. You can hold the Ctrl key while selecting to select multiple non-consecutive tasks.

Step 2: Select the Mark on Track tool from the Task tab on the Ribbon.

Step 3: Select Update Tasks.

In this example, multiple tasks have been selected.

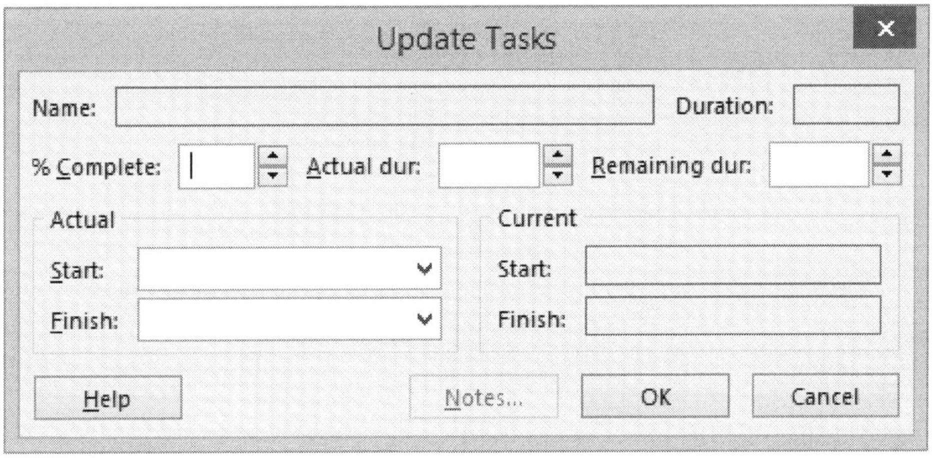

Step 4: Enter (or use the up and down arrows) the % Complete for the selected task(s).

Step 5: Select Ok.

Here is an example where only one task was selected.

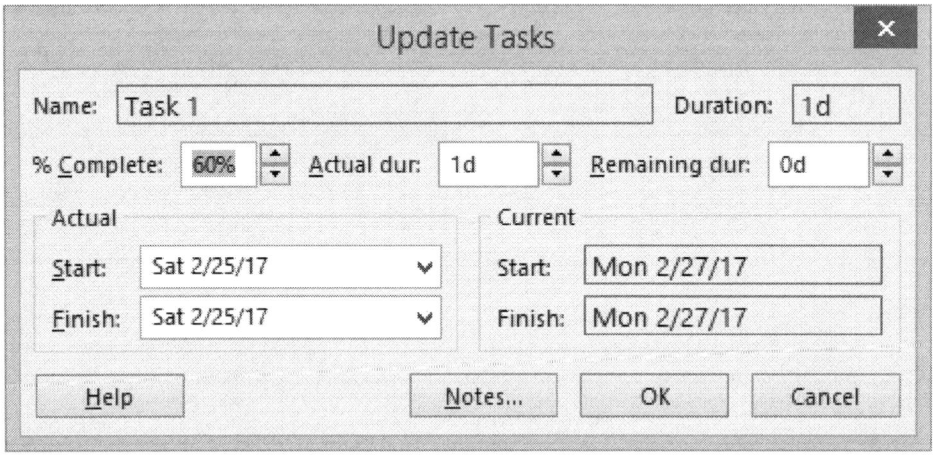

Notice the progress lines on the Gantt chart.

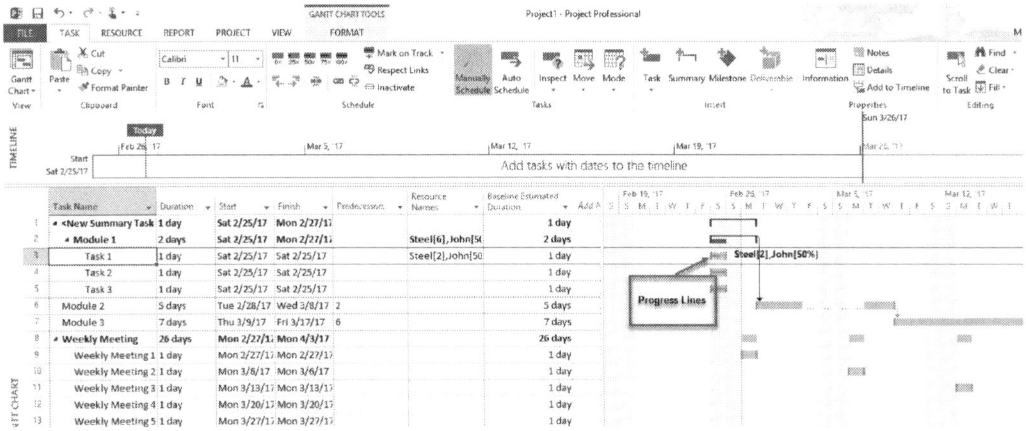

Updating the Project

The Update Project dialog box allows you to update work as complete through a selected date or to reschedule any uncompleted work to start after a selected date.

Use the following procedure to update the project. Select any tasks, if applicable, to update selected tasks.

Step 1: Select the Project tab from the Ribbon.

Step 2: Select Update Project.

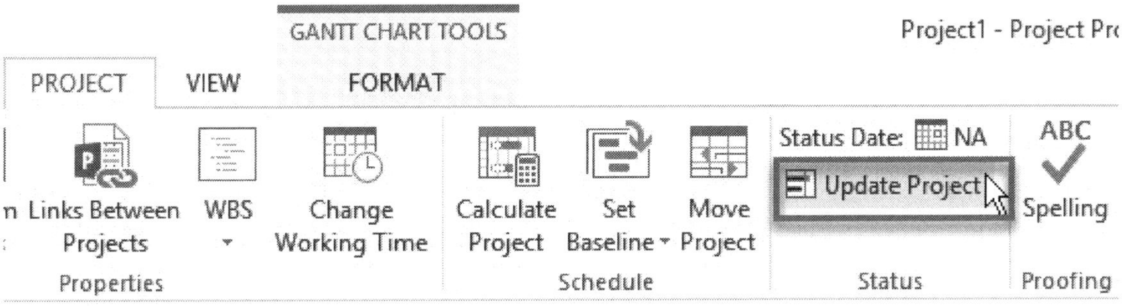

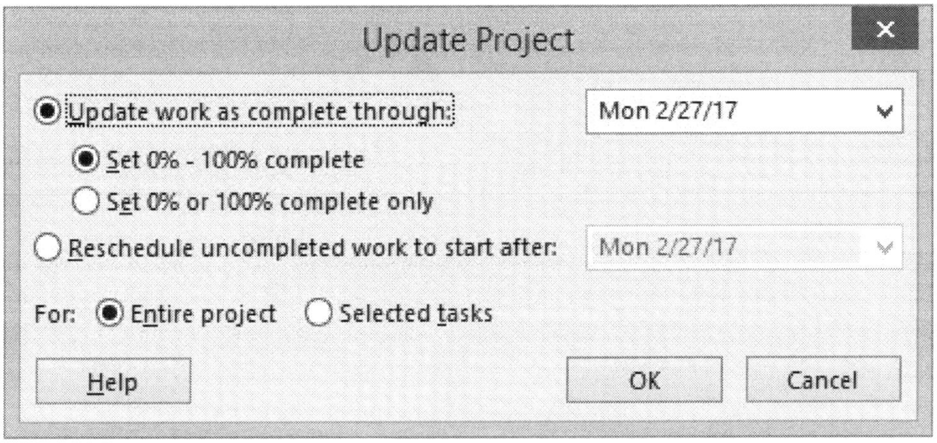

Step 3: Select the desired options and choose the date.

Step 4: Select Ok.

About the Project Status Date

The Project Status date allows you to view the project status and earned valued calculations. The Status Date defaults to the current date, but you can change the date, such as if you want to create reports on Monday for the status at the end of the day Friday.

Use the following procedure to change the status date.

Step 1: Select the Project tab from the Ribbon.

Step 2: Select Status Date.

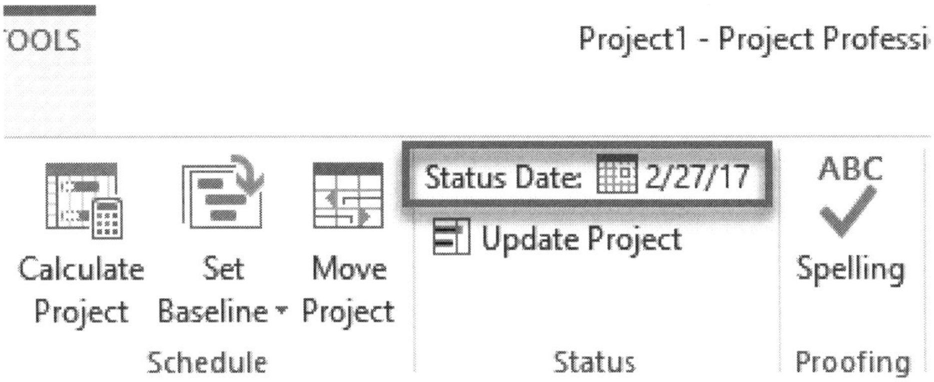

Step 3: Select a new date.

Step 4: Select Ok.

Chapter 8 – Updating and Tracking Your Progress

An important aspect of project management is checking on the project's progress and adjusting in your plan where necessary. This chapter will explain first how to view the critical path. It will explain the use of change highlighting in Project 2016. You will also learn how to use the Task Inspector Pane.

Viewing the Critical Path

The critical path includes tasks that drive the completion date of your project. When there are many tasks in a project that includes task dependencies and other linked tasks, it may be difficult to determine which are the most critical tasks. Viewing the critical path can help the project management team make decisions about adding more resources to a project or changing the tasks in different ways to improve the critical path of the project.

Use the following procedure to view the critical path.

Step 1: Select the Gantt Chart Format Tools tab.

Step 2: Check the Critical Path box.

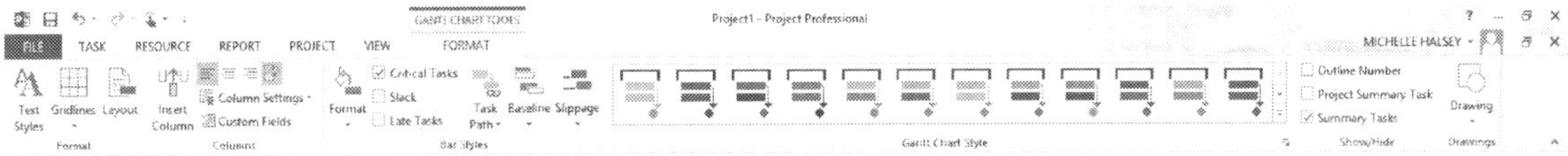

The tasks in the critical path are displayed in red.

Use the following procedure to filter the task list to only include items in the critical path.

Step 1: Select the View tab from the Ribbon.

Step 2: Select the drop-down arrow next to the Filter icon.

Step 3: Select Critical to show the Critical Path.

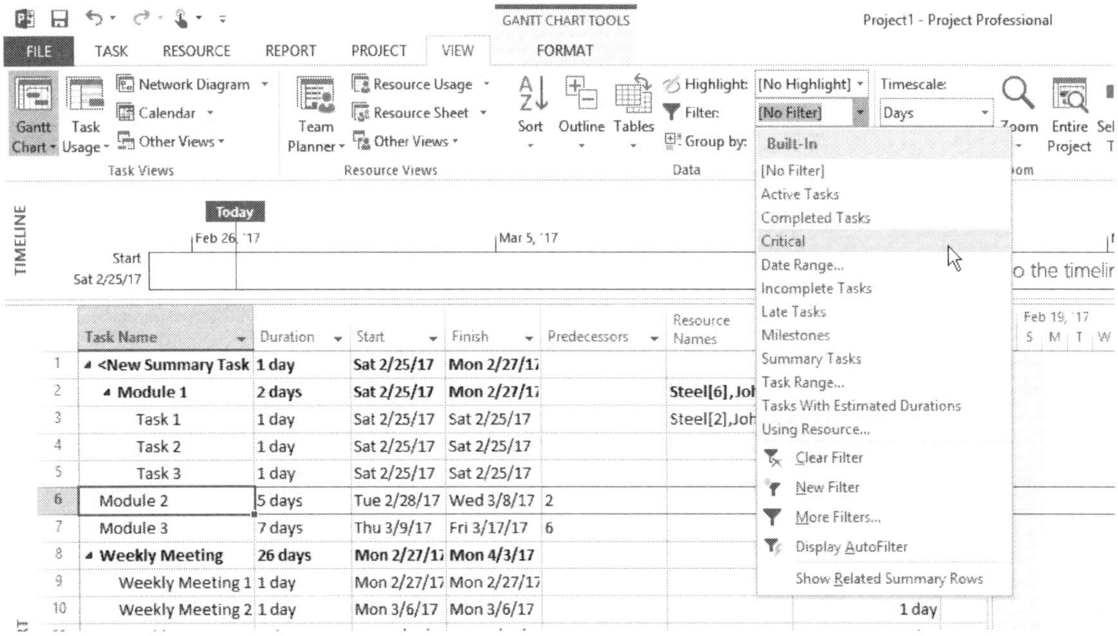

Project only displays the tasks that are part of the critical path.

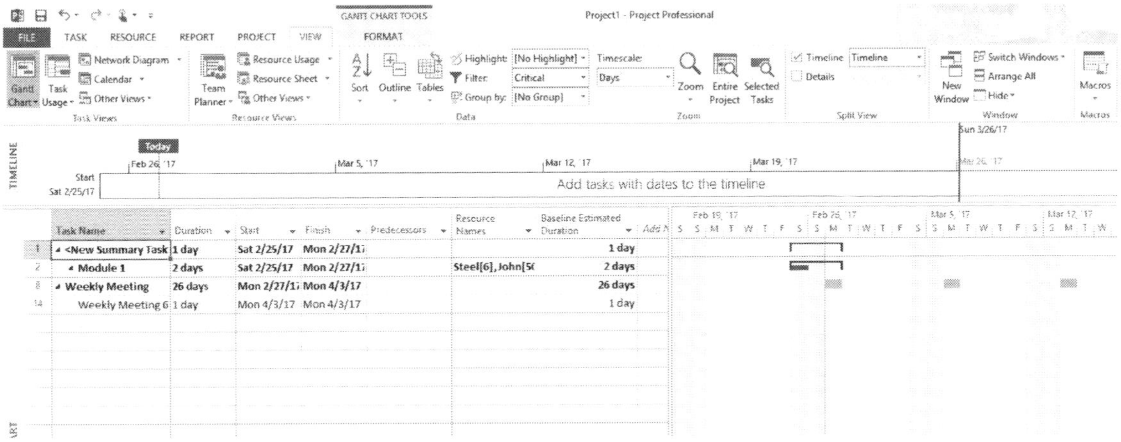

Using Change Highlighting

During your project, you will inevitably need to make changes. Change highlighting can help you see how making changes to items affects the rest of your project.

Use the following procedure to view changed cells.

Step 1: Make a change in the sample file. In this example, the duration for one task was changed from 2 days to 1 day. Press Enter after making the change.

Step 2: Review the highlighted cells to see how the project has changed based on that change.

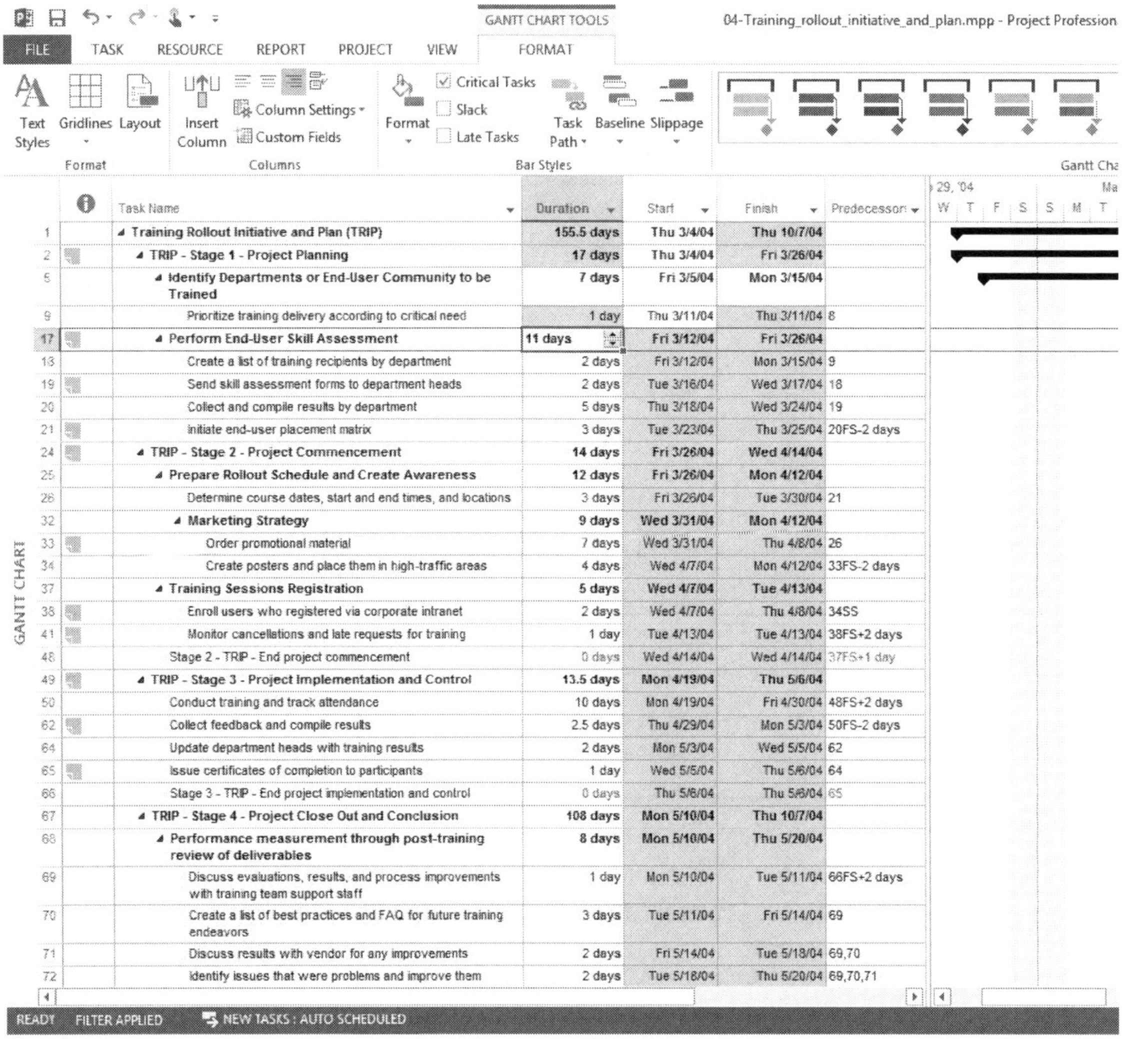

Use the following procedure to change the highlighting color.

Step 1: Select the Format tab from the Ribbon.

Step 2: Select Text Styles.

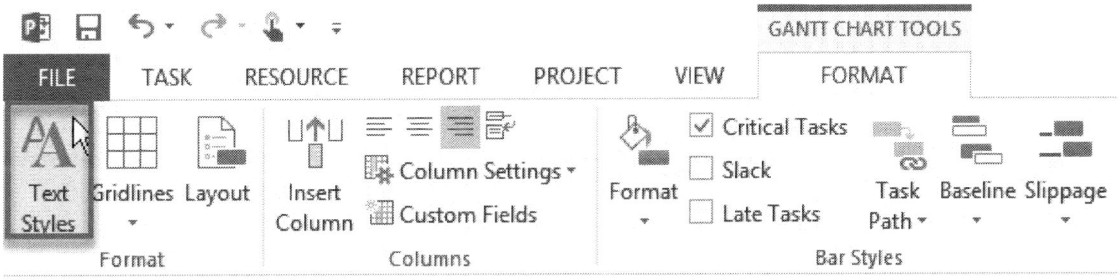

Step 3: Select Changed Cells from the Item to Change drop-down list.

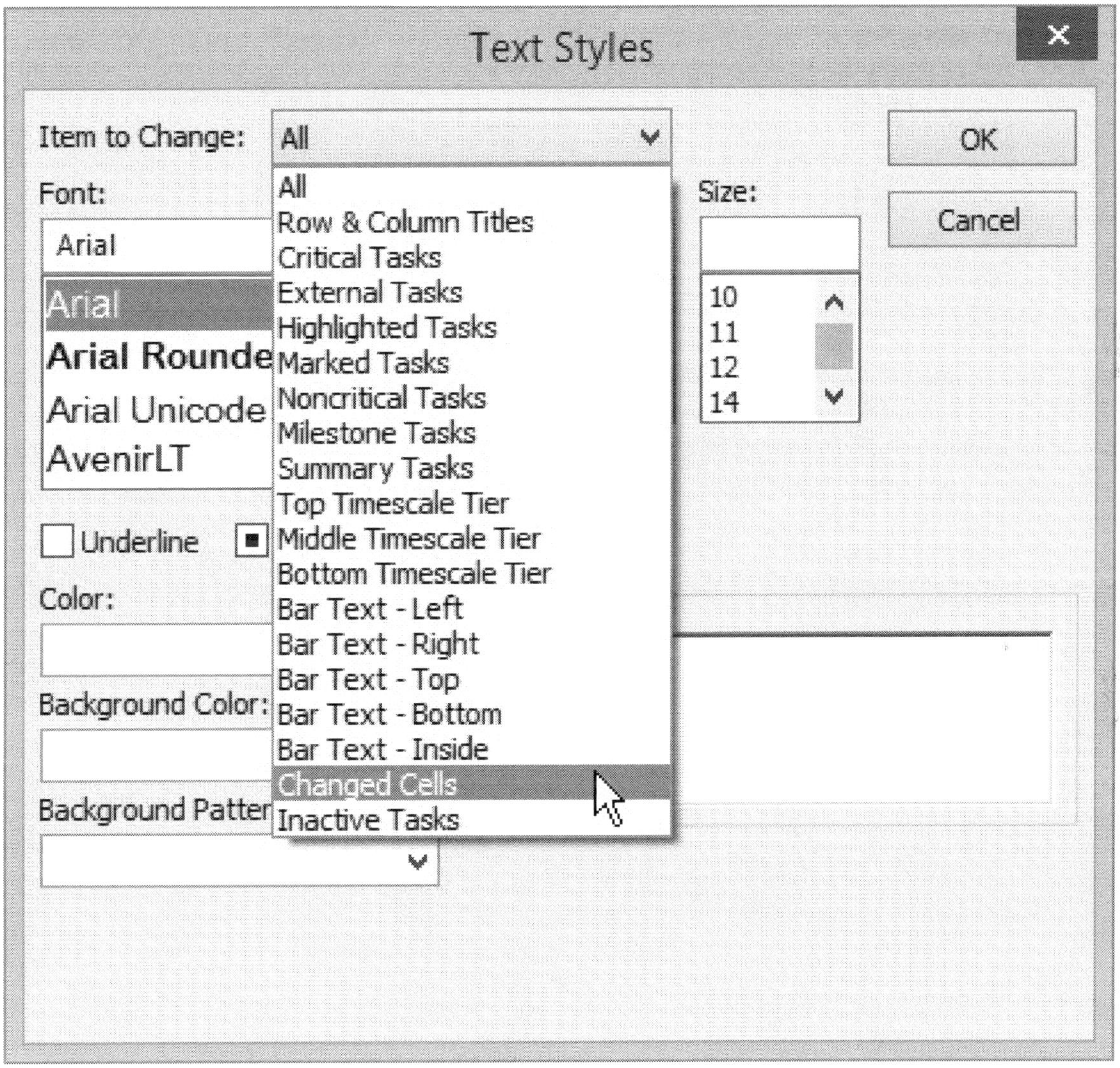

Step 4: Select the desired formatting.

Step 5: Select Ok.

Using the Task Inspector Pane

The Task Inspector Pane can help you track the timing and sequence of tasks. It can also show you how changes to one task affect the rest of the project.

Use the following procedure to display the Task Inspector Pane.

Step 1: Select the Task tab from the Ribbon.

Step 2: Select Inspect.

Project displays the Task Inspector pane on the left side of the screen. In this example, Project makes recommendations for improving the schedule. Click on the button to make the recommended change.

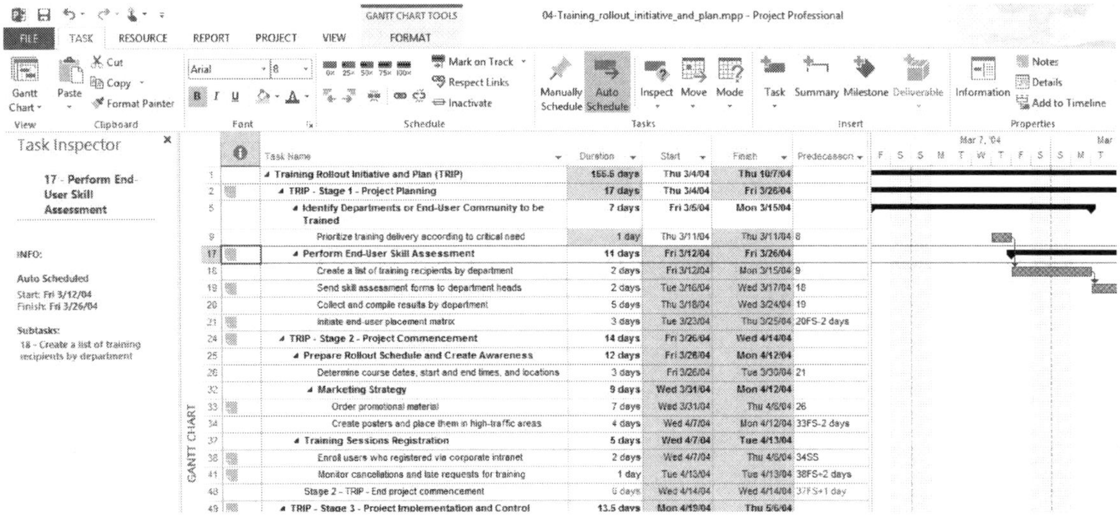

Chapter 9 – Creating Reports

In this chapter, you will learn how to work with reports. Project includes many different default reports, as well as visual reports that work with either Excel or Visio. This chapter will explain how to create basic and visual reports. You will also learn how to compare different versions of a project to see what has changed.

Creating Basic Reports

Project 2016 includes a few basic reports. When you generate a report, you are taken to the Backstage View to preview or print the report.

Use the following procedure to open a report.

Step 1: Select the Report tab from the Ribbon.

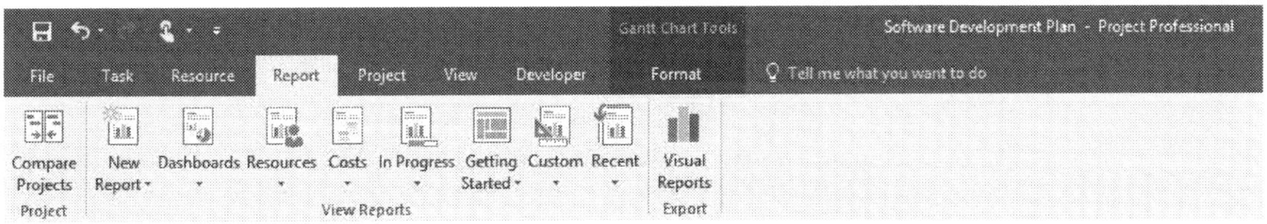

Step 2: Select Custom, More Reports, and then Dashboard.

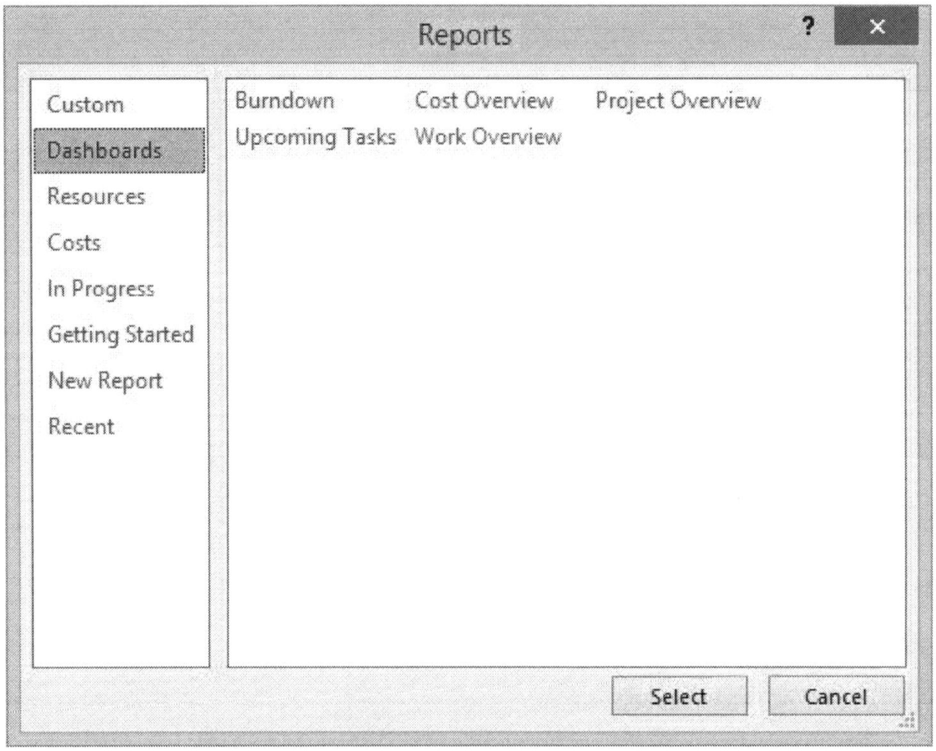

Step 3: Select Project Overview and choose the Select button.

Step 4: Some of the reports require a beginning date range. You would select it from the drop-down calendar and select Ok. Then you would select the end date range from the drop-down calendar and select Ok.

Project displays the report preview on the right side of the Backstage View. You can use the zoom tool (in the bottom right hand corner) to change the view. You can also print the report.

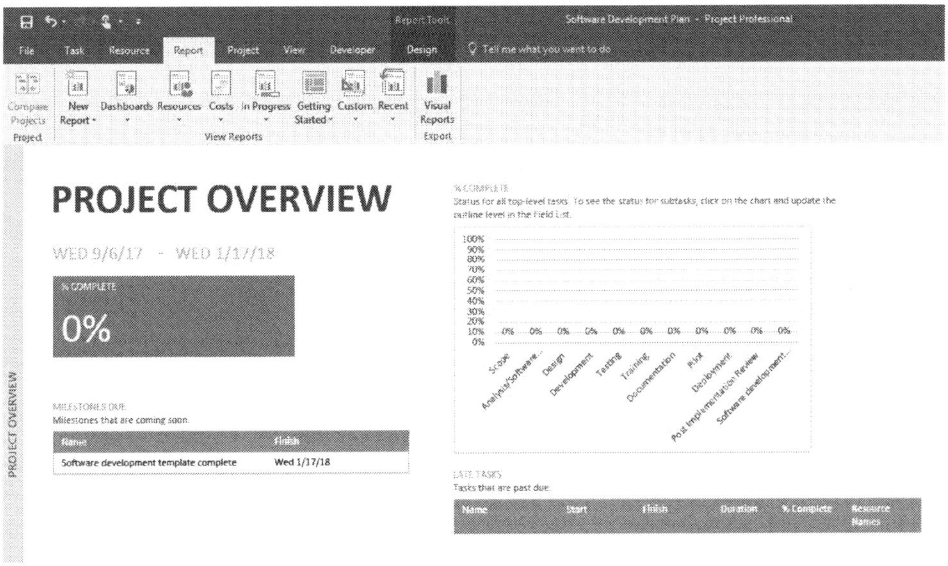

Creating a Visual Report

Project 2016 includes a few visual reports. Visual reports work with either Excel or Visio, depending on the selected report.

Use the following procedure to open a visual report.

Step 1: Select the Project tab from the Ribbon.

Step 2: Select Visual Reports.

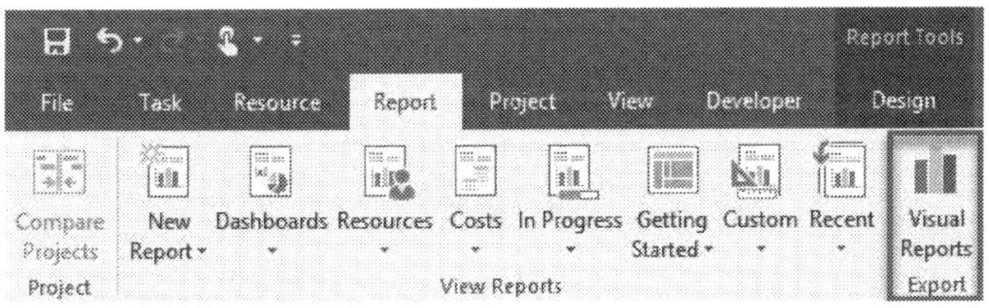

Step 3: For this example, select Resource Work Summary Report.

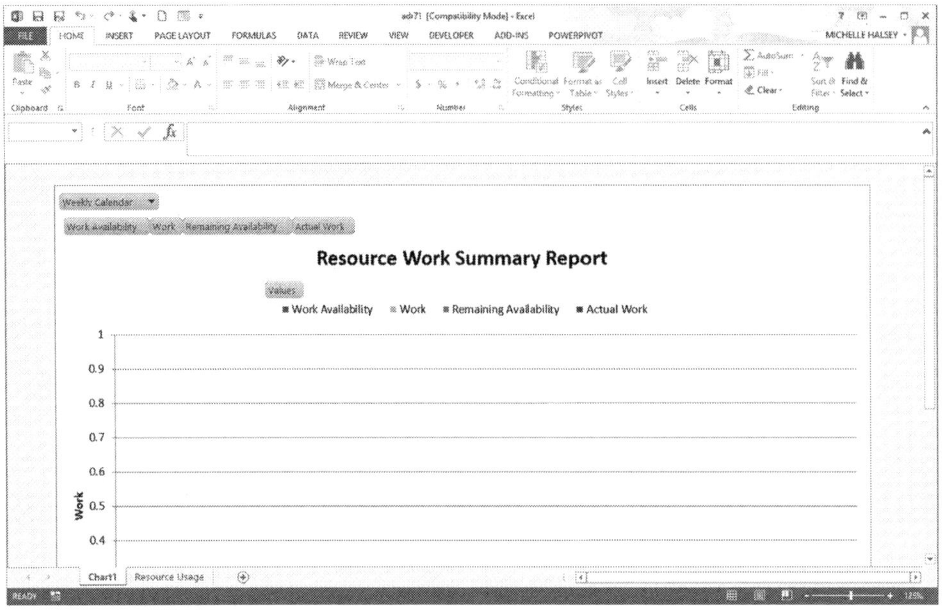

Step 4: Select View.

Visual Reports has the following tabs:

All

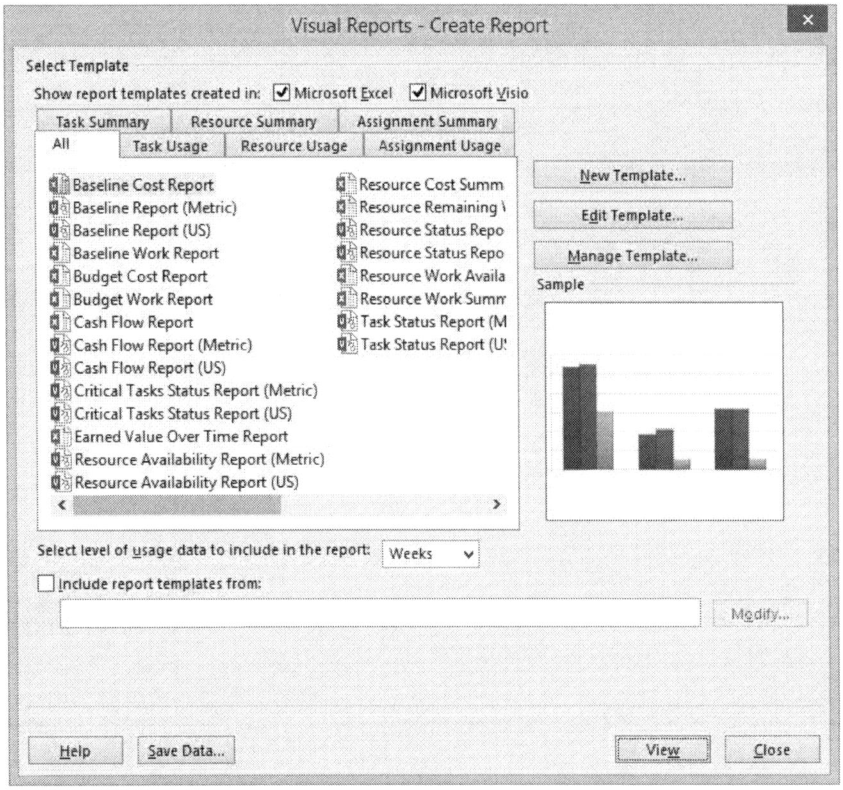

Task Usage

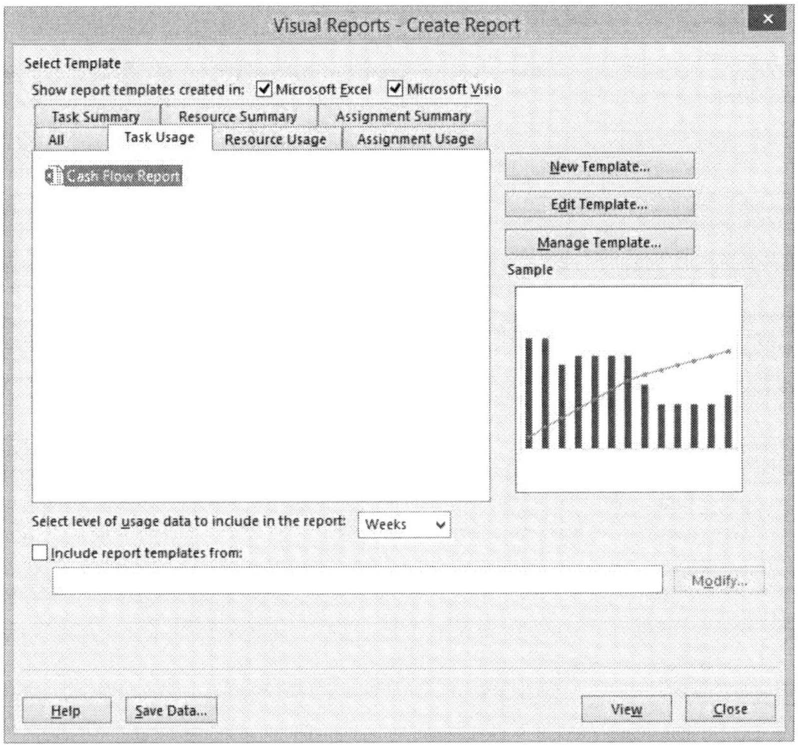

Resource Usage

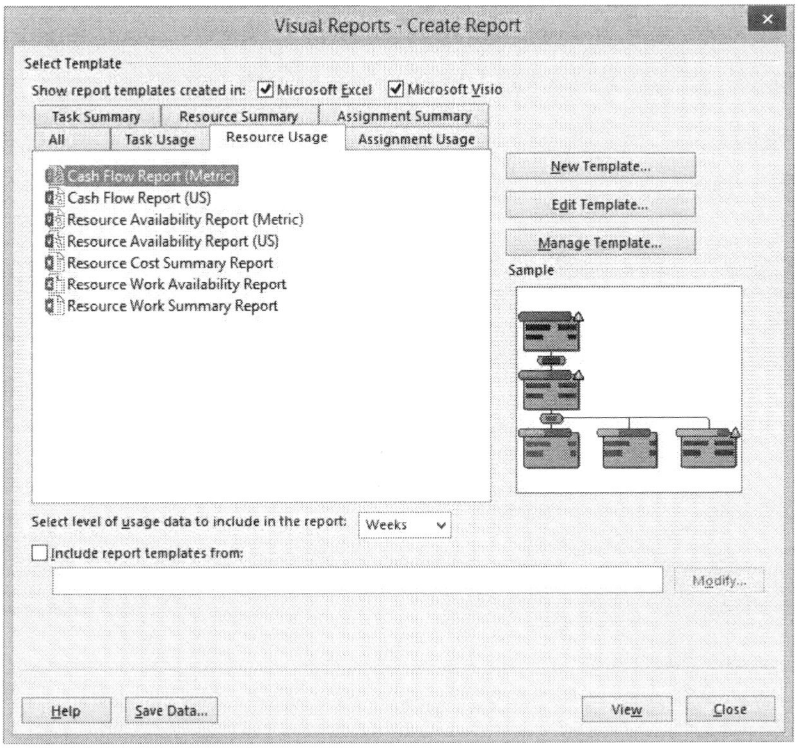

Assignment Usage

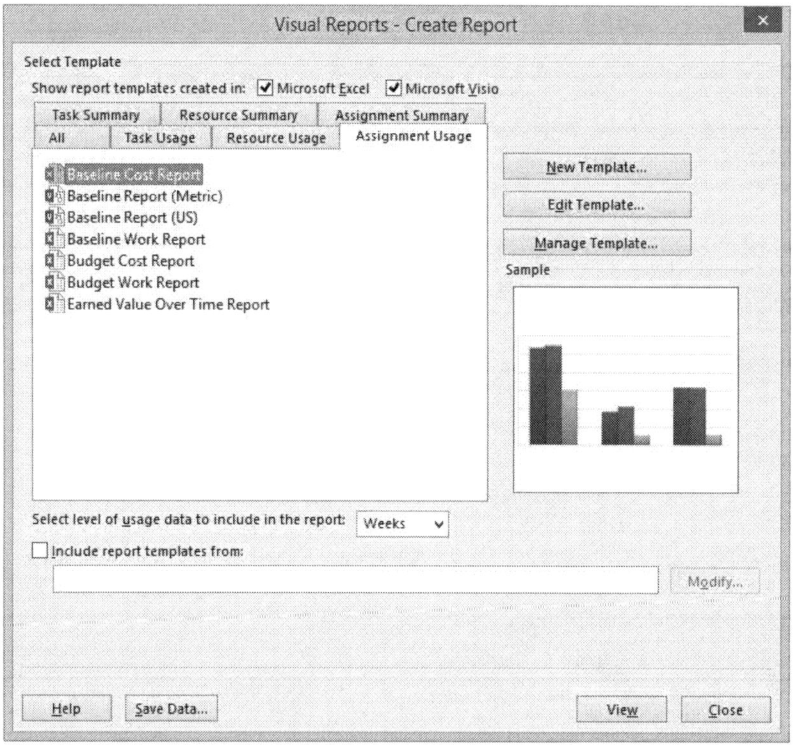

Task Summary

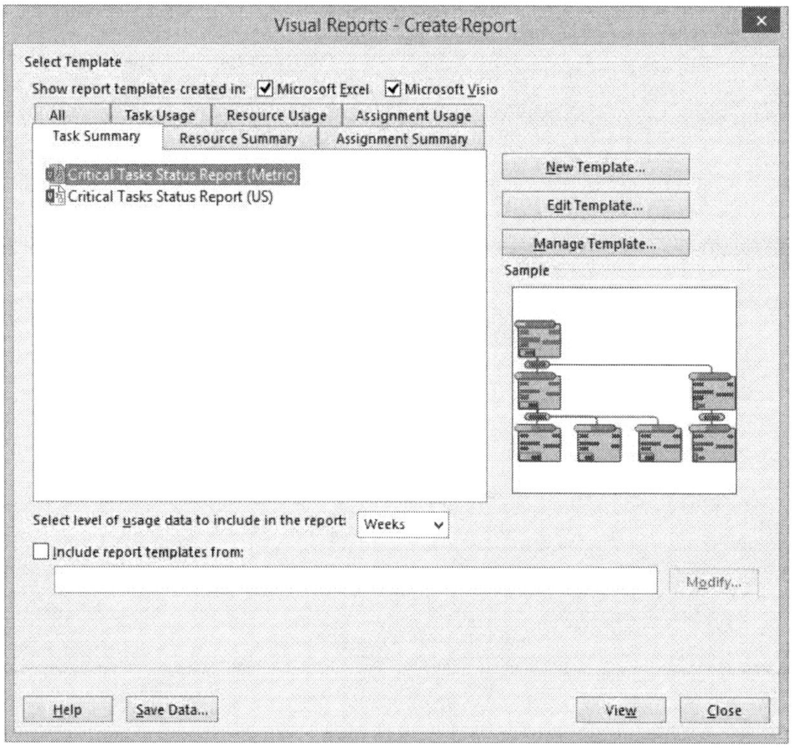

Resource Summary

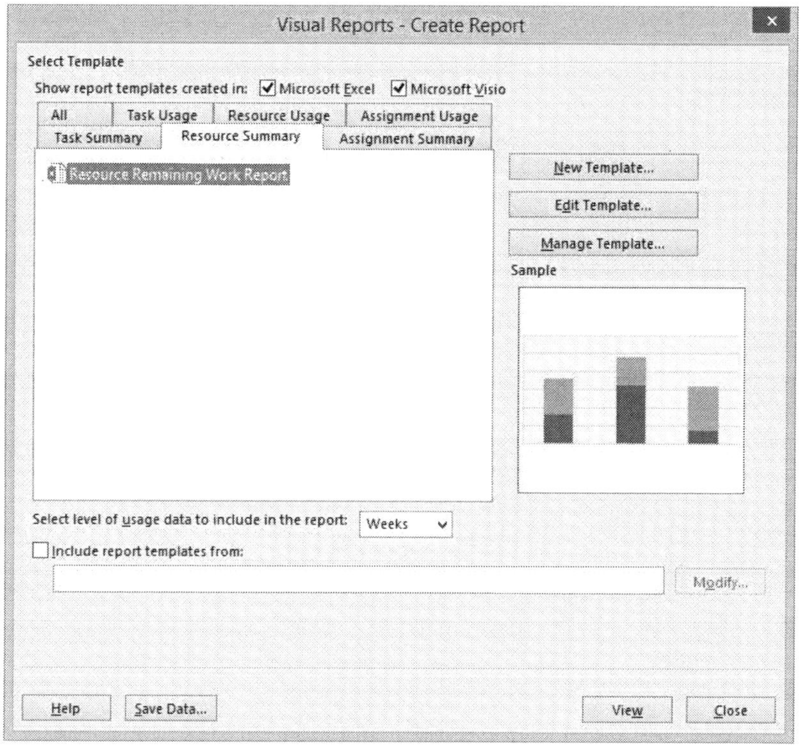

Assignment Summary

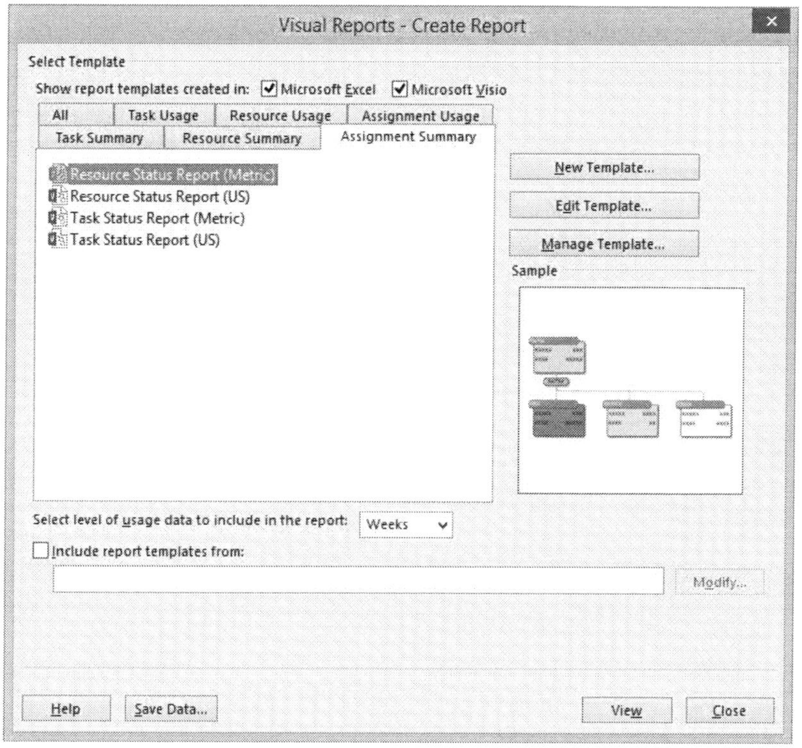

In this example, Project generates the selected report and displays the results as a Microsoft Excel Pivot Chart report.

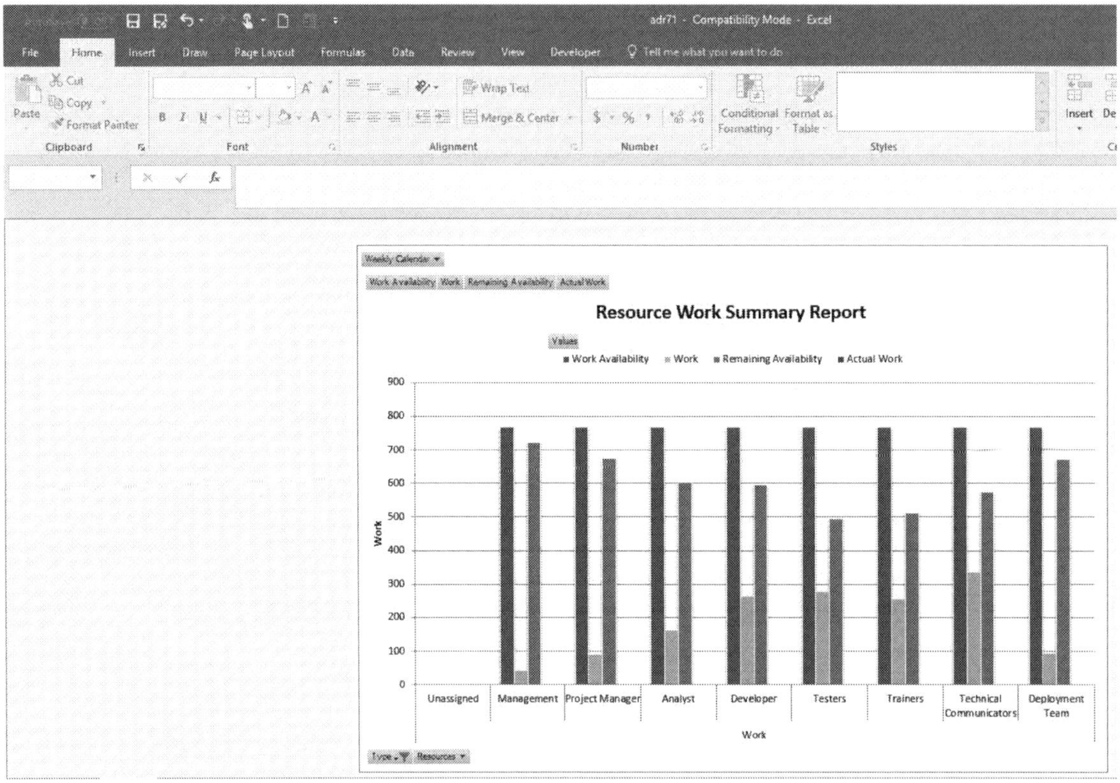

Comparing Projects

Use the following procedure to compare different versions of a project.

Step 1: Select the Report tab from the Ribbon.

Step 2: Select Compare Projects.

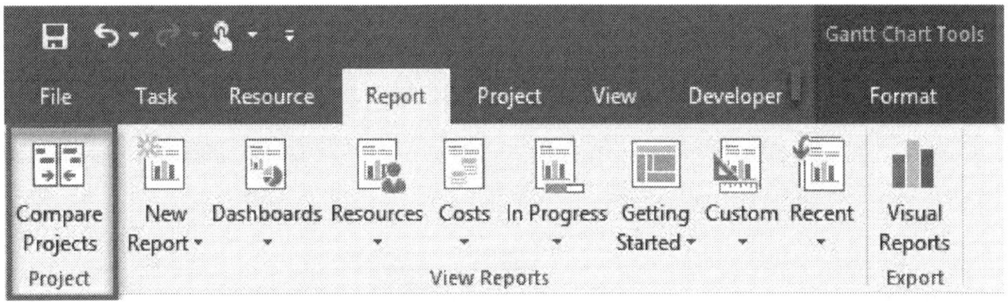

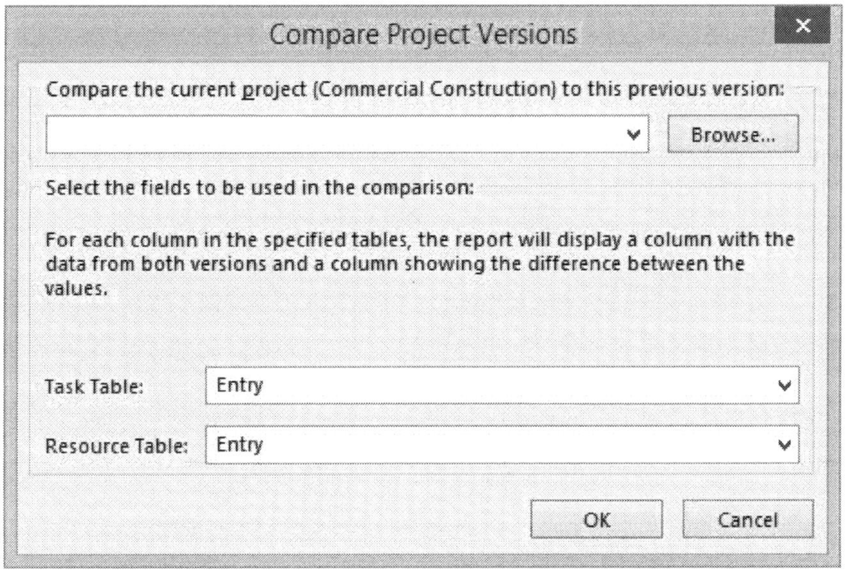

Step 3: If the file you want to use for comparison is open, select it from the drop-down list. Otherwise, select Browse and locate the file on your computer.

Step 4: Select Ok.

Project displays a new project, which is the comparison report. It also displays both projects at the bottom of the screen. The left side of the Comparison Report includes information about differences between the projects.

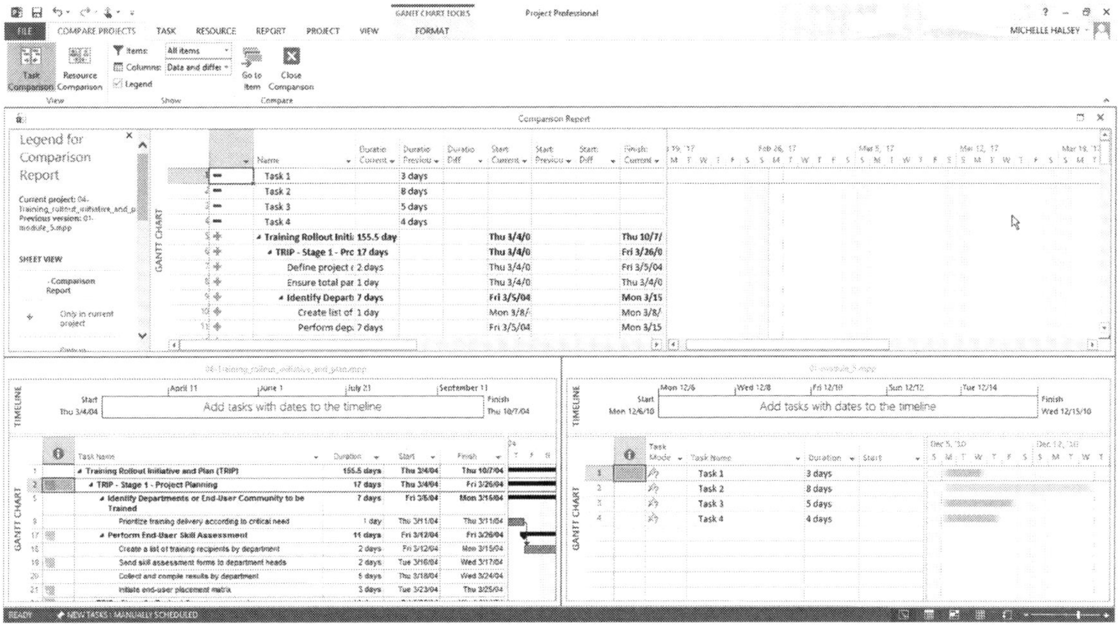

Chapter 10 – Adding the Finishing Touches

You can share your project with others in a few different ways. This chapter will help you prepare your project for other eyes by checking the spelling. You will learn how to use the Page Setup dialog box. Finally, you will learn how to print, e-mail, and create a PDF of your project to share with others.

Checking Your Spelling

It is important to check your spelling in your project so that you communicate clearly with other people who will read the project plan and schedule. The Project 2016 spell checker will find possible errors, suggest corrections, and show where the error is located.

Use the following procedure to check spelling.

Step 1: Select the Project tab from the Ribbon.

Step 2: Select Spelling.

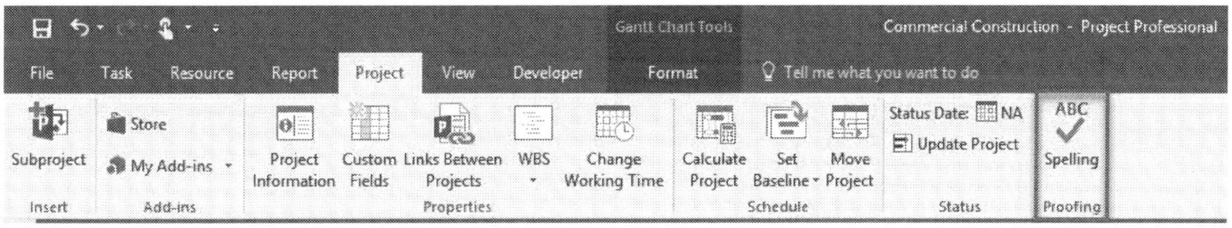

Step 3: Project displays the Spelling dialog box with the first possible spelling error displayed.

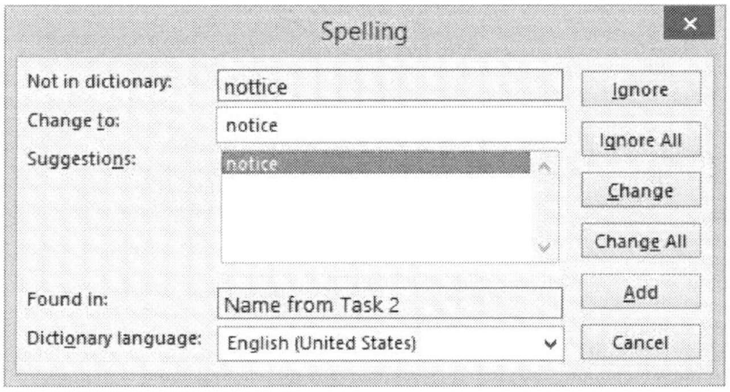

The buttons on the Spelling dialog box.

- The Ignore button allows you to keep the word as the current spelling, but only for the current location.
- The Ignore All button allows you to ignore the misspelling for the whole project.
- The Change button allows you to change the misspelled word to the highlighted choice in the Suggestions area. You can highlight any word in the Suggestions area and select Change.
- The Change All button allows you to notify Project to make this spelling correction any time it encounters this spelling error in this project.
- The Add button allows you to add the word to your dictionary for all Project files.
- The Suggestions area lists possible changes for the misspelling. There may be many choices, just one, or no choices, based on Project's ability to match the error to other possibilities.
- The Found in area displays the location of the spelling error.
- The Dictionary language allows you to select which language to use for the spell check.

Using the Page Setup Dialog

The Page Setup dialog box has several tabs, depending on which type of view you are working with in Project.

Use the following procedure to open the Page Setup dialog box.

Step 1: Select the File tab from the Ribbon to open the Backstage View.

Step 2: Select Print.

Step 3: Select the Page Setup link.

Here is the Page Setup dialog box tab for the default view – Gantt Chart. The Page tab includes the orientation, an option to scale, the paper size, and options for printing.

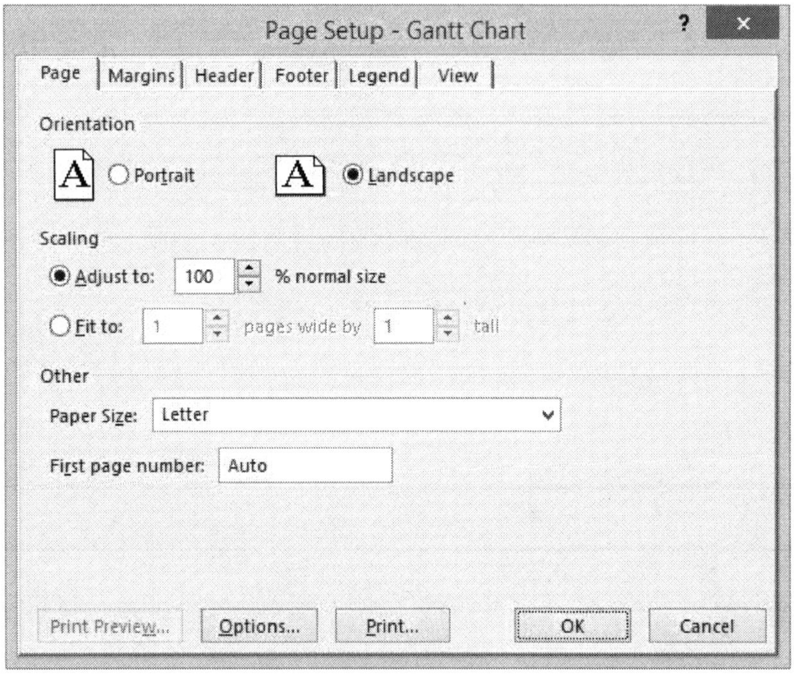

The Margins tab includes options for setting the page margins.

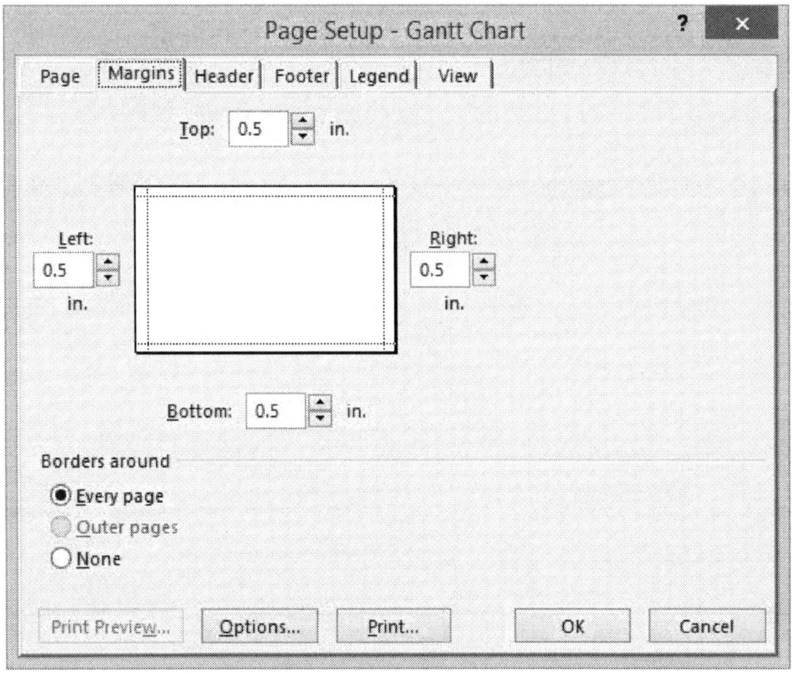

The Header and Footer tabs allows you to create headers and footers on the document.

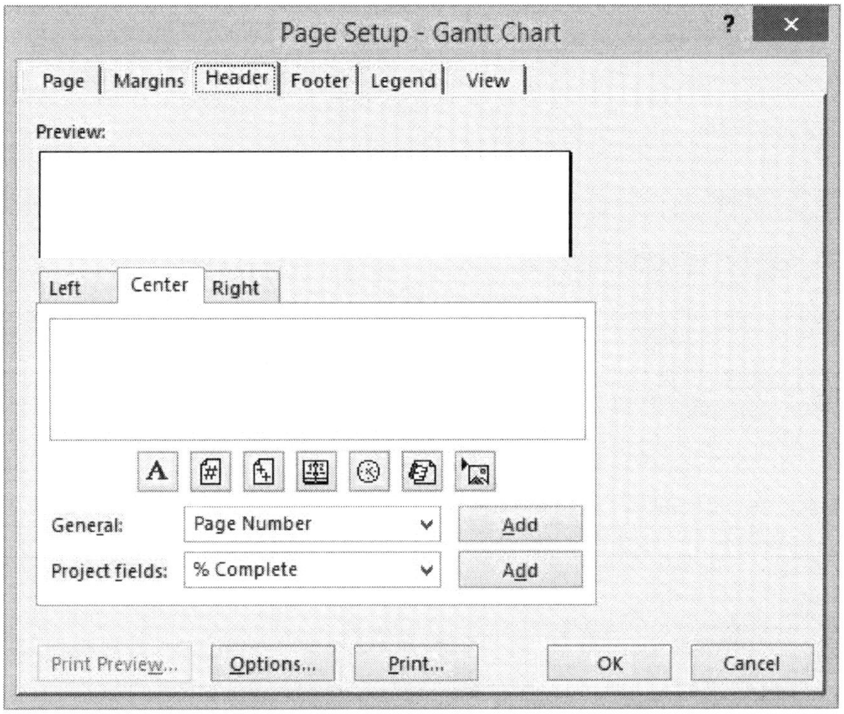

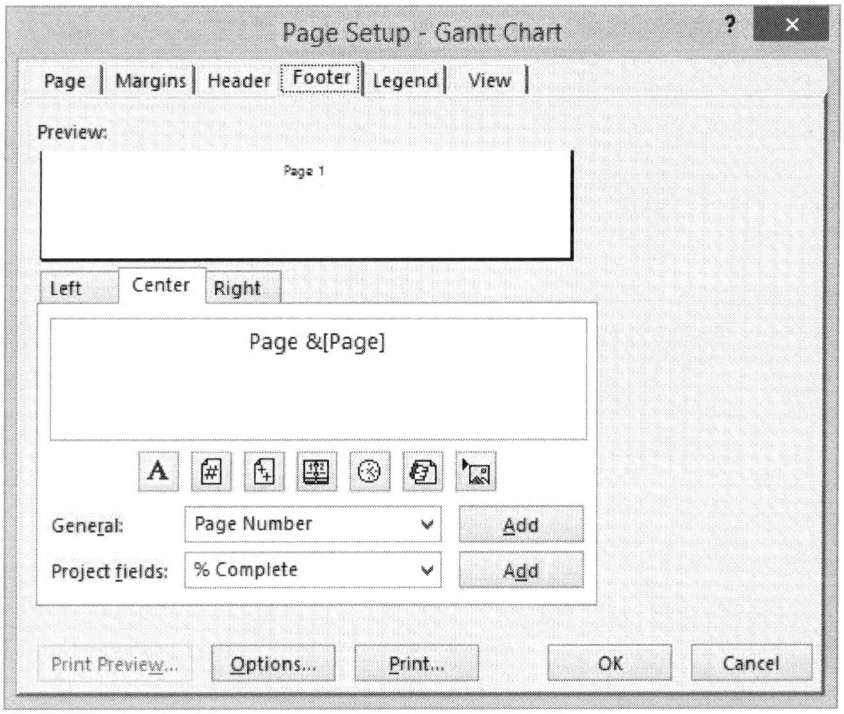

The Legend tab allows you to include Legend information for the project.

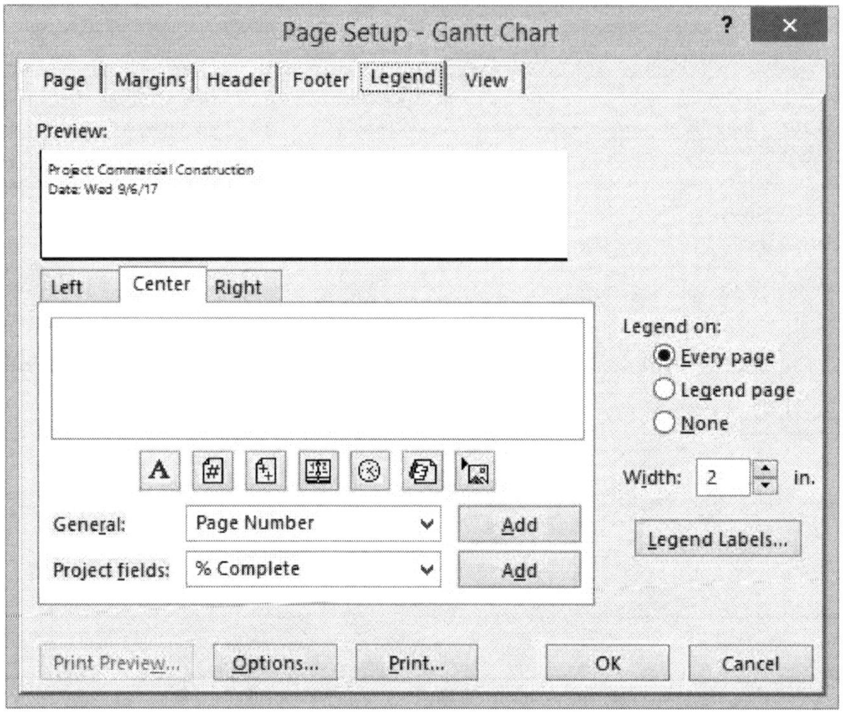

The View tab allows you to include additional information in the page setup.

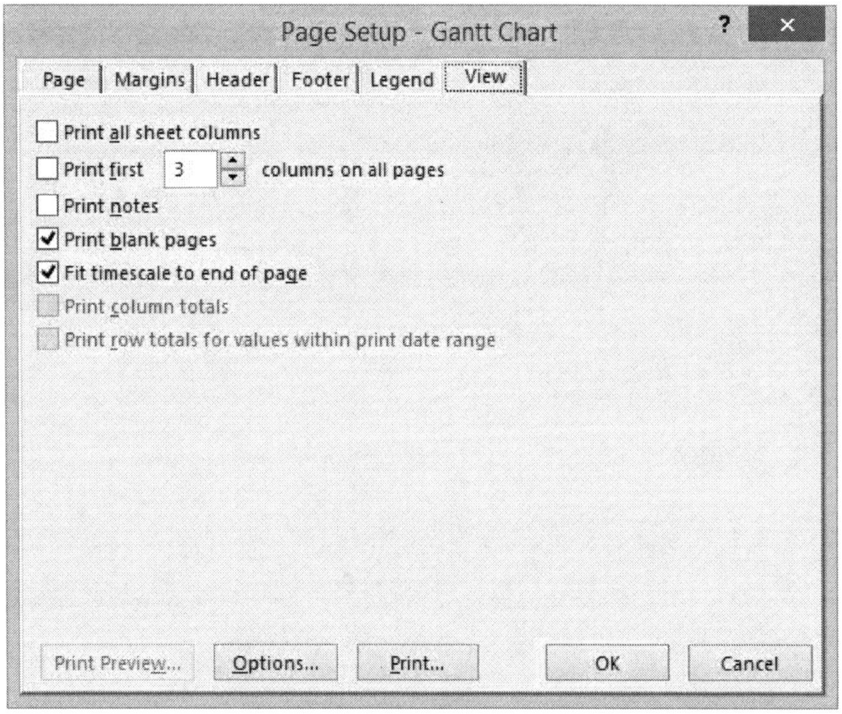

Printing a Project

The new Project 2016 Print tab in the Backstage View allows you to easily set your printing options and print your project.

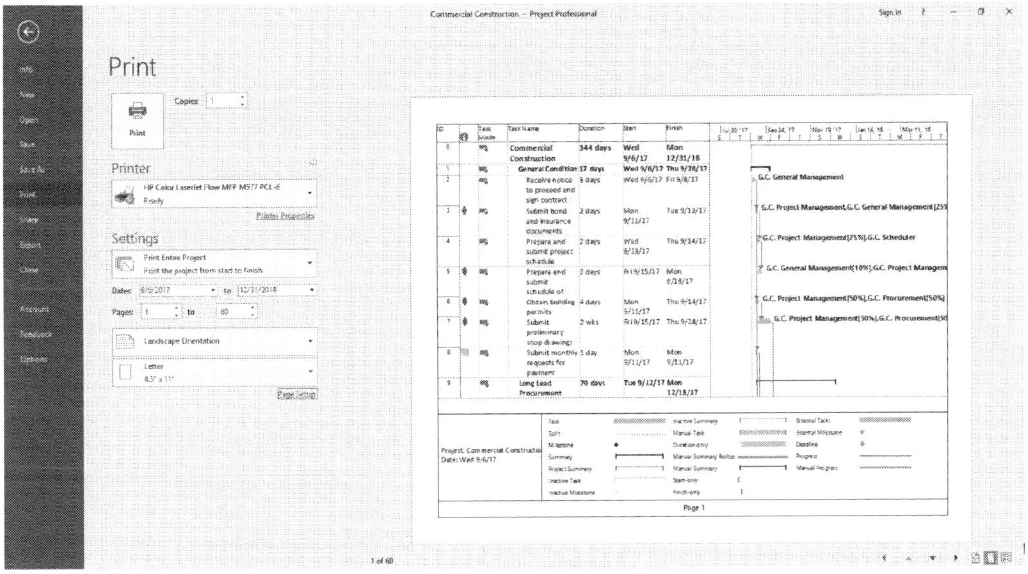

The buttons on the Print tab of the Backstage View.

- The PRINT button allows you to print the document using the current settings.
- The COPIES field allows you to print one or more copies of the project.
- The PRINTER allows you to select a different printer. The printer properties allow you to set the properties for that printer.
- The SETTINGS tool allows you to select different pages of your project or different dates.
- The DATES field allows you to select a date range to include in the print out.
- The PAGES field allows you to specify a custom page range to print.
- The other SETTINGS control additional settings for print, including the orientation and the paper size.

- There is also a link to the PAGE SETUP dialog box.

E-mailing a Project

The new Project 2016 Share tab in the Backstage View allows you to easily send the current project using your default email application. You can send it as an attachment or you can send a link to the file.

Use the following procedure to email a project file.

Step 1: Select the File tab on the Ribbon.

Step 2: Select the Save & Send tab in the Backstage View.

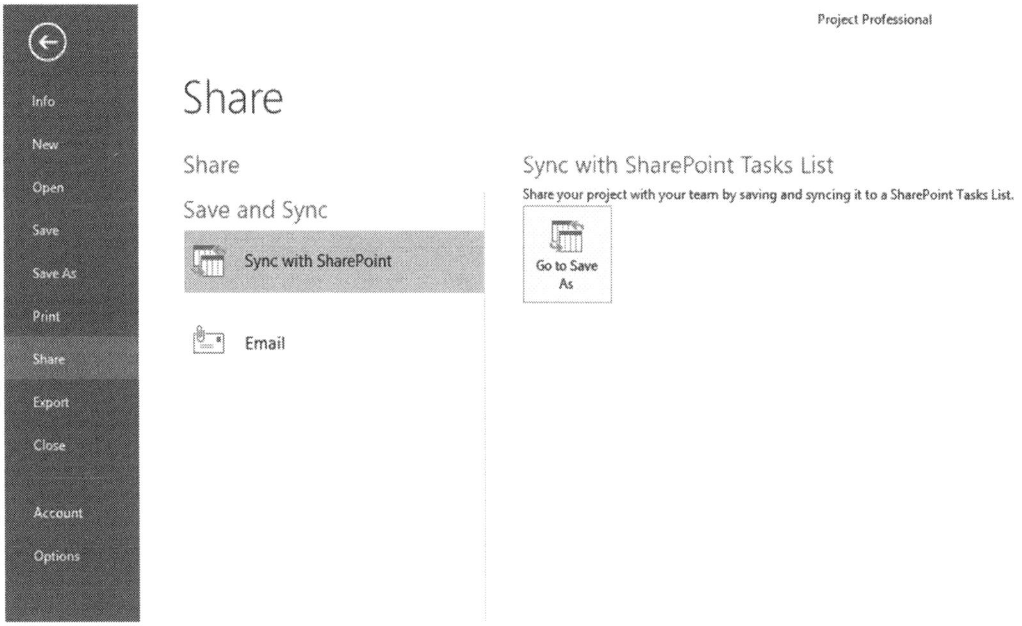

Step 3: Select Send as Attachment under Save & Send.

Step 4: Select Send as Attachment.

Creating a PDF

The new Project 2016 Share tab in the Backstage View allows you to easily convert the project to a PDF file for sharing with viewers who do not have Project installed on their computers.

Use the following procedure to save a project file as a PDF file.

Step 1: Select the File tab on the Ribbon.

Step 2: Select the Save & Send tab in the Backstage View.

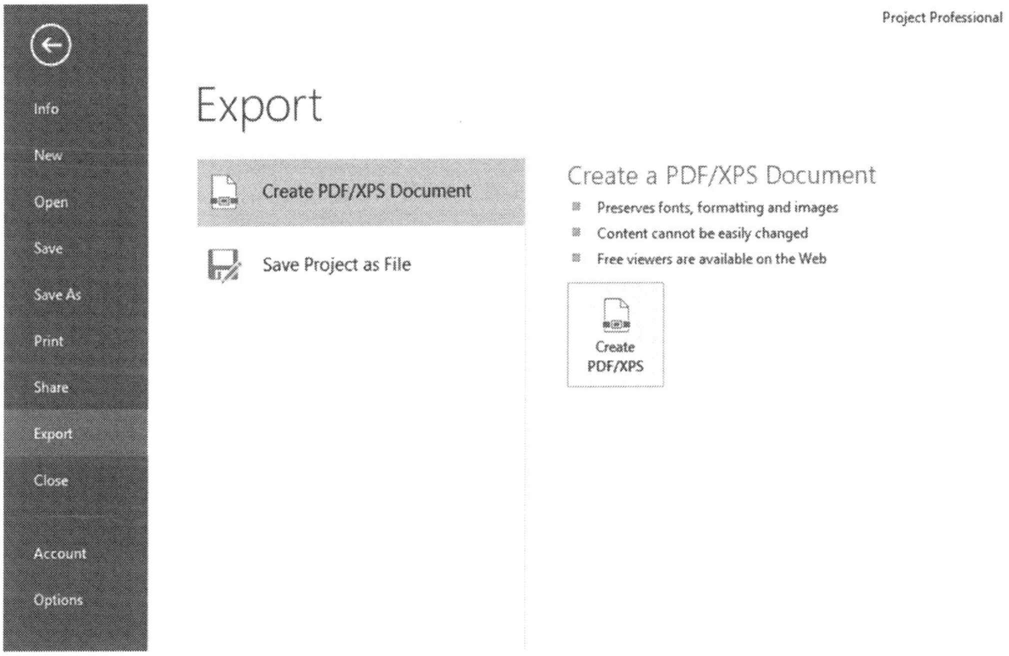

Step 3: Select Create PDF/XPS under Save & Send.

Step 4: Select Create PDF/XPS.

Step 5: Indicate where to save the file on your computer and select OK.

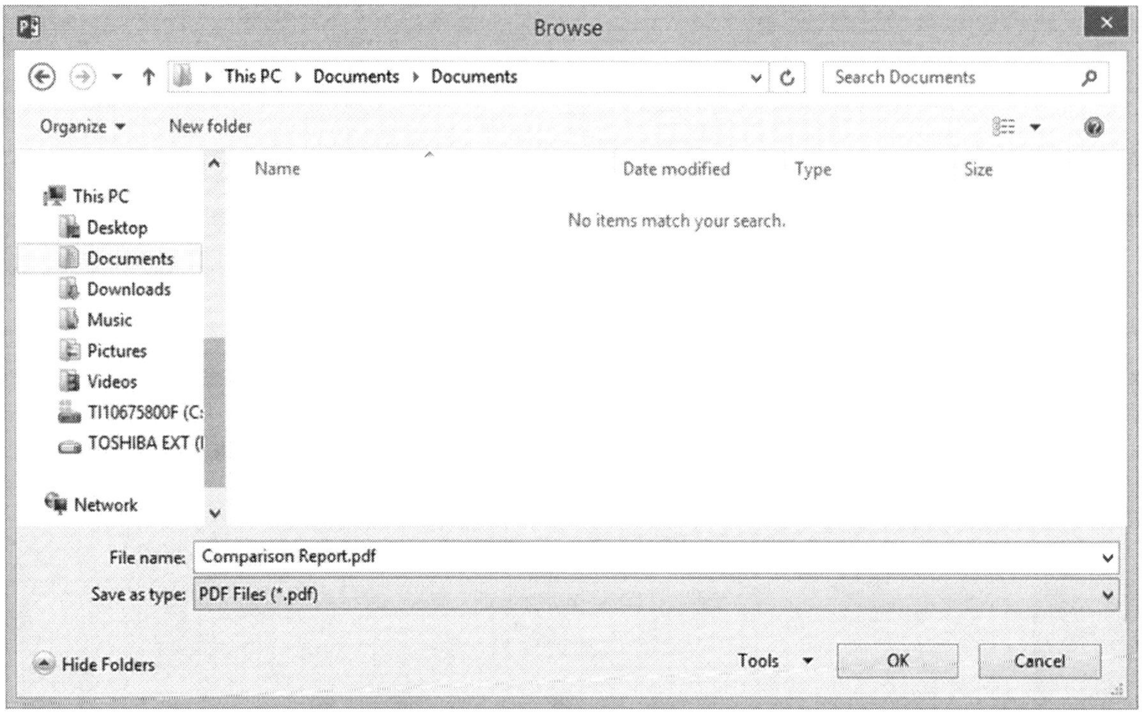
Browse
This PC ▸ Documents ▸ Documents
Search Documents
Organize ▾
New folder
This PC
Desktop
Documents
Downloads
Music
Pictures
Videos
TI10675800F (C:
TOSHIBA EXT (I
Network
Name
Date modified
Type
Size
No items match your search.
File name: Comparison Report.pdf
Save as type: PDF Files (*.pdf)
Hide Folders
Tools
OK
Cancel

Chapter 11 – Viewing the Project

In this chapter, you will learn some advanced ways to view a project. You will learn how to split views to look at different information at the same time or use the Task Form for entering task details. You will learn how to sort, group, and filter information, including using AutoFilters. Finally, you will review how to use Zoom.

Using Split Views

Use the following procedure to setup split views.

Step 1: Select the View tab from the Ribbon.

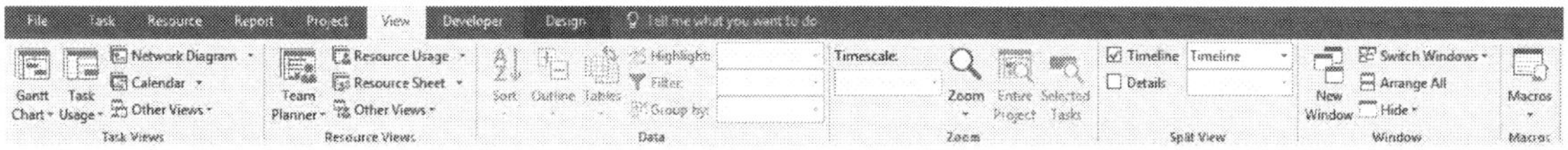

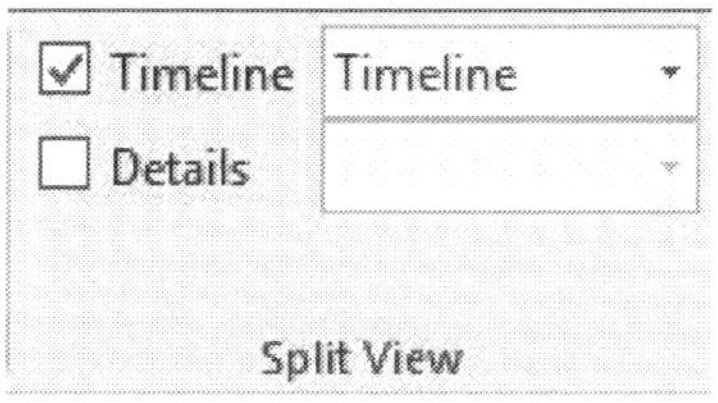

Step 2: You can show the Timeline or another view in the split window. To show the Timeline, check the Timeline box. The Timeline portion of the screen is displayed at the top. You can change the bottom view using the other View tools just as you would if the Timeline is not displayed.

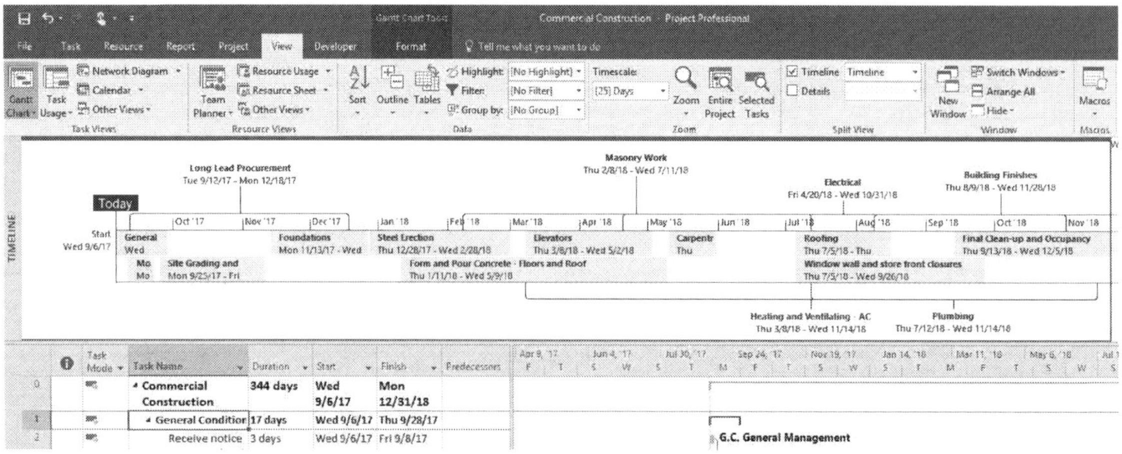

Step 3: To show a detail view in the split window, check the Details box on the View tab of the Ribbon. Select the desired view from the drop-down list.

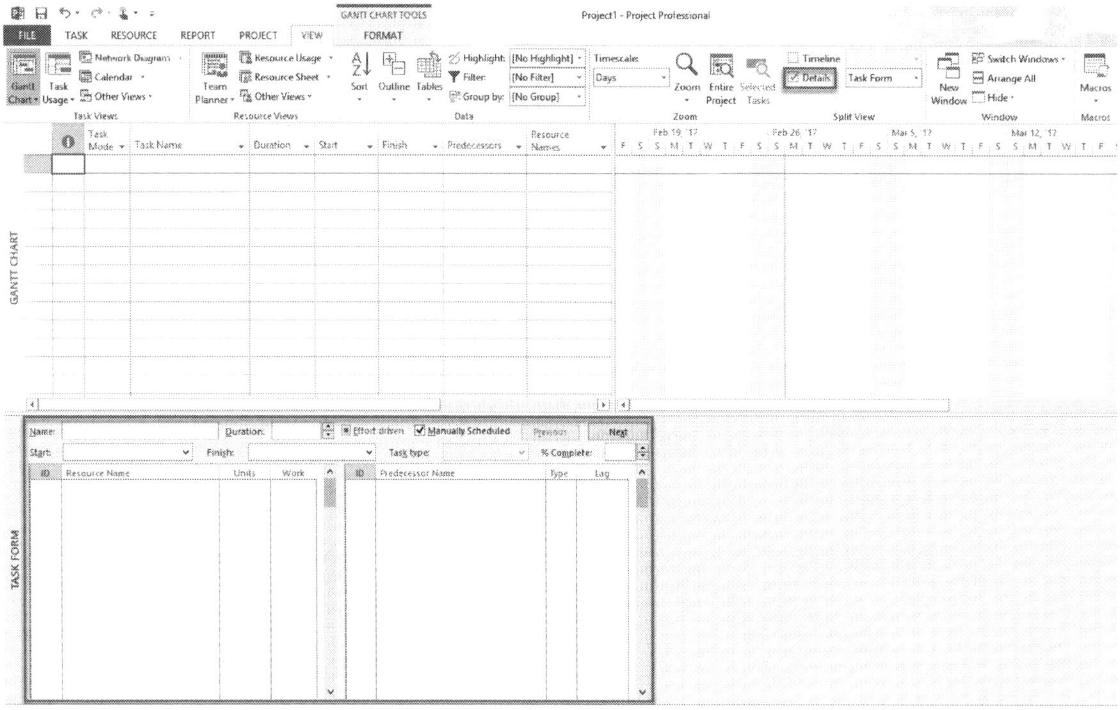

To change the view displayed in one of the split areas, use the following procedure.

Step 1: You can change the top view by selecting the desired view from the View tab on the Ribbon.

Step 2: To change the bottom view, choose a different option from the Details drop-down list in the Split View area of the View tab on the Ribbon. Or you can right-click the title bar on the left and select the desired view from the context menu.

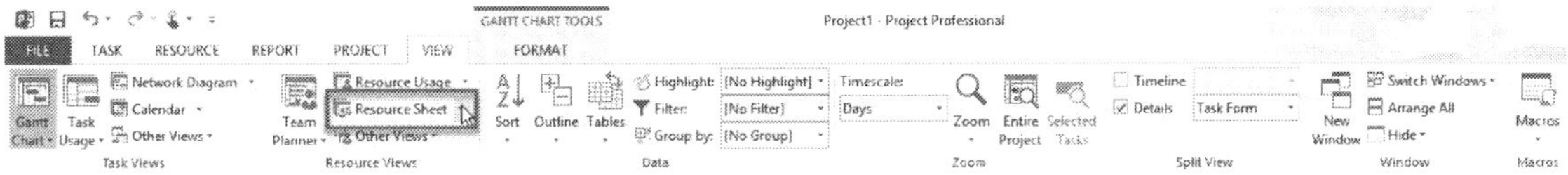

To remove a split, use the following procedure.

Step 1: Uncheck the box in the Split View area on the View tab of the Ribbon to remove the split.

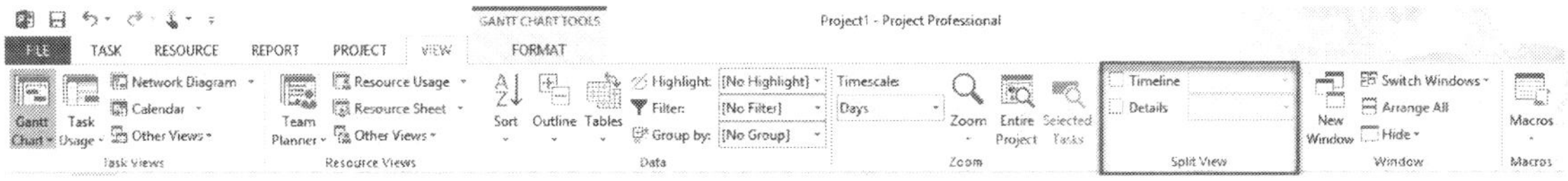

Project removes the bottom view and expands the top view to fill the window.

Sorting Information

Use the following procedure to sort tracks.

Step 1: Highlight the tasks that you want to sort by selecting the row number. You can click and drag tasks that are next to each other or you can press the Ctrl key while you click the different rows.

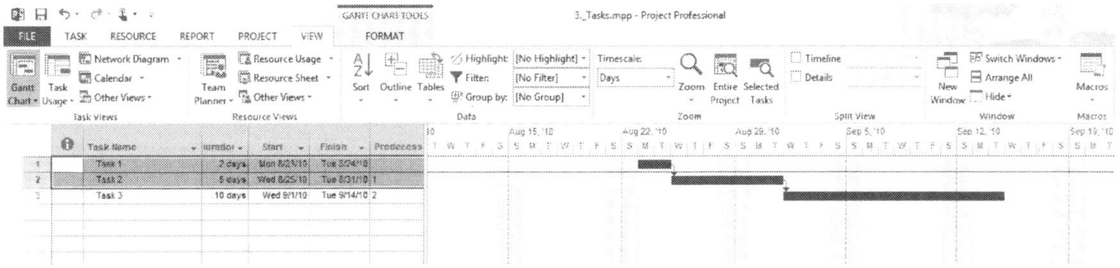

Step 2: Select the View tab from the Ribbon.

Step 3: Select Sort.

Step 4: Select one of the sorting options.

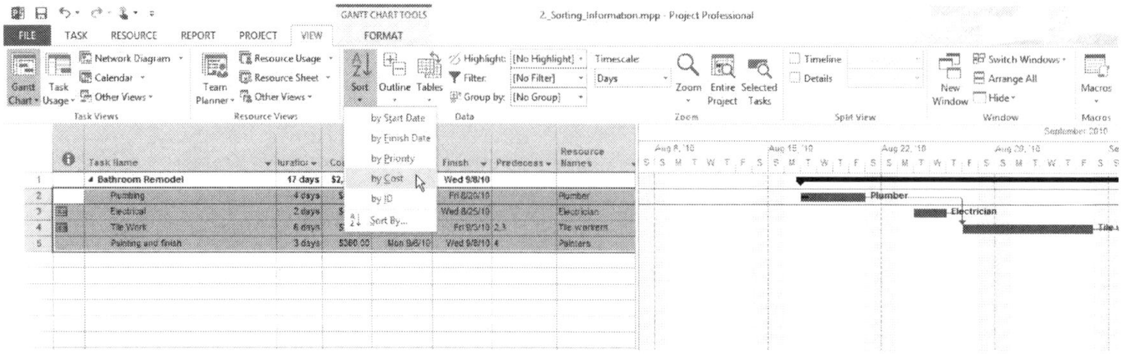

Project rearranges the tasks according to the sort option you selected.

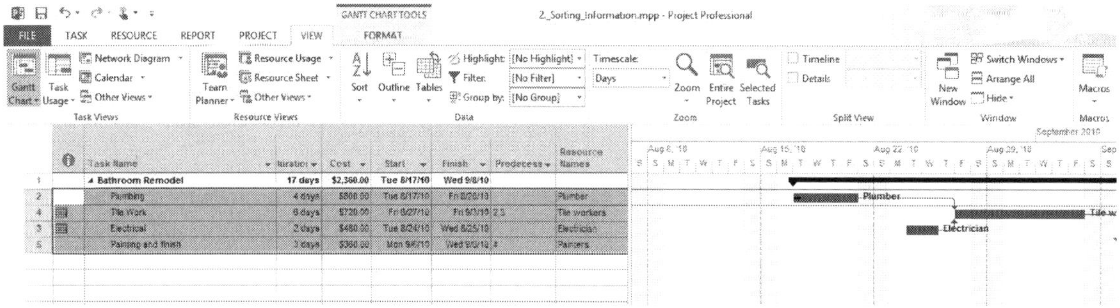

The Sort dialog box can be accessed by using the following steps.

Step 1: Select Sort from the Project menu. Select Sort by.

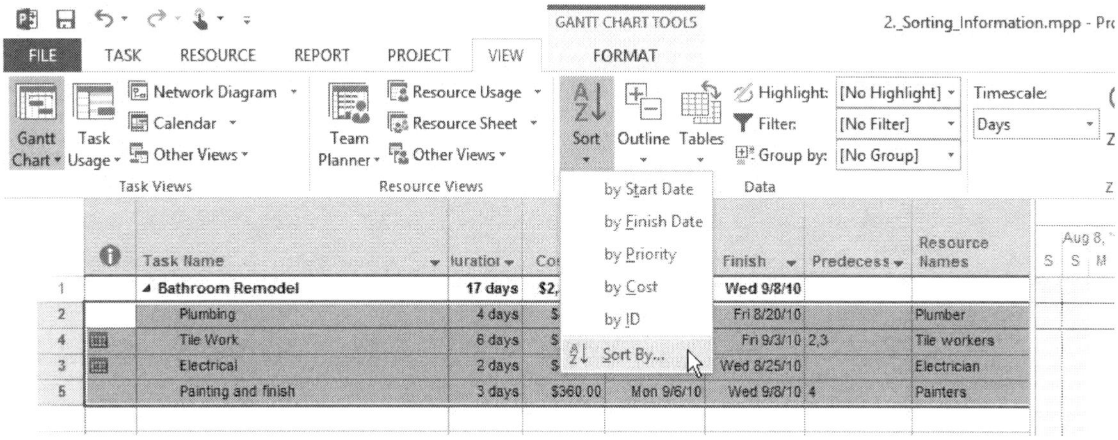

Project opens the Sort dialog box.

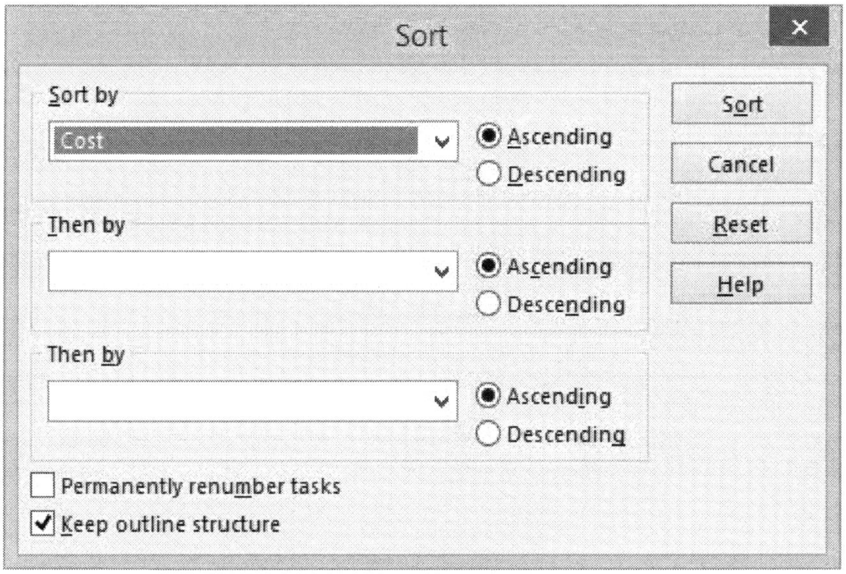

Step 1: Select an option from the first drop-down list to indicate the first sorting option. Select Ascending or Descending.

Step 2: Select an option from the second drop-down list to indicate the second sorting option. Select Ascending or Descending.

Step 3: Select an option from the third drop-down list to indicate the third sorting option. Select Ascending or Descending.

Step 4: Check the Permanently renumber tasks box to change the task ID to the new sorting order.

Step 5: Check the Keep outline structure box to keep your current outline structure.

Step 6: Select Sort. To start over, select Reset. Or select Cancel to close the Sort dialog box without sorting.

Grouping Information

Use the following procedure to group tasks.

Step 1: Select the View tab from the Ribbon.

Step 2: Select the Group By drop-down list.

Step 3: Select one of the grouping options.

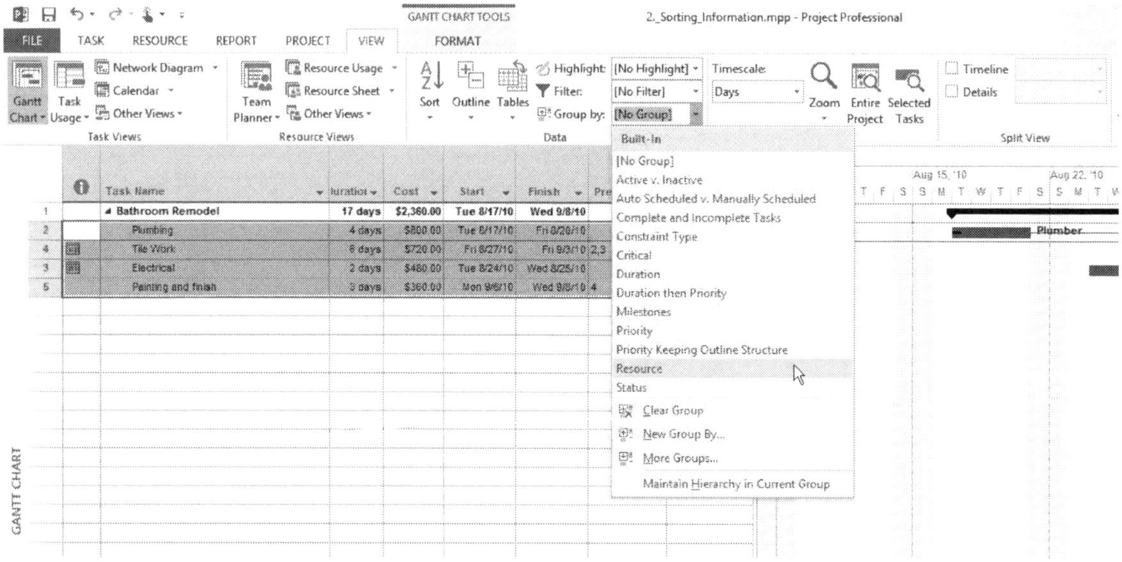

Project rearranges the tasks according to the grouping option you selected. It also displays a colored header for each group.

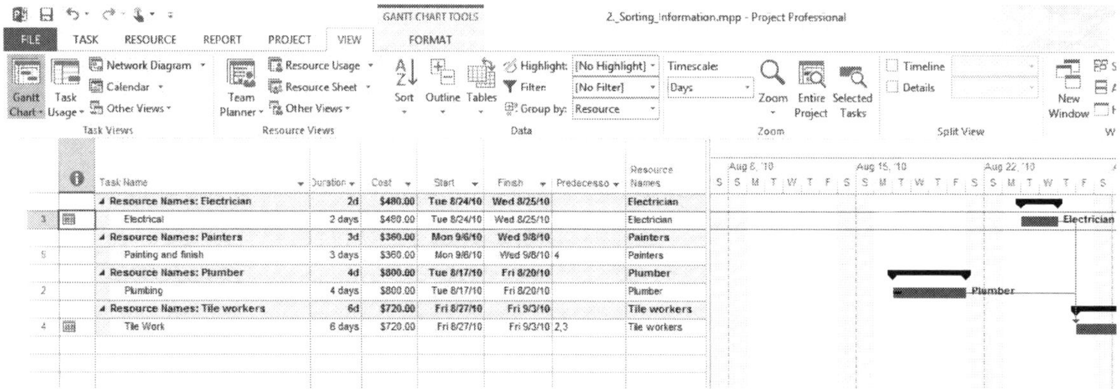

Accessing the More Groups dialog box.

Step 1: Select the View tab from the Ribbon.

Step 2: Select the Group By drop-down list.

Step 3: Select More Groups.

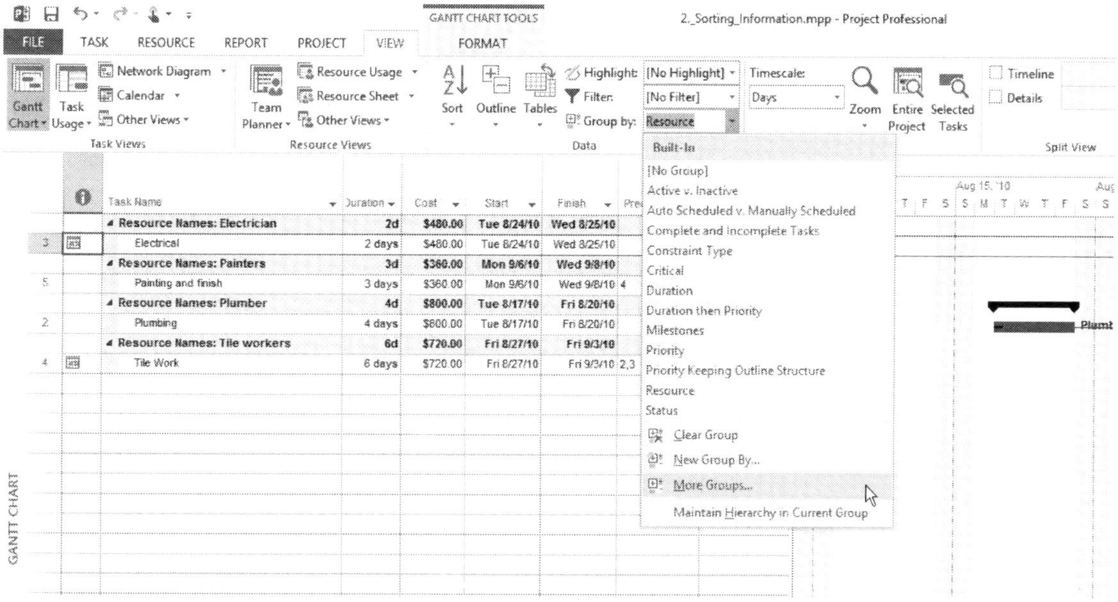

Project opens the More Groups dialog box.

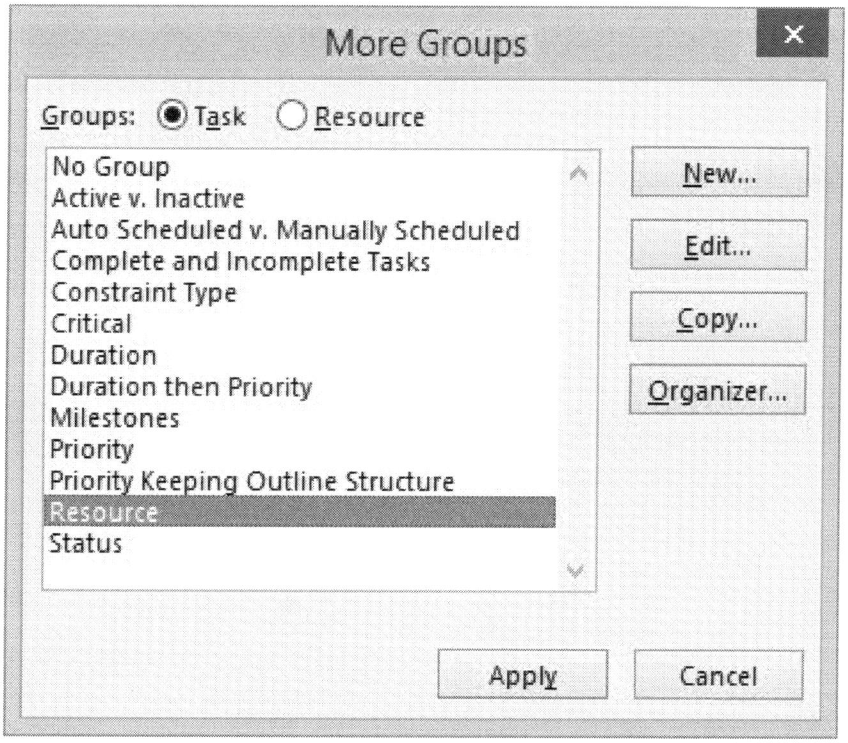

Step 4: Select either Task or Resource to narrow the list of groups.

Step 5: You can copy or create a new grouping by selecting New or Copy. You can edit a grouping by selecting Edit. These options open the Group Definition dialog box to allow you to define the grouping.

Step 6: Select a grouping from the list and select Apply. Or select Cancel to close the More Groups dialog box without applying a grouping.

The Group Definition dialog box.

Step 1: You can open the Group Definition dialog box by selecting New, Edit, or Copy from the More Groups dialog box. You can also select Group by from the Project menu and select Customize Group By.

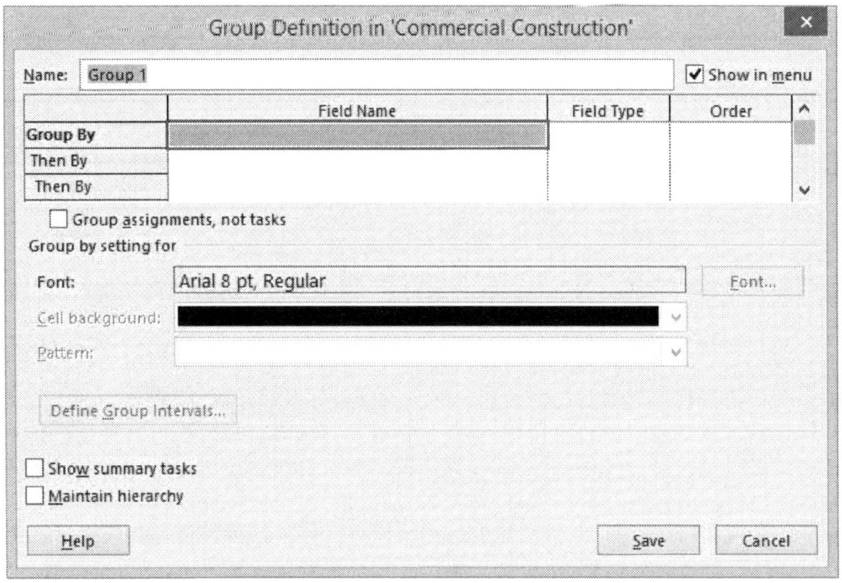

Step 2: The top area allows you to select multiple methods of grouping. For each level of grouping, select the following:

- Field Name
- Field Type
- Order

Step 3: The Group assignments, not tasks box allows you to group by assignments instead of tasks.

Step 4: The Group By settings area allows you to change the formatting for the group headings.

- Select the Font button to select a new font.
- Select a new Cell background from the drop-down list.
- Select a Pattern for the cell background, if desired.

Step 5: The Define Group Intervals button allows you to define how groups are determined.

Step 6: Check the Show summary tasks box if you want to show the summary for each grouping.

Step 7: Check the Maintain hierarchy box to make sure your hierarchy is not changed with the grouping.

Step 8: Select OK to apply the group format. Or select Cancel to close the Group Definition dialog box without changing the group definition.

Filtering Information

Use the following procedures to filter tasks.

Step 1: Select the View tab from the Ribbon.

Step 2: Select the Filter drop-down list.

Step 3: Select one of the filtering options.

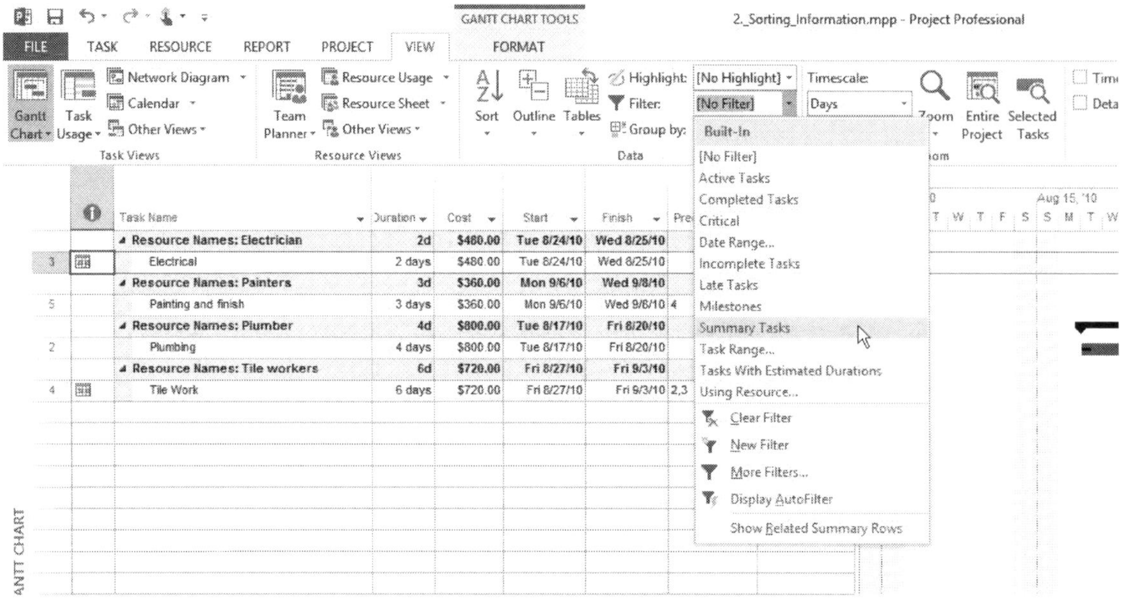

Step 4: Some filters require you to provide additional information. Select the required information and select OK.

Project only shows the tasks that meet the filter criteria.

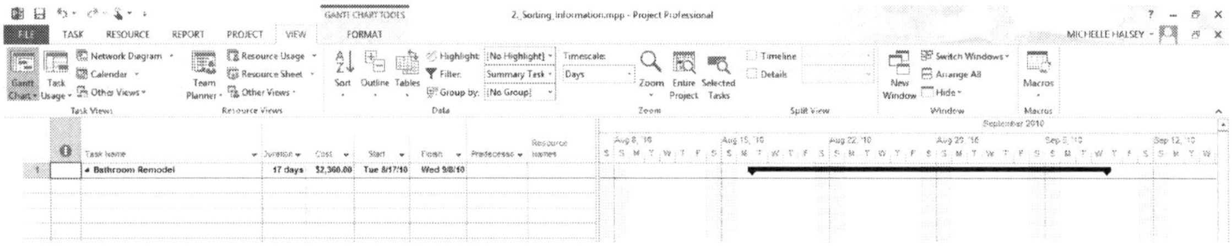

The More Filters dialog box.

Step 1: Select Filtered For from the View menu. Select More Filters.

Project opens the More Filters dialog box.

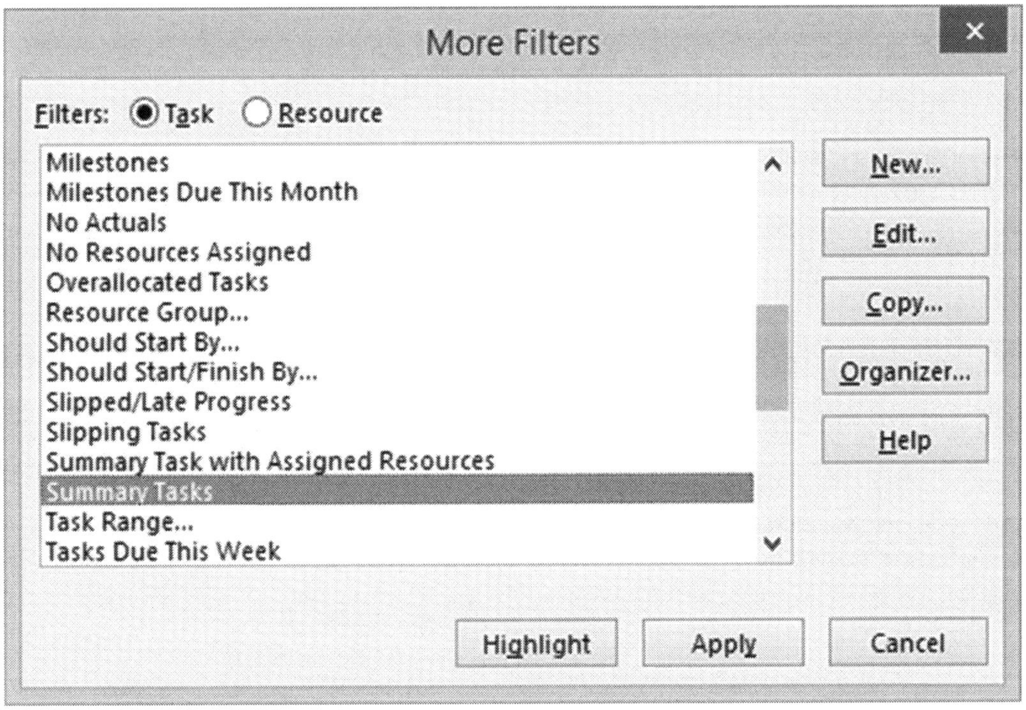

Step 2: Select either Task or Resource to narrow the list of filters.

Step 3: You can copy or create a new filter by selecting New or Copy. You can edit a filter by selecting Edit. These options open the Filter Definition dialog box to allow you to define the filter.

Step 4: Select a filter from the list and select Apply. Or select Cancel to close the More Filters dialog box without applying a filter.

The Filter Definition dialog box can be accessed by using the following procedure.

Step 1: You can open the Filter Definition dialog box by selecting New, Edit, or Copy from the More Filters dialog box.

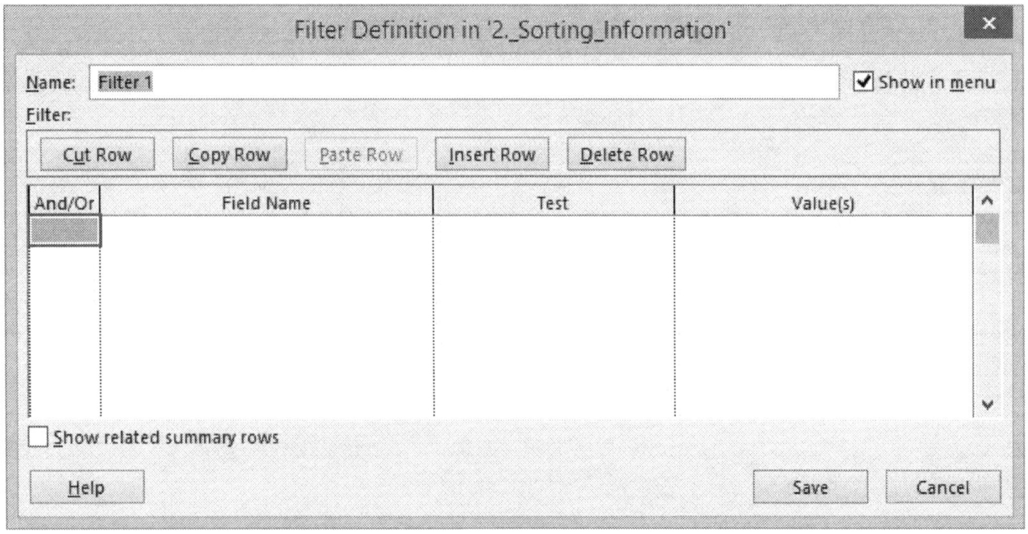

Step 2: The bottom area displays the filter definitions. For each line, select the following:

- And/Or
- Field Name
- Test
- Value

Step 3: Check the Show Related summary tasks box if you want to show the summary tasks.

Step 4: You can insert or delete rows, as well as cut, copy, or paste rows using the buttons above the filter definition rows.

Step 5: Select OK to apply the filter. Or select Cancel to close the Filter Definition dialog box without changing the filter definition.

Using AutoFilters

Using the AutoFilter feature.

Step 1: Select the View tab from the Ribbon.

Step 2: Select the Filter drop-down list.

Step 3: Select Display AutoFilter.

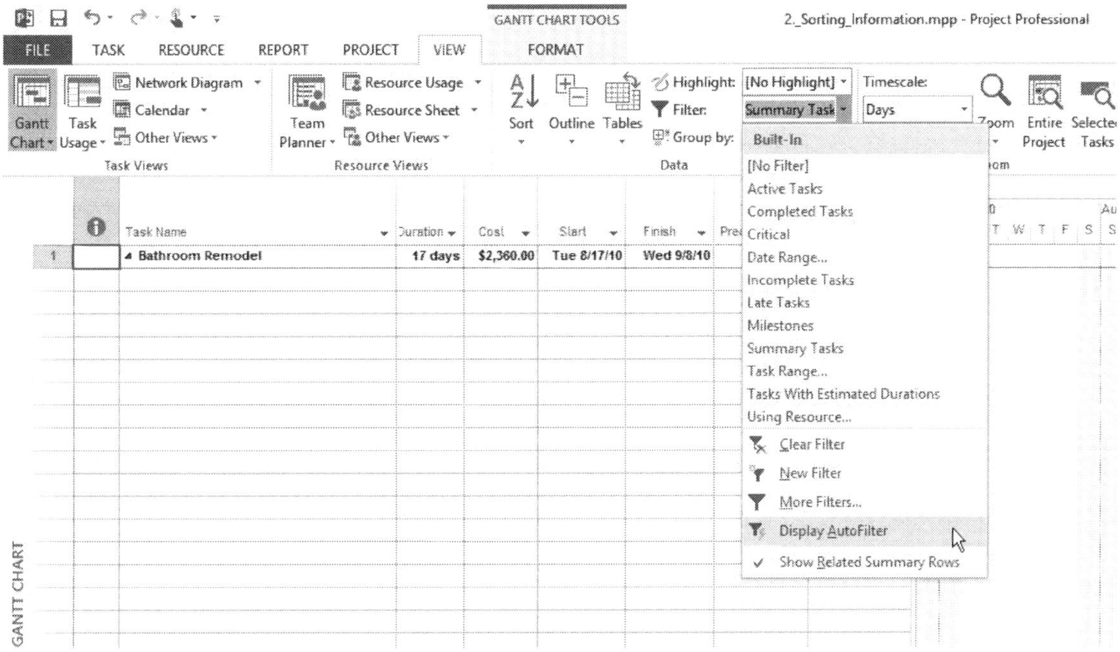

Project displays an arrow on the column headings.

Step 4: Select the arrow next to the column you want to use for filtering to see the filtering options.

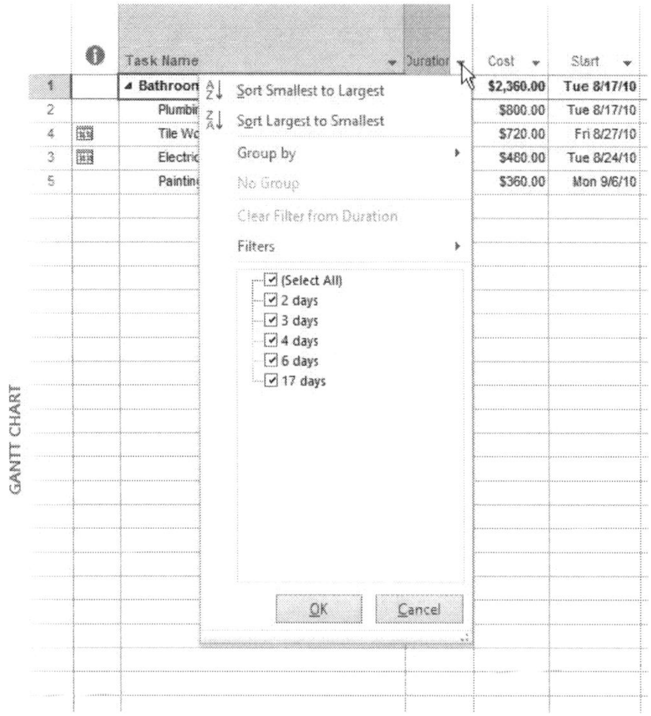

Step 5: Check the AutoFilter option(s) you want to display. Or check Select All to show all tasks.

Step 6: The AutoFilter menu includes custom options for grouping the filtered information. Select a group if desired.

Step 7: Select OK.

Step 8: Repeat with multiple column headings, if desired. Project applies the filters in the order you select them.

Using Zoom

To open the Zoom dialog box, use the following procedure.

Step 1: Select Zoom from the View tab on the Ribbon.

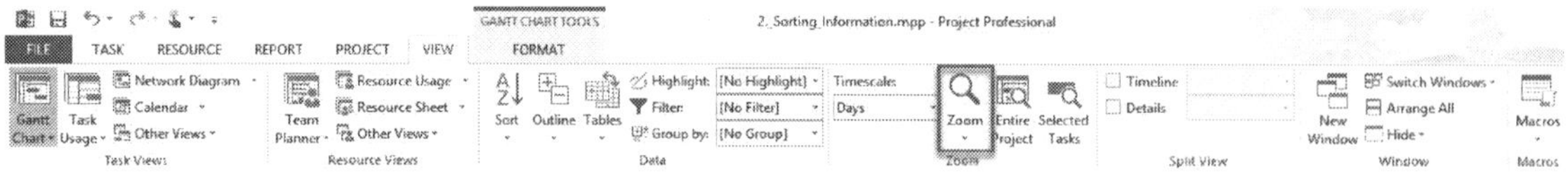

Step 2: Select Zoom.

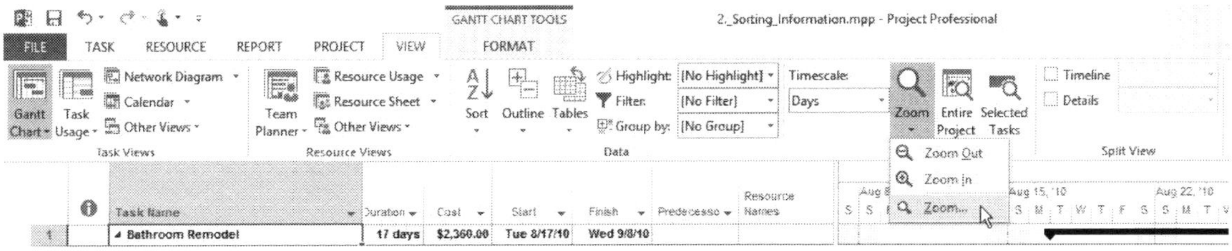

Project displays the Zoom dialog box.

Step 3: Select a Zoom option. If you select Custom, enter the number and select the duration to indicate the custom zoom.

Step 4: Select OK. To start over, select Reset. To close the Zoom dialog box without changing the view, select Cancel.

Chapter 12 – Working with Tasks

This chapter will teach you some ways to work with tasks. You will learn how to display the Project Summary task to get a high-level view of the duration, start and end dates for your project (as well as any other columns of information you may be viewing). You will also learn how to create milestones. If your plan needs a little tweaking, you may want to rearrange your tasks or cancel them. This chapter also explains how to create manually scheduled tasks. Finally, you will learn how to create a recurring task.

More advanced features for working with tasks will be covered as well. You can overlap or delay tasks to shorten the project duration or adjust a schedule to allow for time between linked tasks. Learn how to set task deadlines and constraints, and how to split tasks to accommodate a task interruption, such as when a resource is temporarily reassigned to another task. Finally, we will look at task types, task calendars, and task indicators.

Displaying the Project Summary Task on a New Project

To display the project summary task, use the following procedure.

Step 1: Select the Format tab from the Ribbon.

Step 2: Check the Project Summary Task box.

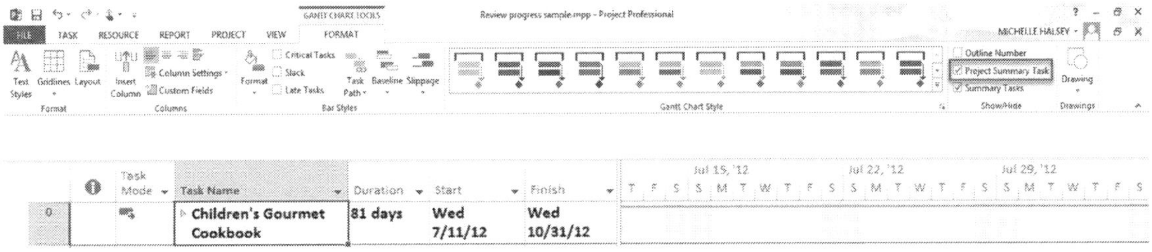

Creating Milestones

To create a milestone, use the following procedure.

Step 1: On the empty row in the Gantt Chart view, enter the name of the milestone.

Step 2: Enter 0 as the duration.

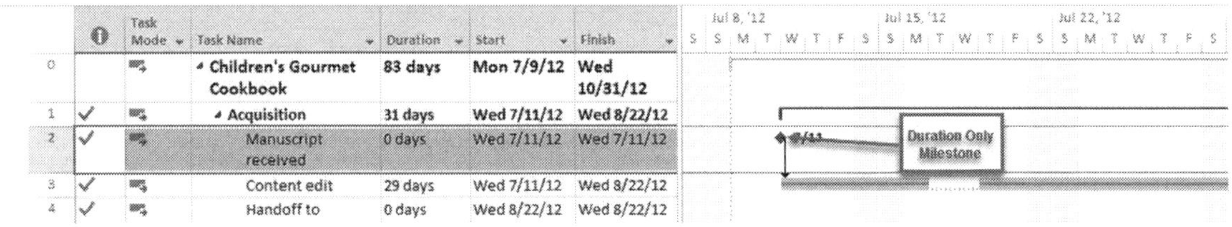

To create a milestone for a task with a duration longer than 0, use the following procedure.

Step 1: Enter the task name and duration as you would for any new task.

Step 2: Double-click the task to open the Task information dialog box.

Step 3: Select the Advanced tab.

Step 4: Check the Mark Task as Milestone box.

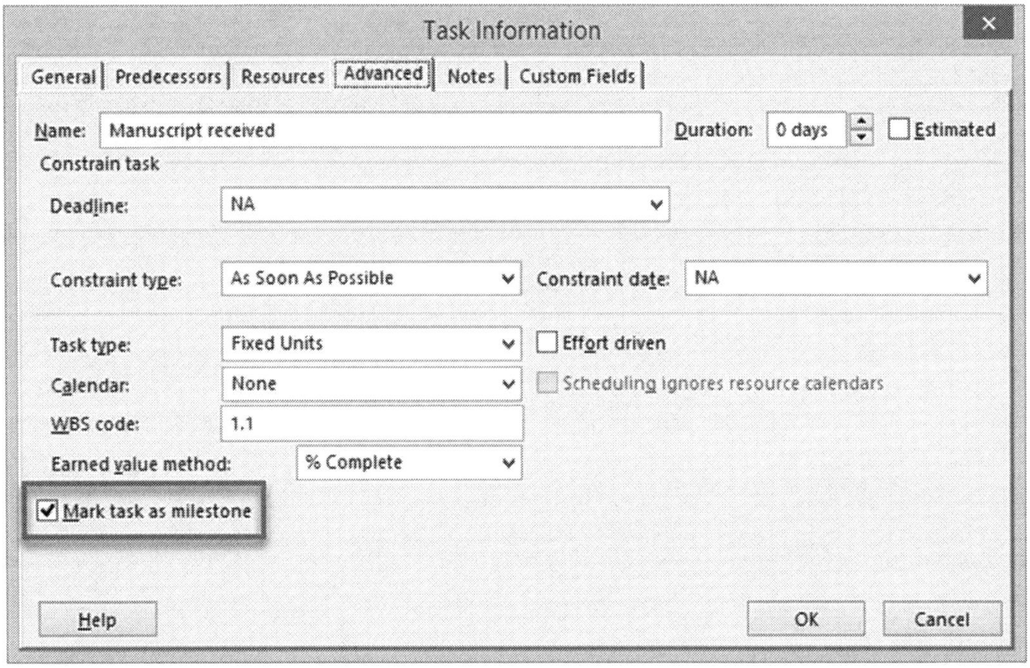

Step 5: Select OK.

Rearranging Tasks

To rearrange a single task, use the following procedure.

Step 1: Select the task that you want to rearrange.

Step 2: Drag it to the new location. The task is shown as a horizontal bar until you release the mouse button.

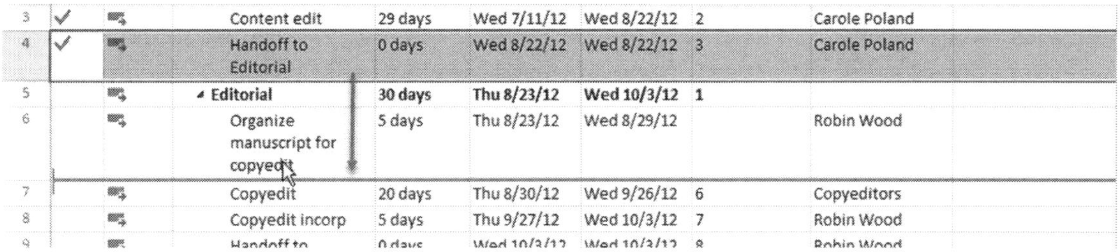

3	✓	Content edit	29 days	Wed 7/11/12	Wed 8/22/12	2	Carole Poland
4	✓	Handoff to Editorial	0 days	Wed 8/22/12	Wed 8/22/12	3	Carole Poland
5		**Editorial**	**30 days**	**Thu 8/23/12**	**Wed 10/3/12**	**1**	
6		Organize manuscript for copyedit	5 days	Thu 8/23/12	Wed 8/29/12		Robin Wood
7		Copyedit	20 days	Thu 8/30/12	Wed 9/26/12	6	Copyeditors
8		Copyedit incorp	5 days	Thu 9/27/12	Wed 10/3/12	7	Robin Wood

To rearrange a summary task, use the following procedure.

Step 1: Select the summary task that you want to rearrange.

Step 2: Drag it to the new location. The group is shown as a horizontal bar until you release the mouse button.

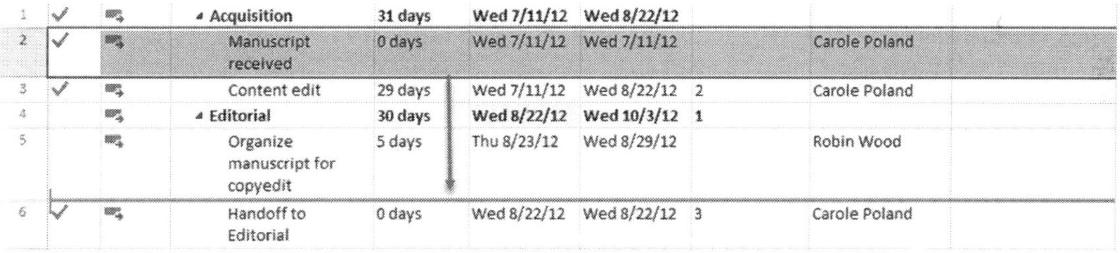

1	✓	**Acquisition**	**31 days**	**Wed 7/11/12**	**Wed 8/22/12**		
2	✓	Manuscript received	0 days	Wed 7/11/12	Wed 7/11/12		Carole Poland
3	✓	Content edit	29 days	Wed 7/11/12	Wed 8/22/12	2	Carole Poland
4		**Editorial**	**30 days**	**Wed 8/22/12**	**Wed 10/3/12**	**1**	
5		Organize manuscript for copyedit	5 days	Thu 8/23/12	Wed 8/29/12		Robin Wood
6	✓	Handoff to Editorial	0 days	Wed 8/22/12	Wed 8/22/12	3	Carole Poland

Canceling an Unneeded Task

To delete a task, use the following procedure.

Step 1: Right click on the task you want to remove from the project plan.

Step 2: Select Delete from the context menu.

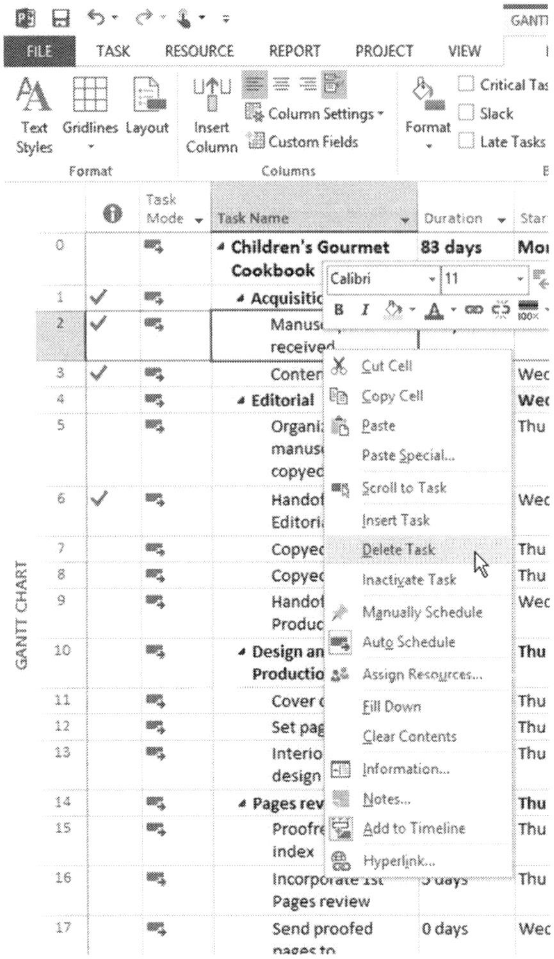

The task is removed.

To inactivate a task, use the following procedure.

Step 1: Highlight the task in the project plan.

Step 2: Select Inactivate from the Task tab on the Ribbon.

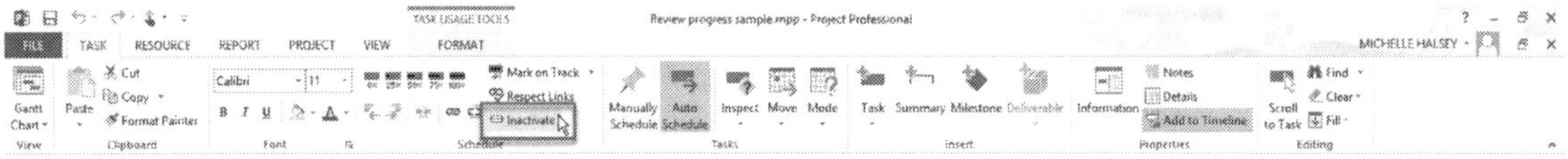

The task is crossed out to show that it is inactive.

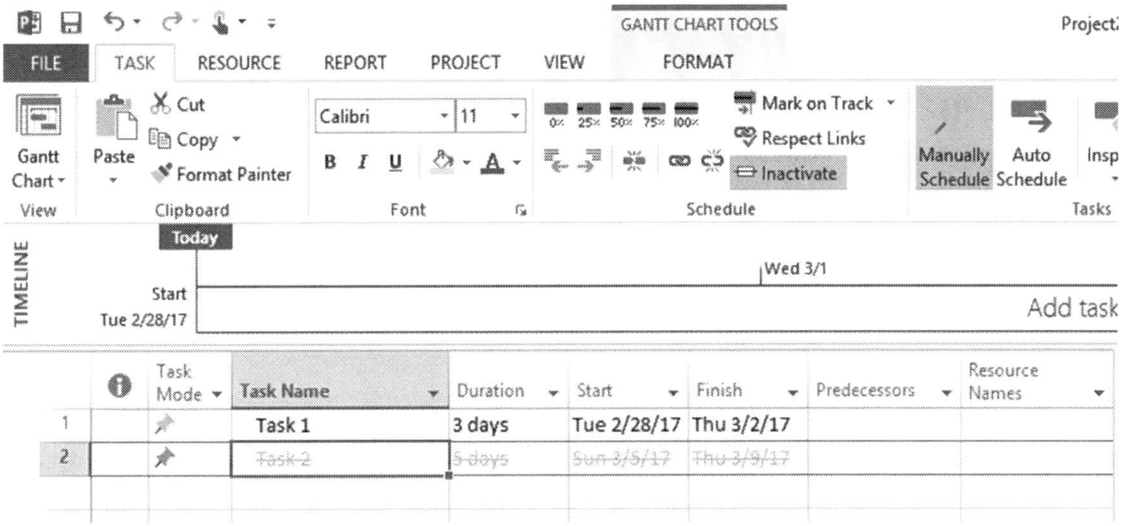

Creating Manually Scheduled Tasks

To change an existing task to be manually scheduled, use the following procedure.

Step 1: Select the task that you want to change in the Gant chart view of the Project plan.

Step 2: Select Manually Schedule.

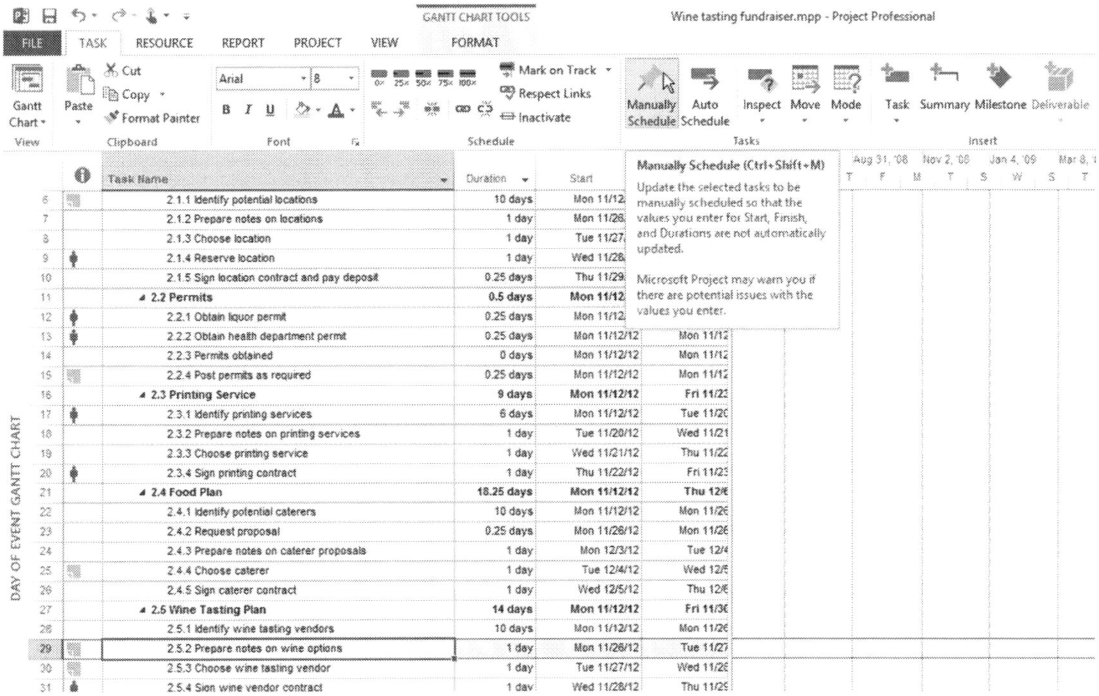

To change the default to manually schedule all new tasks, use the following procedure.

Step 1: On the Status Bar at the bottom of the screen, select the New Tasks: Auto Scheduled notification.

Step 2: Select Manually Scheduled – Task dates are not automatically updated.

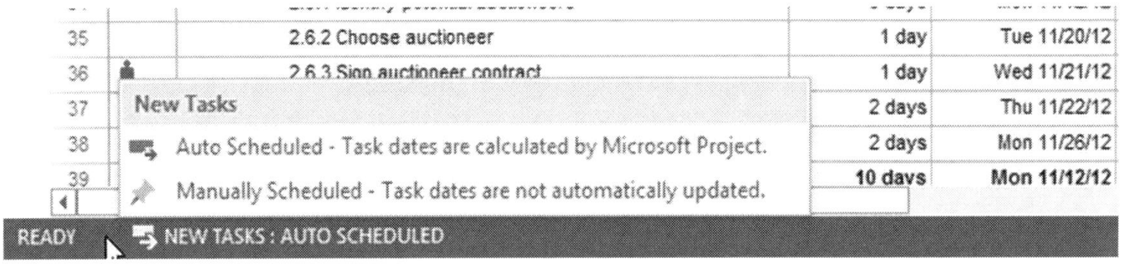

Creating a Recurring Task

To create a recurring task, use the following procedure.

Step 1: Select the row on the Gantt chart view of the Project plan where you want the recurring task to appear.

Step 2: On the Task tab of the Ribbon, select the arrow below the Task tool in the Insert group.

Step 3: Select Recurring Task.

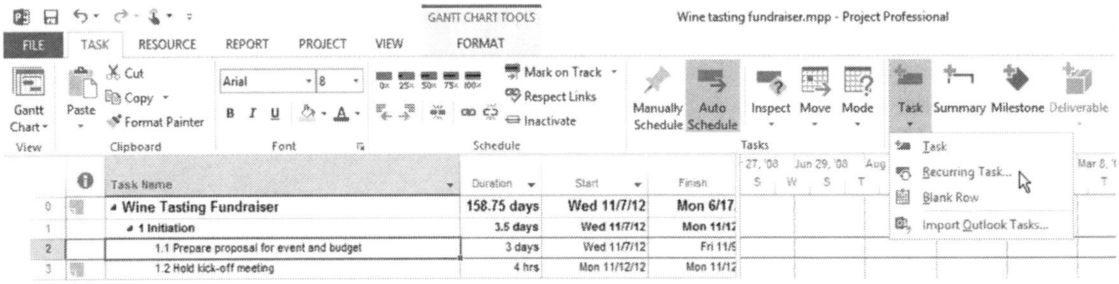

Step 4: In the Recurring Task Information dialog box, enter a Task Name.

Step 5: Enter the Duration for the task.

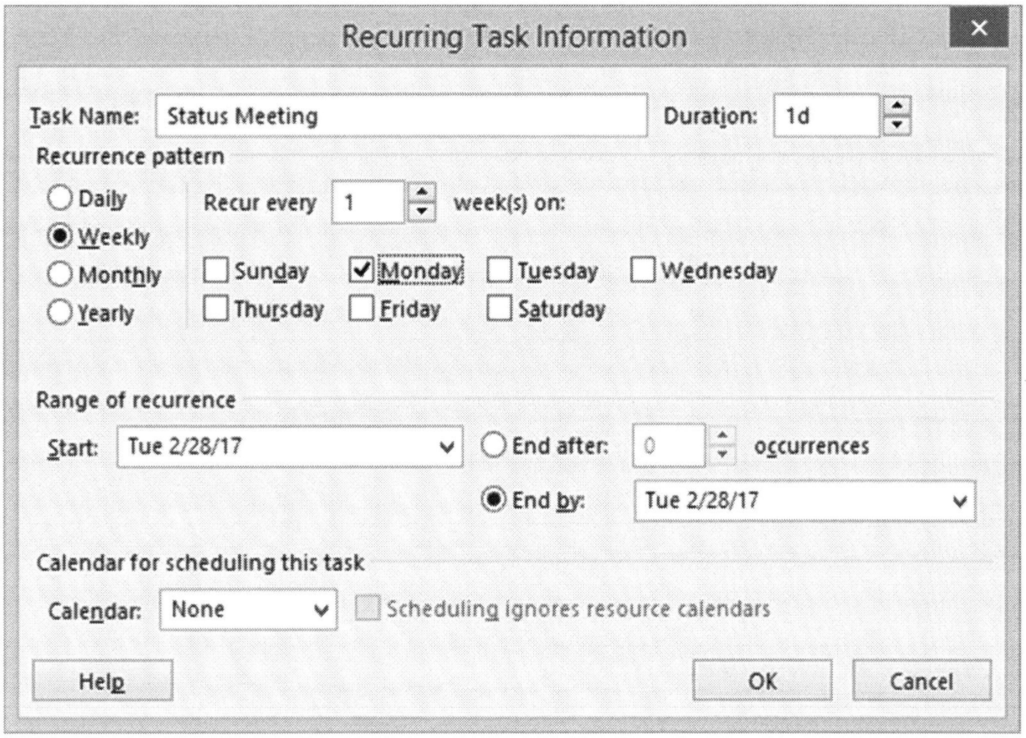

Step 6: Select which interval the task will repeat (Daily, Weekly, Monthly or Yearly). Depending on your selection, complete additional details for the recurrence pattern:

If you selected Daily, enter the number of days and whether it is every day or just work days.

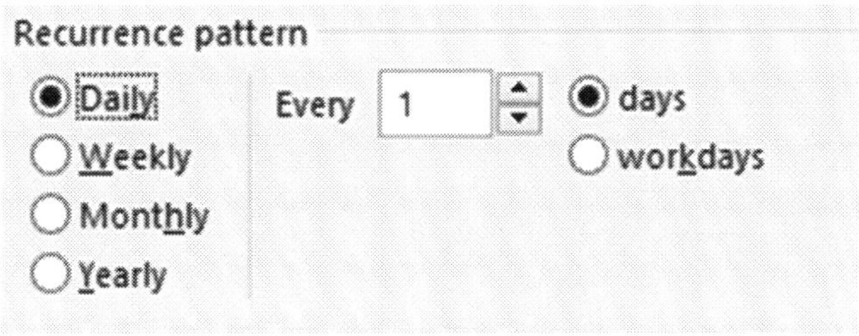

If you selected Weekly, enter the number of weeks and check the boxes for the days of the weeks on which the task recurs.

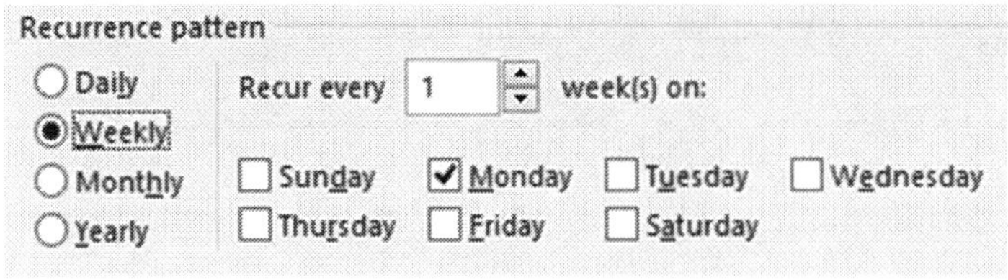

If you selected Monthly, select the pattern by date or by day of the week and which week.

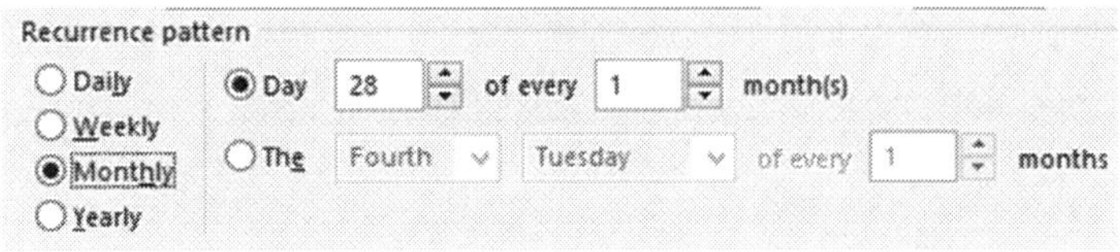

If you selected Yearly, select the pattern by date or by day of the week and which week and which month.

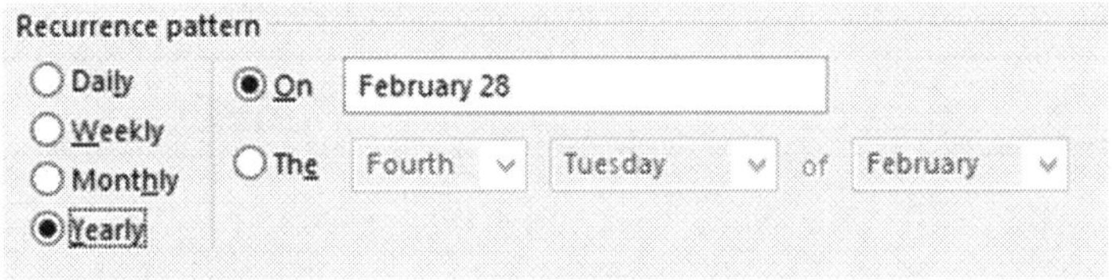

Step 1: For the Range of Recurrence, enter the Start Date and either the End Date or the number of occurrences to end after.

Step 2: Select a Resource Calendar from the drop-down list.

Step 3: Select OK.

Overlapping Tasks

To add lead time, use the following procedure.

Step 1: Select the task you want to overlap. It should be a task that has a predecessor. Double-click to open the Task Information dialog box.

Step 2: Select the Predecessors tab.

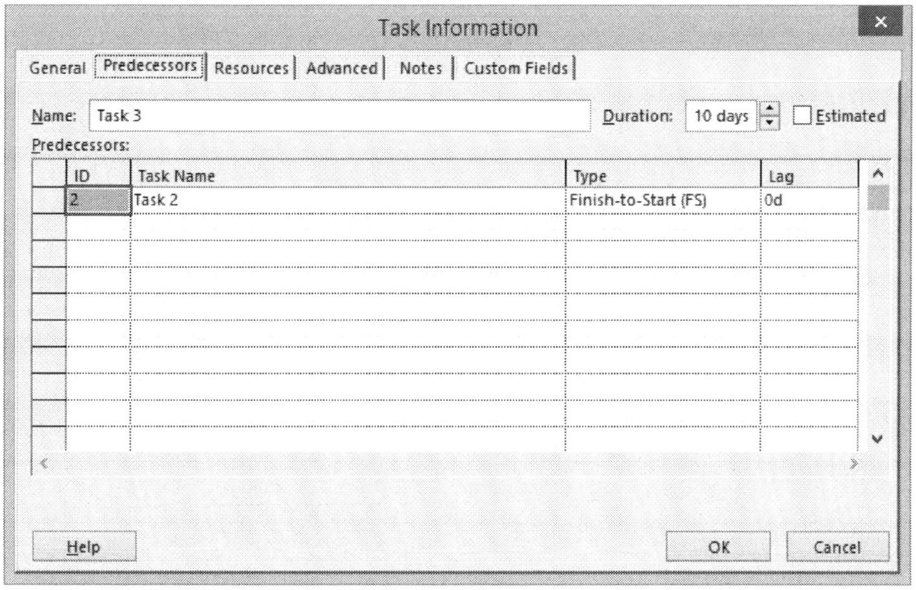

Step 3: In the LAG field, enter -50%.

Step 4: Select OK to close the Task Information dialog box.

Project displays the overlapping tasks on the Gantt chart.

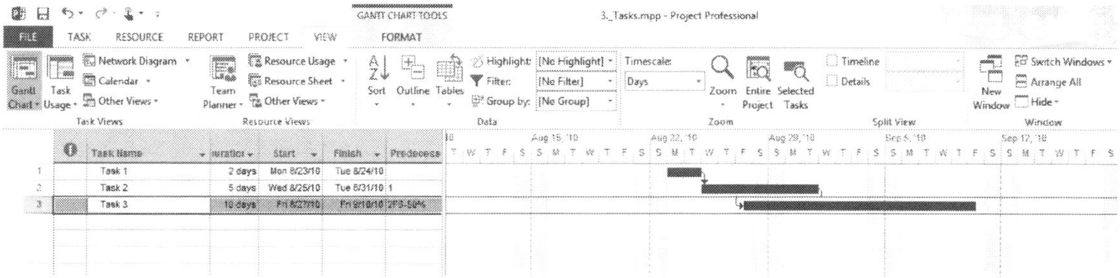

Delaying Tasks

Use the following procedure to add lag time to a task.

Step 1: Select the task you want to delay. It should be a task that has a predecessor. Double-click to open the Task Information dialog box.

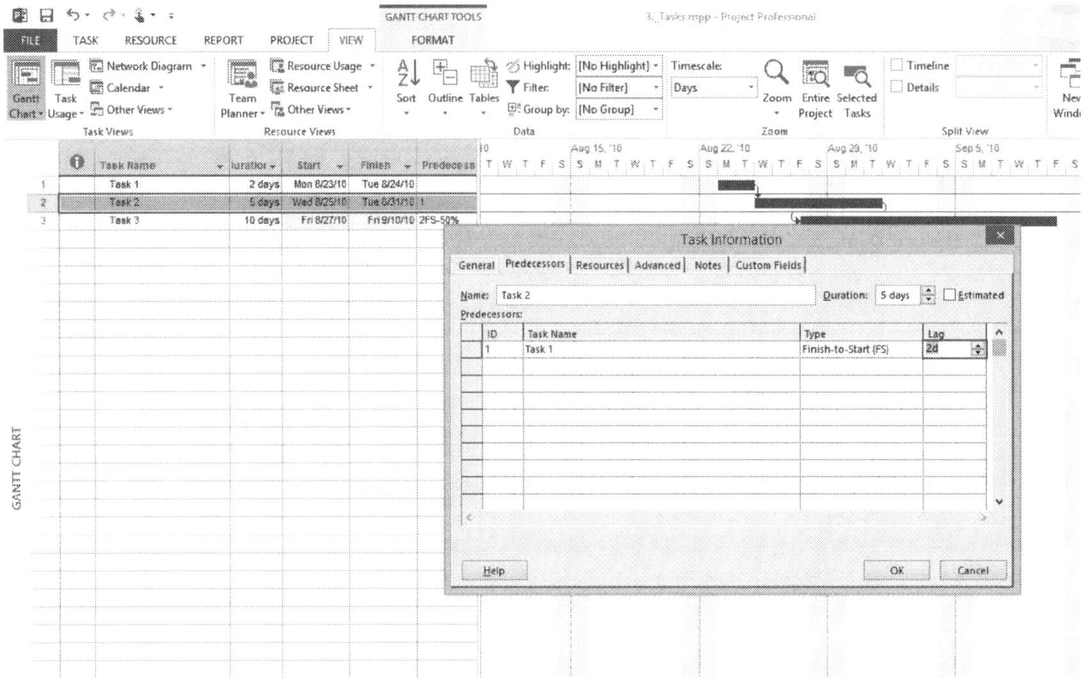

Step 2: In the LAG field, enter 2d.

Step 3: Select OK to close the Task Information dialog box.

Project displays the delayed task on the Gantt chart.

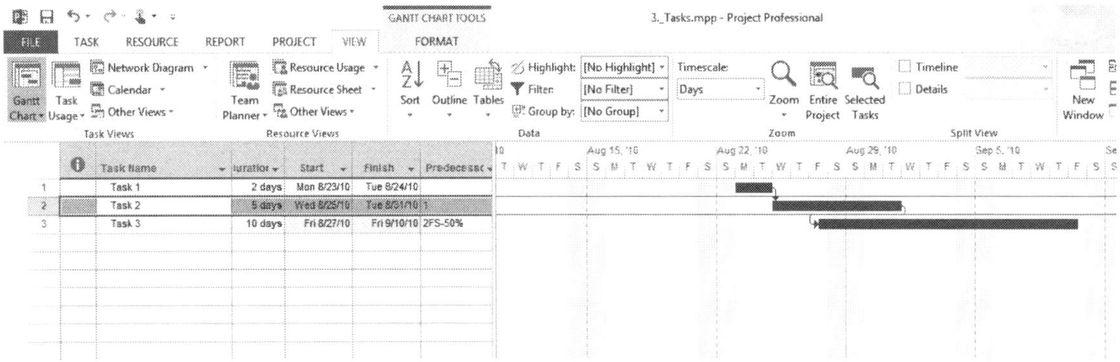

Notice the area shaded in yellow. The task for which Task 2 is a predecessor has changed, because of the delay. Also notice the Predecessors field.

Setting Task Deadlines

To setup a deadline, use the following steps.

Step 1: Select the task for which you want to set a deadline. Double-click to open the Task Information dialog box.

Step 2: Select the Advanced Tab.

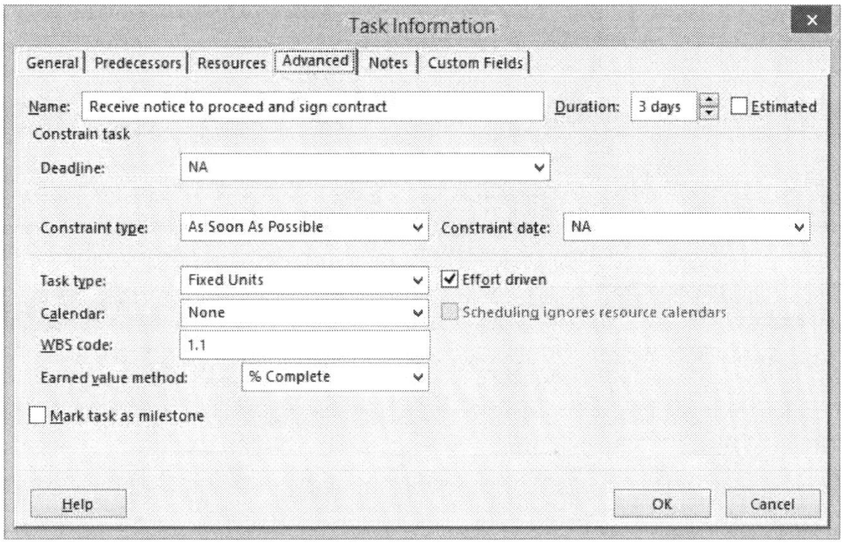

Step 3: In the Deadline field, click the down arrow and select the deadline from the calendar.

Step 4: Select OK to close the Task Information dialog box.

The following example shows a task whose deadline has passed, and the work has not yet been shown complete, so the Indicator column displays a warning.

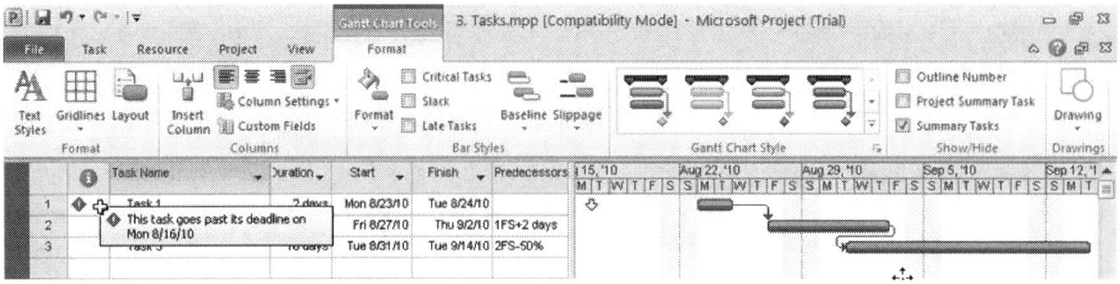

Setting Task Constraints

Use the following procedure to set a constraint type and date.

Step 1: Select the task for which you want to set a constraint. Double-click to open the Task Information dialog box.

Step 2: Select the Advanced Tab.

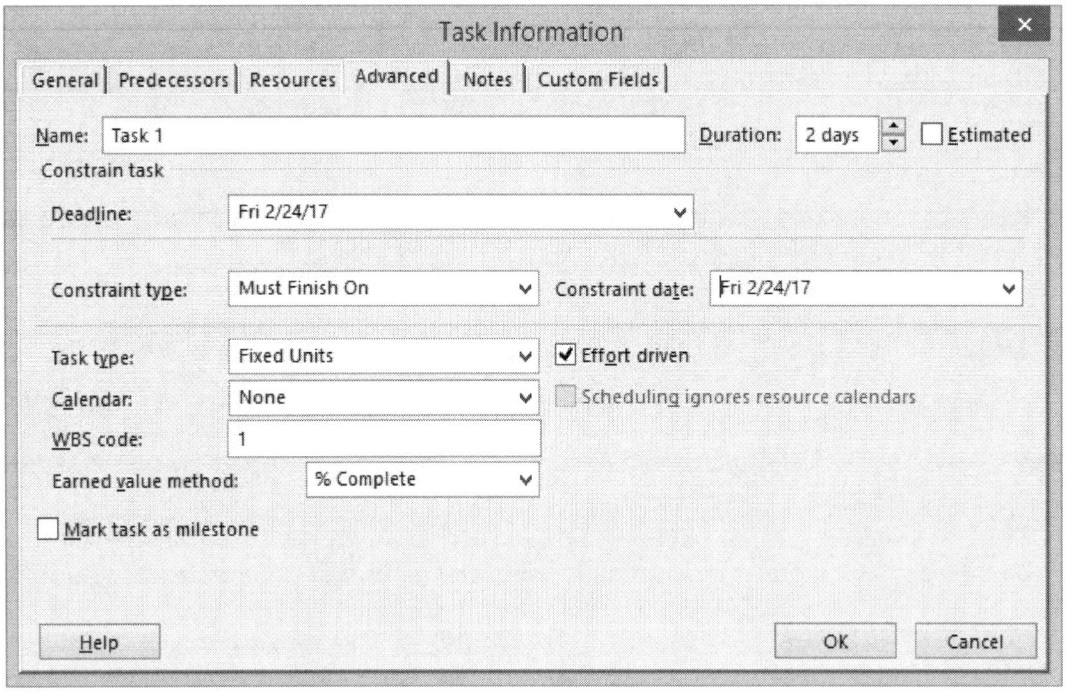

Step 3: Select the Constraint type from the drop-down list.

Step 4: Select the arrow next to the Constraint date and select the constraint date from the calendar.

Step 5: Select OK to close the Task Information dialog box.

Project displays an icon in the Indicator column to indicate that the task includes a constraint. The text for the indicator explains the constraint.

How to open the Constraint Dates table.

Step 1: Select the View tab from the Ribbon.

Step 2: Select the Tables drop-down list.

Step 3: Select More Tables.

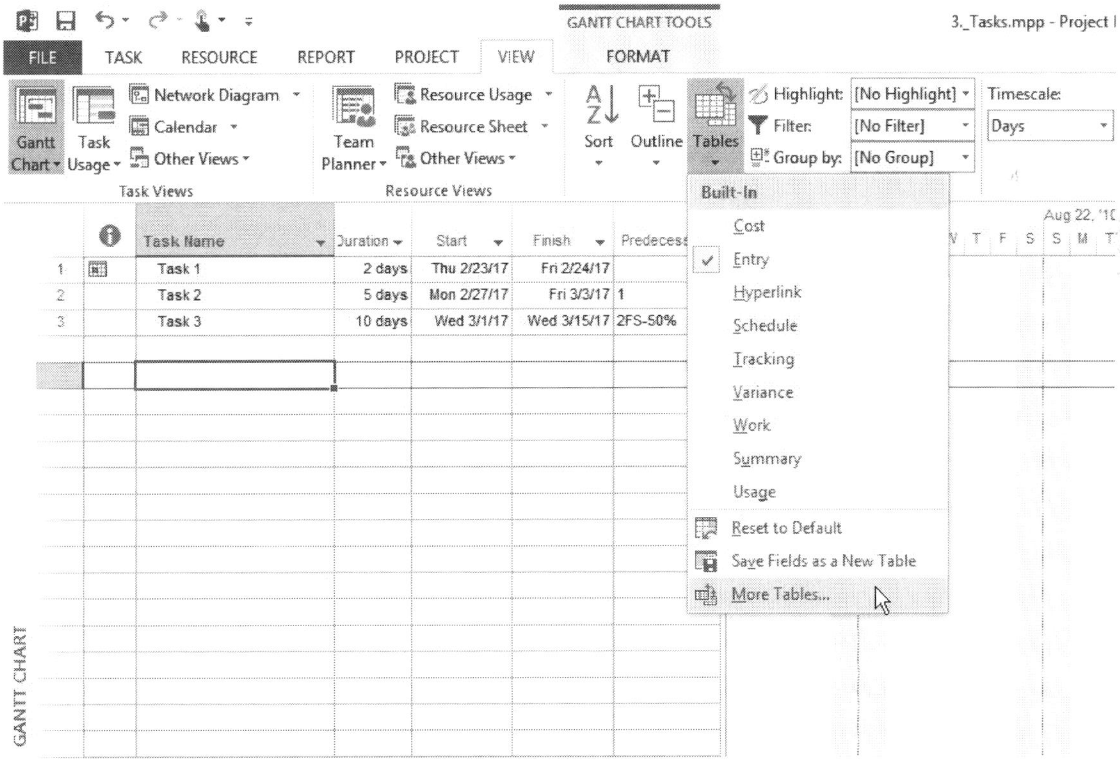

Step 4: Select Constraint Dates.

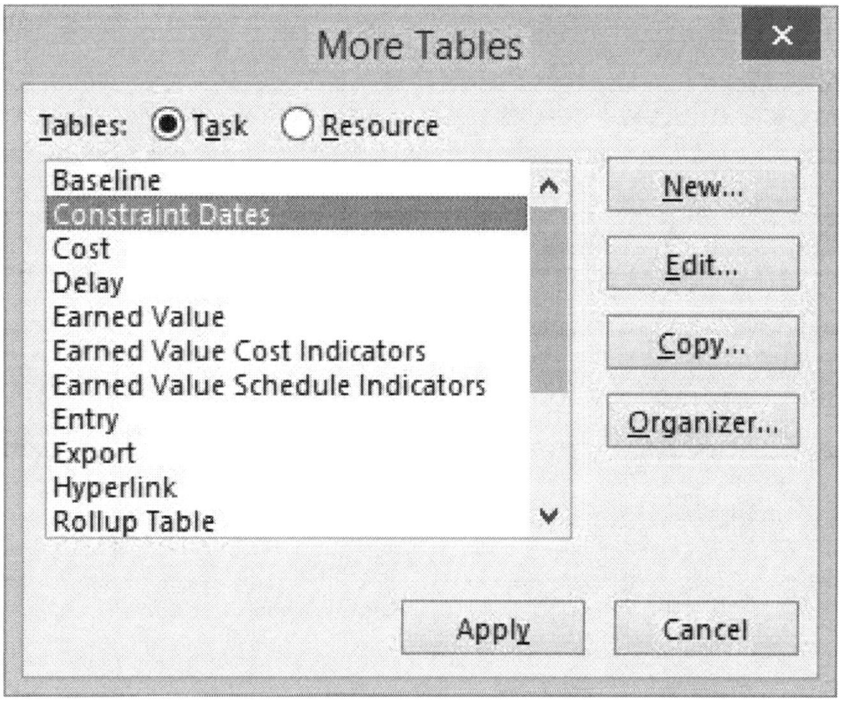

Step 5: Select Apply.

Project displays the Constraint Dates table in the left side of the window. The Gantt chart is still displayed on the right side of the window, allowing you to easily refer to the schedule while setting constraints for multiple tasks.

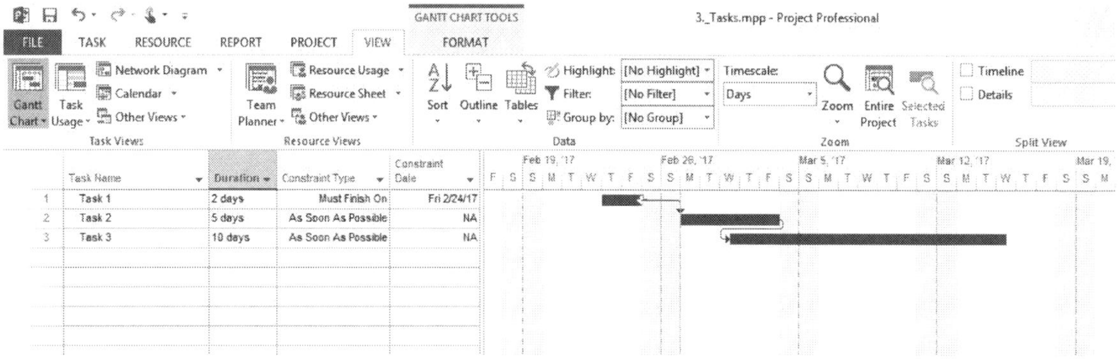

Splitting Tasks

Use the following procedure to split up tasks.

Step 1: Select the task you want to split.

Step 2: Select the Split Tasks tool from the Task tab on the Ribbon.

Project displays the Split Tasks dialog.

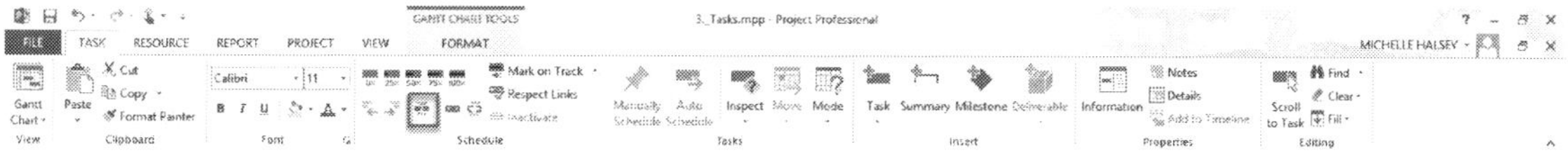

Step 3: Hover your mouse over the task and Project displays the corresponding date in the Split Tasks dialog.

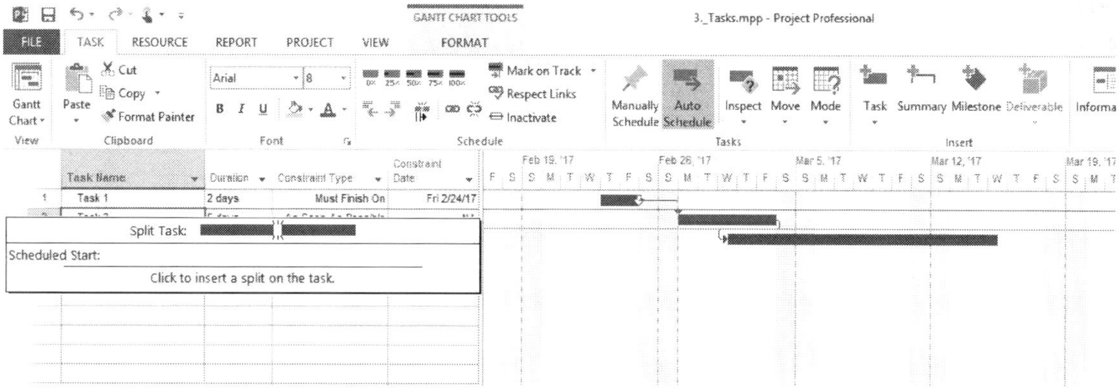

Step 4: When the date corresponds to the date where you need to split the task, click the mouse.

Project inserts a break in the task, as illustrated below.

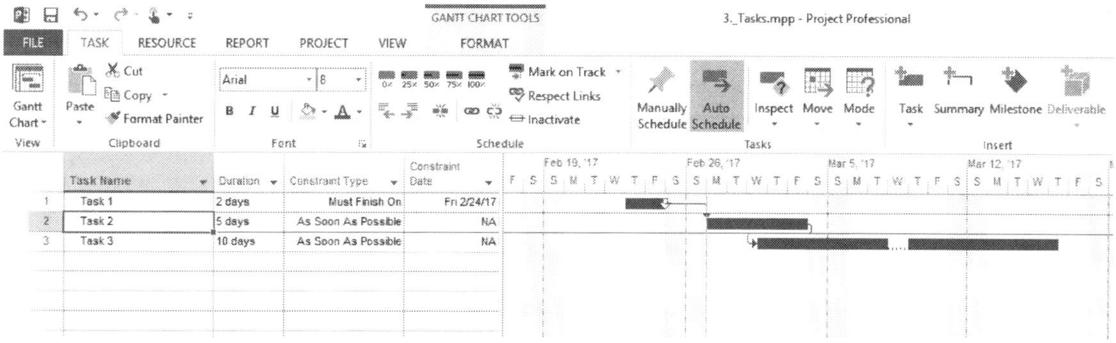

Step 5: You can drag the second part of the task to any start date.

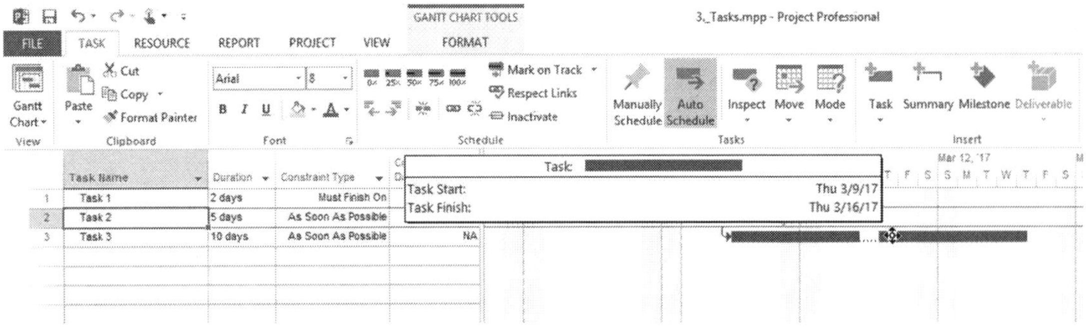

The following illustration shows a task that has been split, and then resumed on the following Monday.

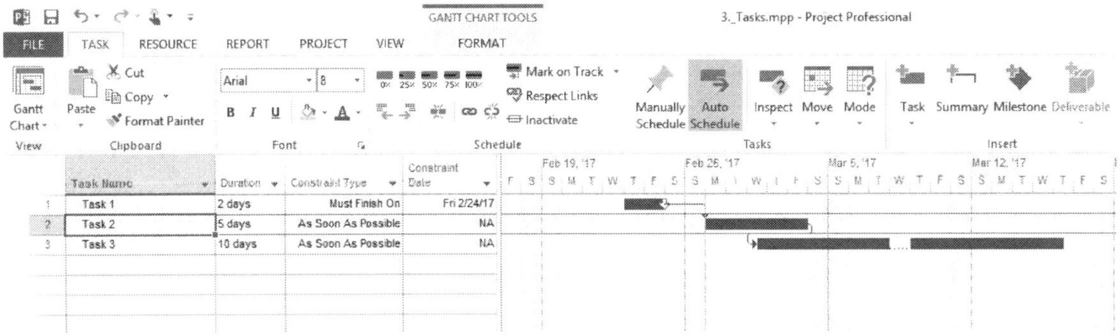

Understanding Task Type

There are three types of tasks that affect scheduling in a project. Project recalculates any changes made differently, based on which type of task you select. Let's look at a fixed unit task first.

Fixed Unit Task

When you revise a fixed unit task, Project makes the recalculations as follows:

- Revisions to units creates a recalculated duration.
- Revisions to duration creates a recalculated work value.
- Revisions to the work value creates a recalculated duration.

Let's look at an example to show how this works. Open the sample file 4. Tasks 2.mpp. The sample task has a duration of 10 days for one full-time resource. Project calculates the work to be 80 hours.

Imagine that another full-time resource is now available to help with the task. Assign the second resource, and see the new duration of 5 days.

Use the following procedure.

Step 1: Double-click the Task to open the Task Information dialog box.

Step 2: Select the Resources tab.

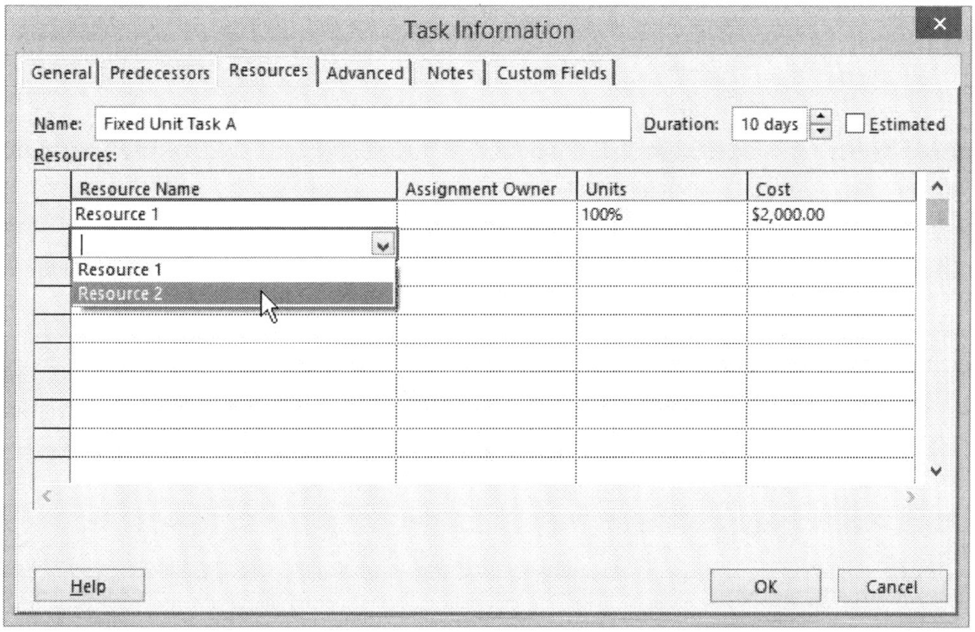

Step 3: Select the down arrow under Resource Name and select Resource 2.

Step 4: Select OK to close the Task Information dialog box.

Project recalculates the duration. Now the task only takes 5 days.

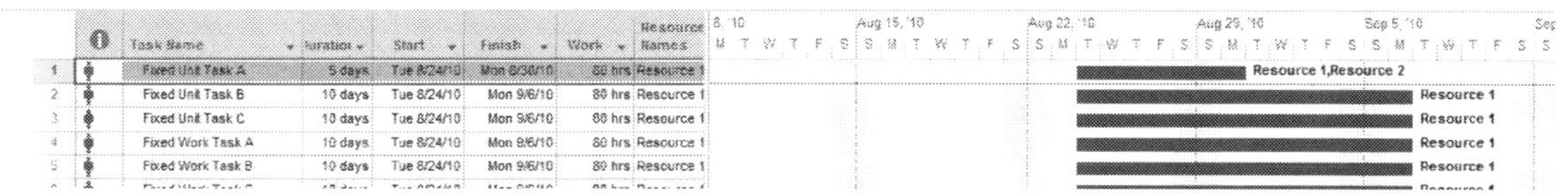

Now imagine that you must complete the task in eight days. Change the duration and see the recalculated work value.

Use the following procedure.

Step 1: Enter 8 as the duration for Fixed Unit Task B.

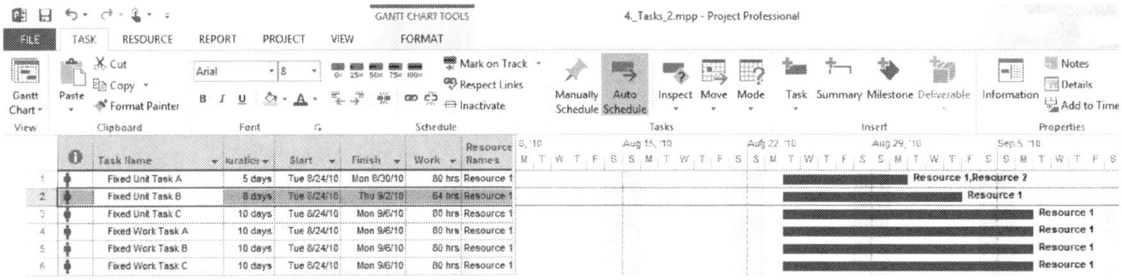

Project recalculates the Work.

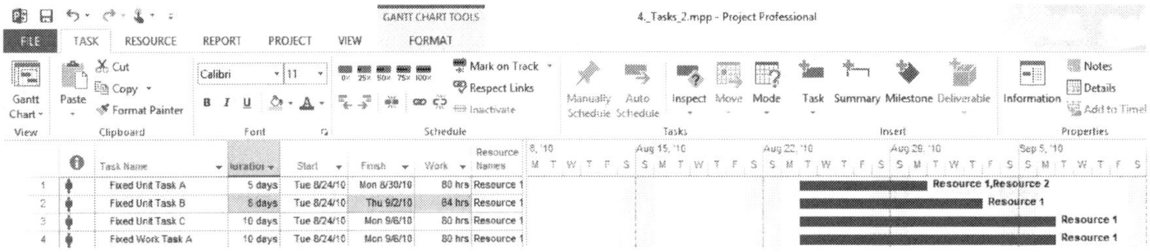

Finally, imagine that you determine the task will take 100 hours of work. You change the work and see the recalculated duration.

Use the following procedure.

Step 1: Enter 100 as the Work for Fixed Unit Task C.

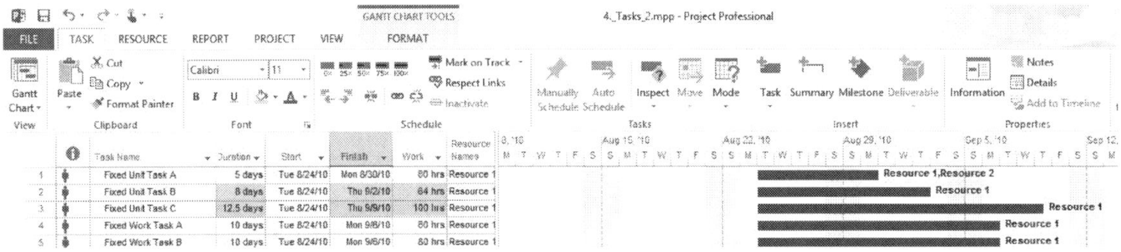

Project recalculates the duration.

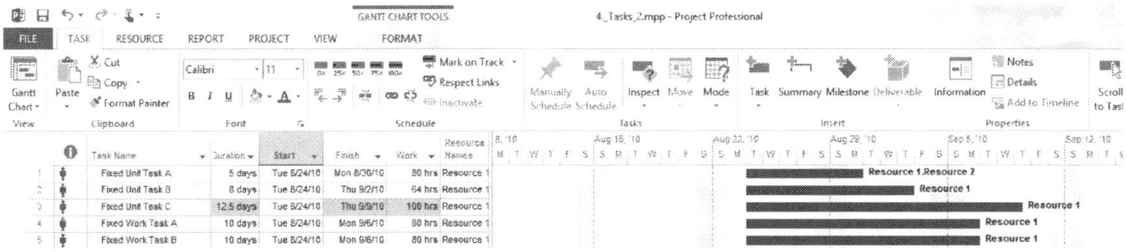

Fixed Work Task

When you revise a fixed work task, Project makes the recalculations as follows:

- Revisions to units creates a recalculated duration.
- Revisions to duration recalculates units.
- Revisions to the work value creates a recalculated duration.

The sample task is the same as the previous example, except the task type is Fixed work.

Imagine that another full-time resource is now available to help with the task. For Fixed Work Task A, assign the second resource, and see the new duration of 5 days. Use the procedure in the previous example to change the resource.

Project recalculates the duration. Now the task only takes 5 days.

Now imagine that you must complete the task in eight days. For Fixed Work Task B, change the duration and see the recalculated work value. Use the procedure in the previous example to change the duration.

Project recalculates the resource units. Resource 1 is now overallocated.

Finally, imagine that you determine the task will take 100 hours of work. For Fixed Work Task C, change the work and see the recalculated duration. Use the procedure in the previous example to change the work.

Project recalculates the duration.

Fixed Duration Task

When you revise a fixed duration task, Project makes the recalculations as follows:

- Revisions to units creates a recalculated work value.
- Revisions to duration creates a recalculated work value.
- Revisions to the work value recalculates the resource units.

The sample task is the same as the previous example, except the task type is Fixed duration.

Imagine that another full-time resource is now available to help with the task. For Fixed Duration Task A, assign the second resource. Project has reallocated each resource to 50%, making each resource available to work on other tasks 50% of the time for the duration of the task. Use the procedure in the previous example to change the resource.

Project recalculates the duration. The duration has not changed. The work has not changed. However, now each resource is allocated at 50%.

Now imagine that you must complete the task in eight days. For Fixed Duration Task B, change the duration and see the recalculated work value. Use the procedure in the previous example to change the duration.

Project recalculates the work. Notice that the work value is 64 hours.

Finally, imagine that you determine the task will take 100 hours of work. For Fixed Duration Task C, change the work. The resource is now over allocated. Use the procedure in the previous example to change the work.

Assigning a Task Calendar

Use the following procedure to create a new calendar specific to a task

Step 1: Select the Project tab from the Ribbon.

Step 2: Select Change Working Time.

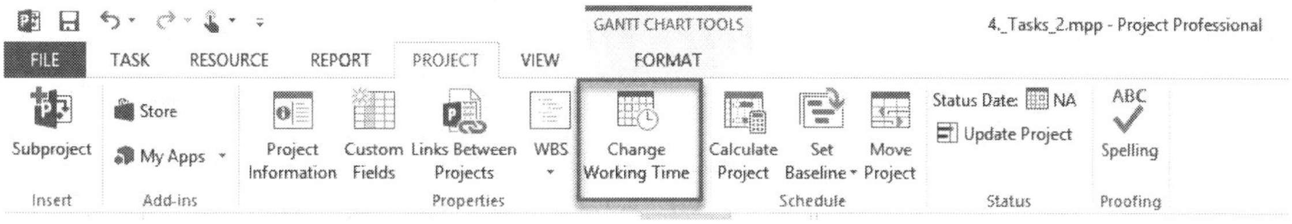

Step 3: Select Create New Calendar.

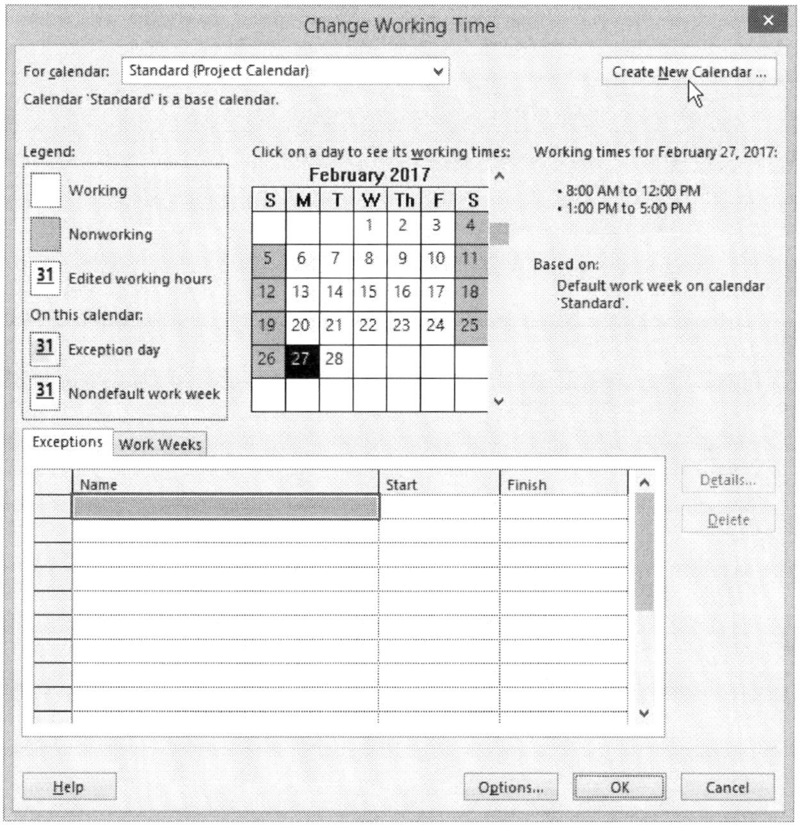

Project displays the Create New Base Calendar dialog box.

Step 4: Enter a name for the calendar.

Step 5: Select an option for the starting point for the base calendar and select OK.

Step 6: In the Change Working Time dialog box, select the Work Weeks tab.

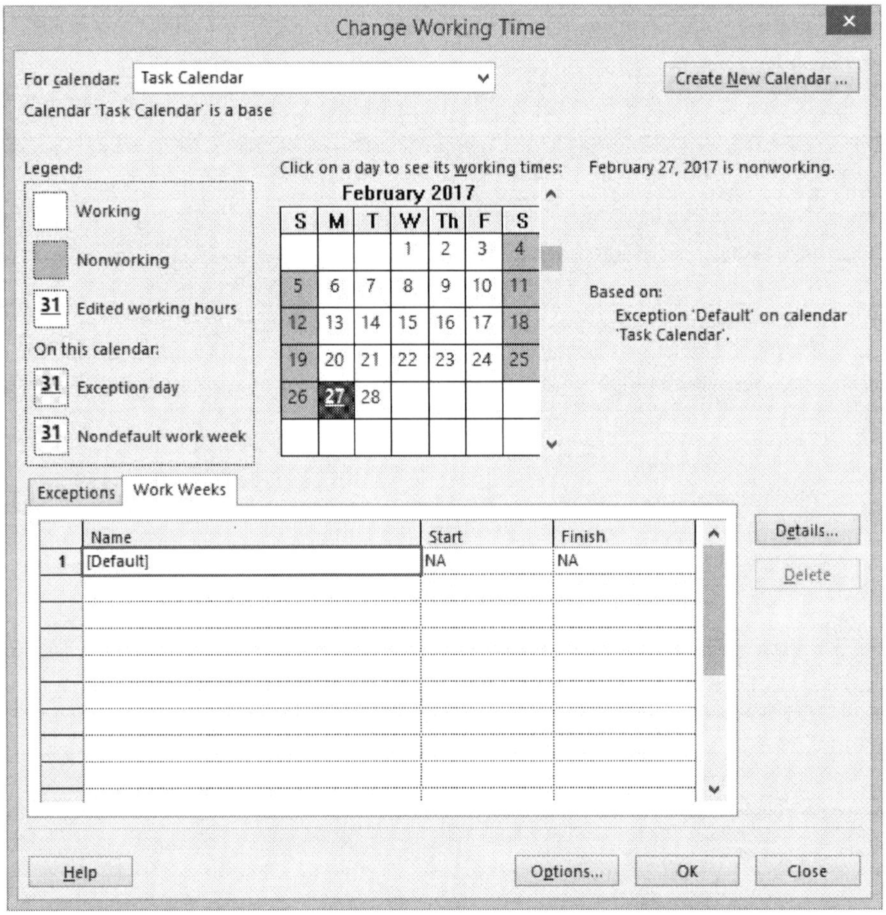

Step 7: Select Details to set the work week.

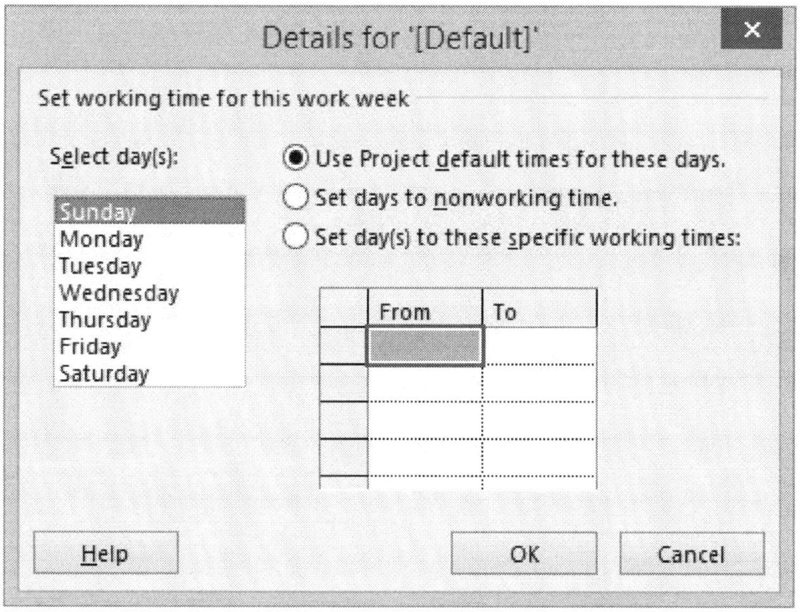

Step 8: For each day of the week, select the day (or press Ctrl while selecting multiple days that are the same) and select an option to use the default working times, set the days to nonworking time, or enter the specific working hours for those days. In this example, we will set everything to nonworking, except Fridays, which have working hours from 4pm to 12 am.

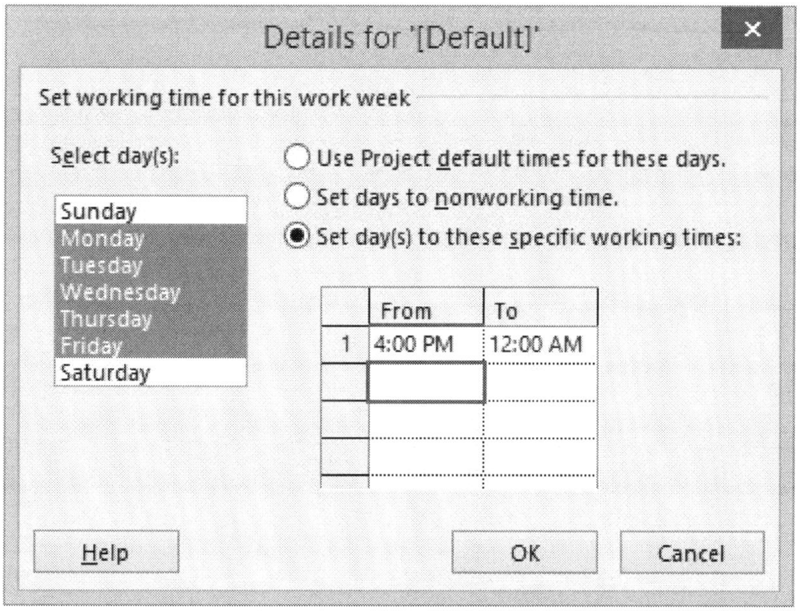

Step 9: Select OK to close the Details dialog box. Select OK to close the Change Working Time dialog box.

Applying the New Calendar to a Task

Step 1: Select the task for which you want to set a task calendar. Double-click to open the Task Information dialog box.

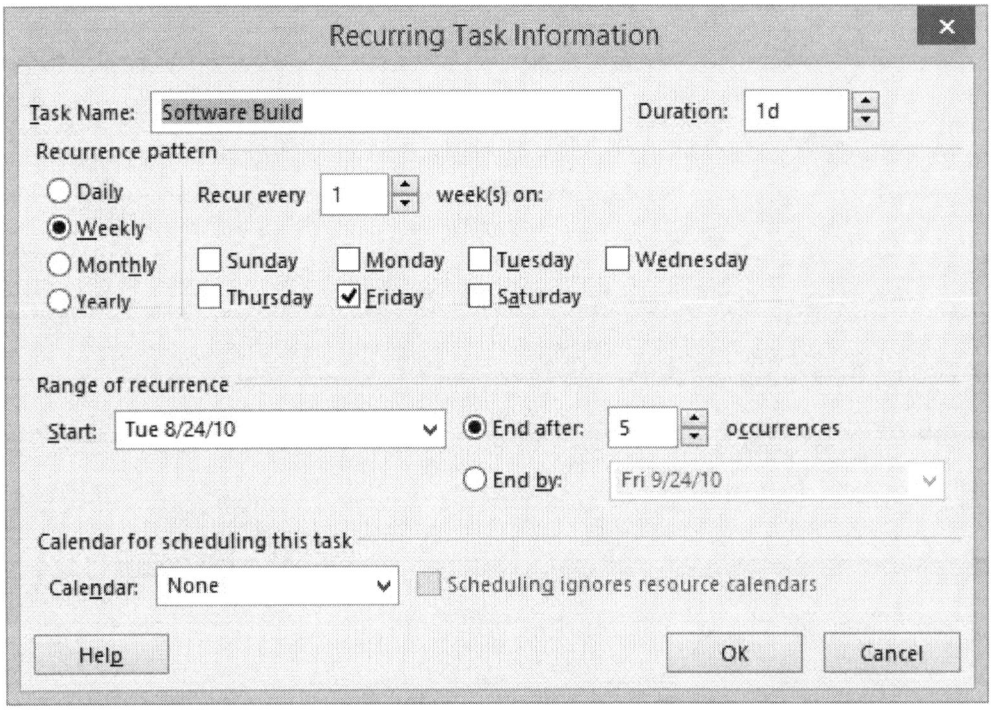

Step 2: Select the calendar you created from the Calendar drop-down list.

Step 3: Check the Scheduling ignore resource calendars, if applicable.

Step 4: Select OK.

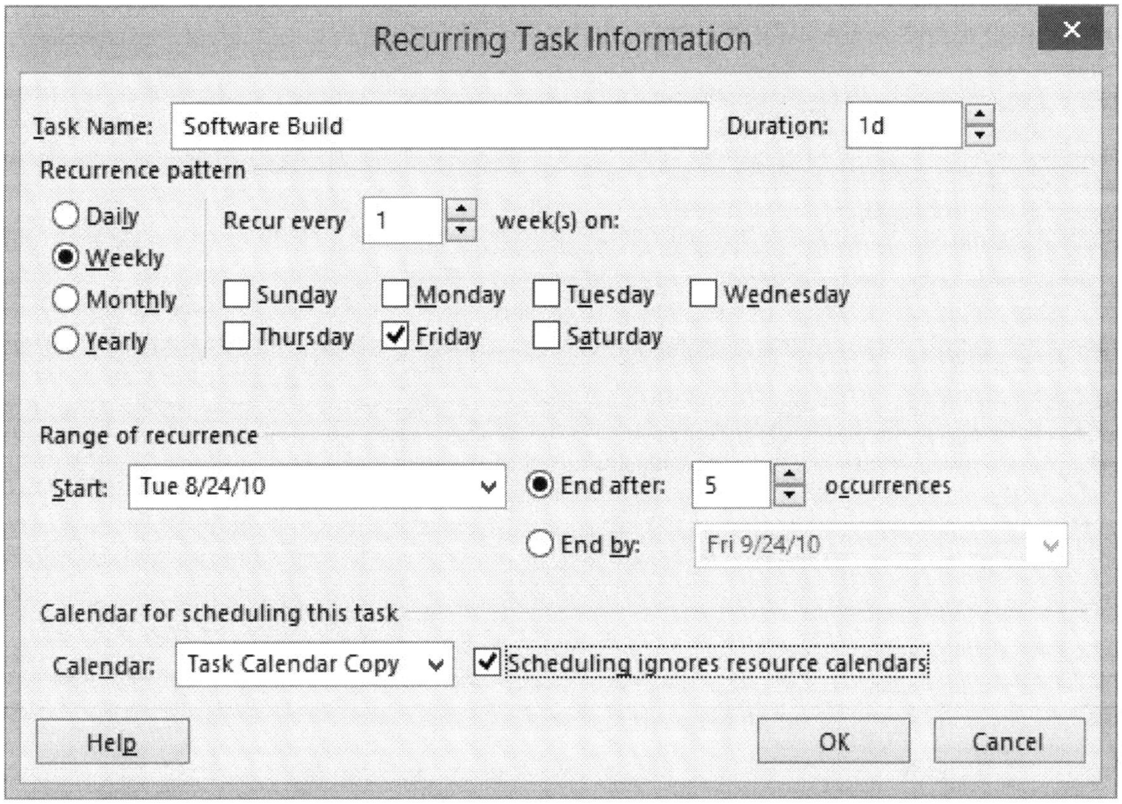

Step 5: In this example, Project displays a warning dialog box. Select Yes to schedule the tasks during nonworking times, as that is when this task occurs.

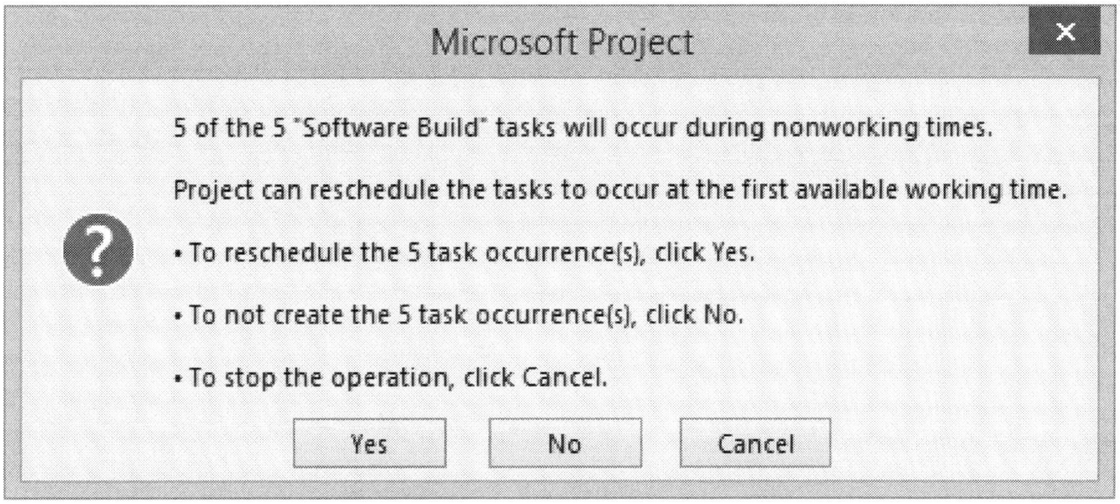

Notice the new duration and task indicators.

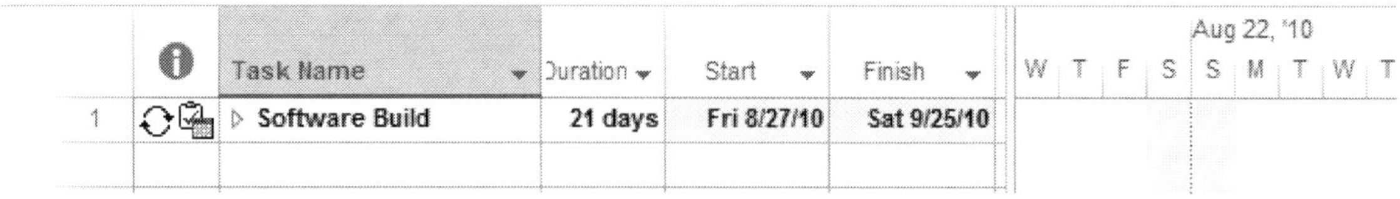

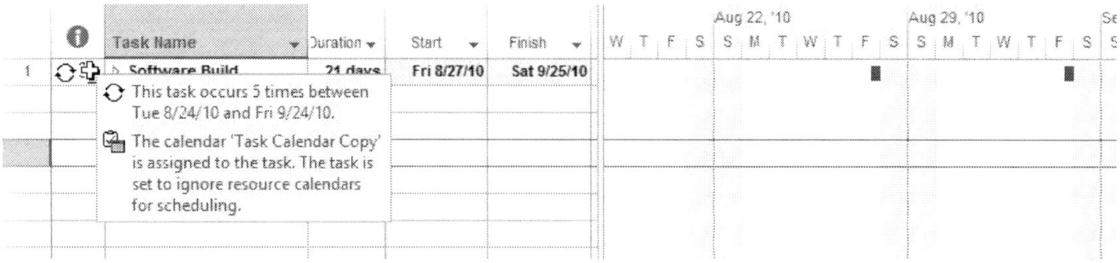

Understanding Task Indicators

The different types of indicators.

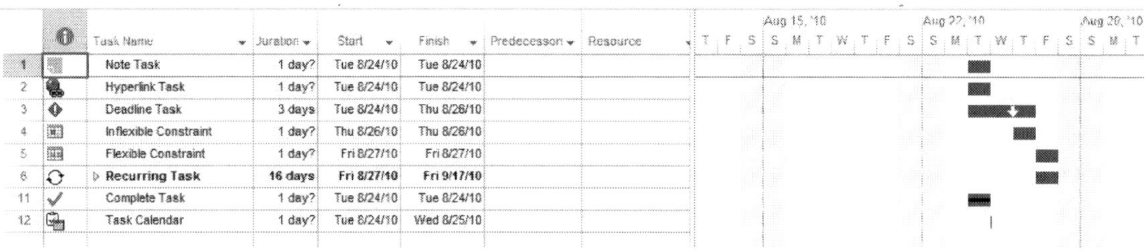

Chapter 13 – Working with Resources

In this chapter, you will learn some advanced techniques for working with resources. You will need to be able to adjust some resource details to have Project develop an accurate schedule. For example, if one of your resources is not available for the first two weeks of a project, you will need to enter that information in Project to adjust the tasks for that resource accordingly. You will learn about resource contours, resource calendars, availability dates, and how to group resources.

Learn how to handle resources in your projects using the Assign Resources dialog box to both remove and replace a resource assignment. You will also learn how to manage unassigned tasks. Then, the chapter covers some options for resolving resource conflicts. You will also learn how to print a view or report of resource information.

We will talk about adding individual and multiple resources to the Enterprise Resource Pool so that you can share resources from your project with the entire organization. You will also learn how to export and import resource information to or from other file formats, like Microsoft Excel. Finally, you will learn how to view resource availability across multiple projects.

Assigning a Resource Calendar

How to assign a new base calendar to the resource calendar.

Step 1: Change the view to the Resource Sheet by selecting Resource Sheet from the View tab.

Step 2: Double-click the resource for which you want to adjust the calendar to open the Resource Information dialog box.

Step 3: In the Resource Information dialog box, select Change Working Time button.

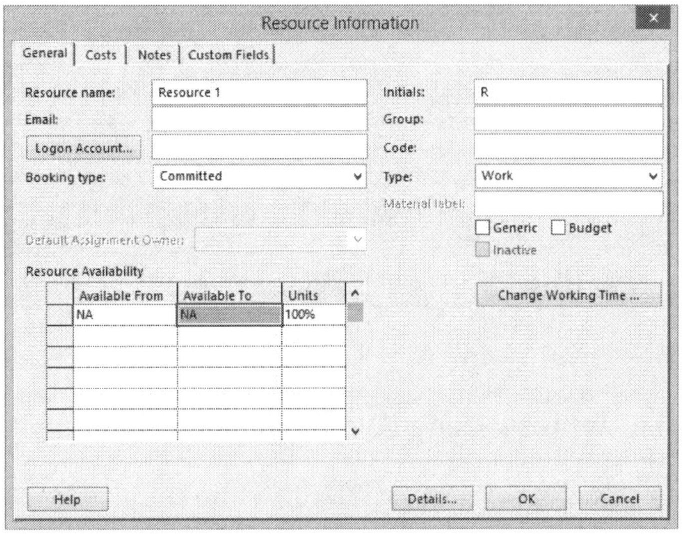

Step 4: Select a new Base Calendar from the drop-down list.

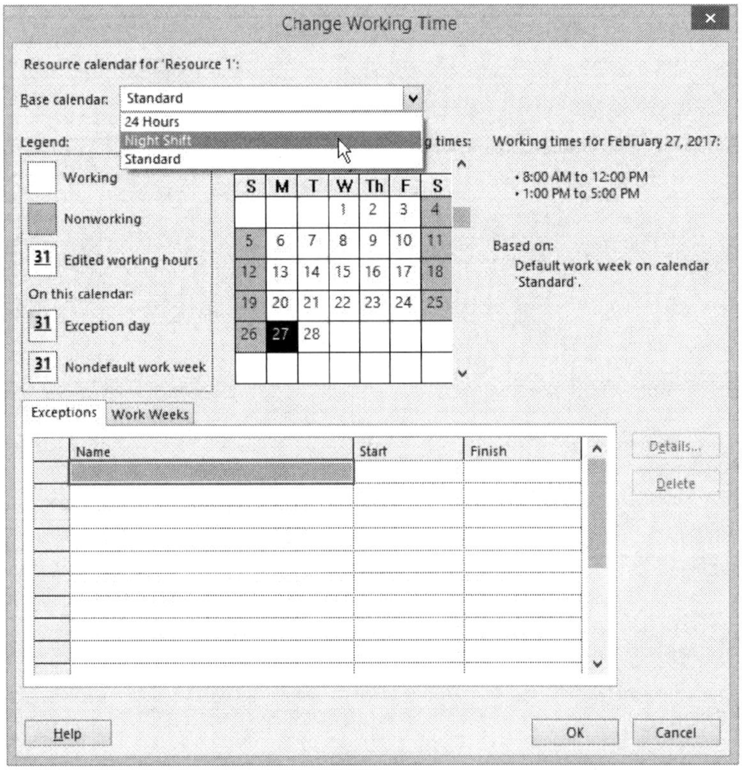

Step 5: Select OK to close the Change Working Time dialog box.

Step 6: Select OK to close the Resource Information dialog box.

Customizing a Resource Calendar

How to enter a non-default work week for a resource calendar.

Step 1: Change the view to the Resource Sheet by selecting Resource Sheet from the View tab.

Step 2: Highlight the resource for which you want to adjust the calendar and double-click to open the Resource Information dialog box.

Step 3: In the Resource Information dialog box, select Change Working Time button.

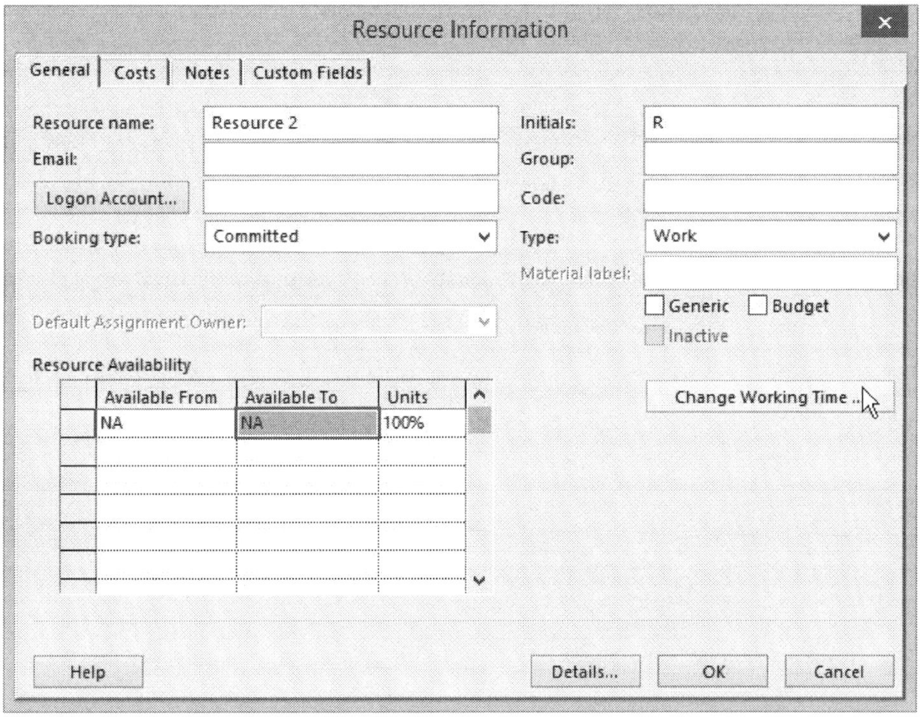

Step 4: Select the Work Weeks tab.

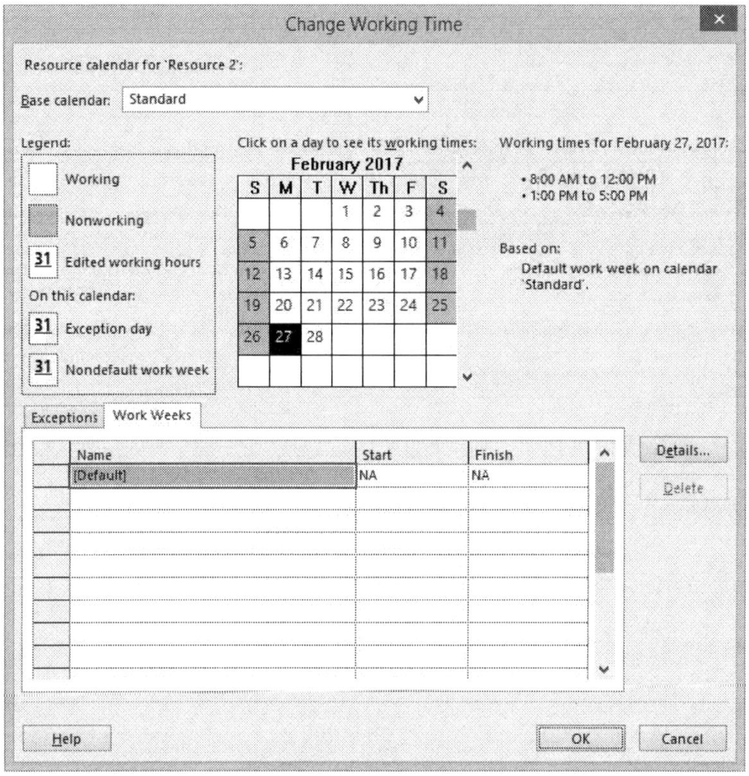

Step 5: Enter a Name for your reference to identify the work week. For example, you might call it "Finishing Project A."

Step 6: Select the Start column to select a Start Date for the non-default time period from the calendar.

Step 7: Select the Finish column to select an End Date for the non-default time period from the calendar.

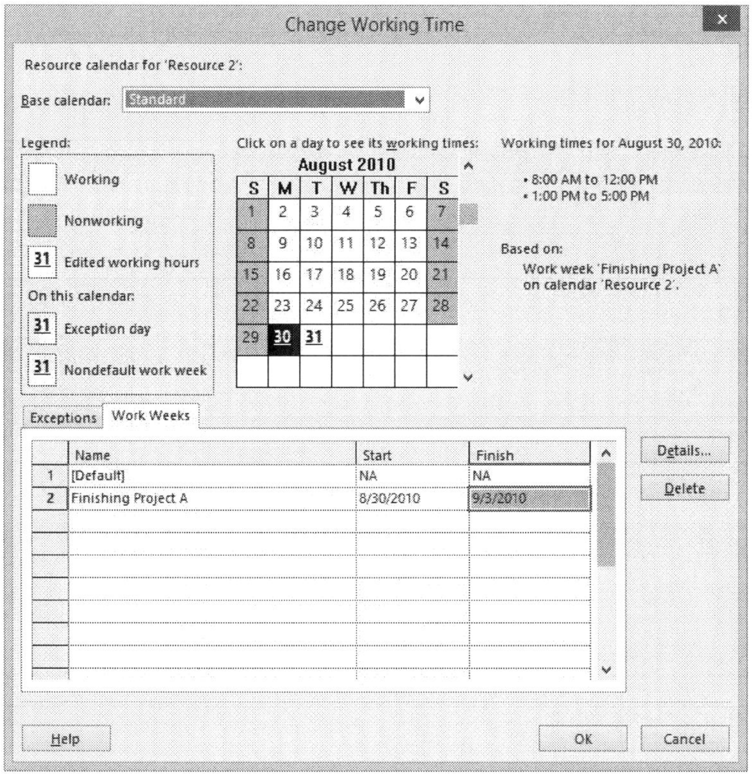

Step 8: Select Details to define the working hours for the selected work week.

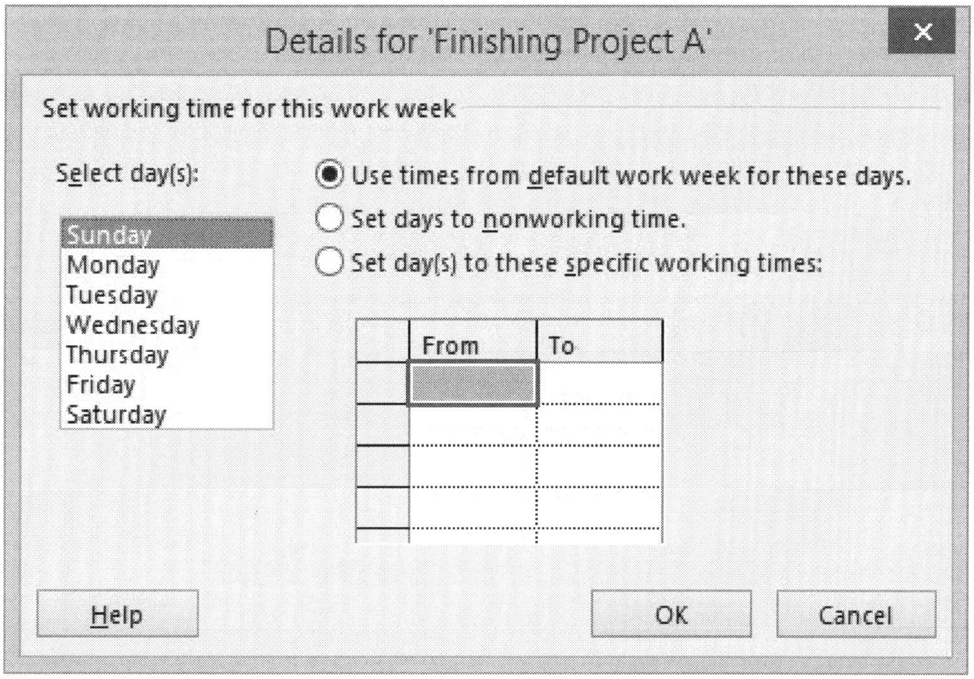

Step 9: Highlight the day or days (hold down the SHIFT or CTRL keys to select multiple days) that you want to define.

Step 10: Select either Use times from default work week for these days, set days to nonworking time, or Set day(s) to these specific working times. If you select the third option, also enter the Start and Finish time for each working period on the selected day(s). Repeat this step to define each day of the week for the selected work week.

Step 11: Select OK to close the Details dialog box.

Step 12: Select OK to close the Change Working Time dialog box.

Step 13: Select OK to close the Resource Information dialog box.

Applying Predefined Resource Contours

Use the following procedure to apply a contour.

Step 1: Select Task Usage from the View tab.

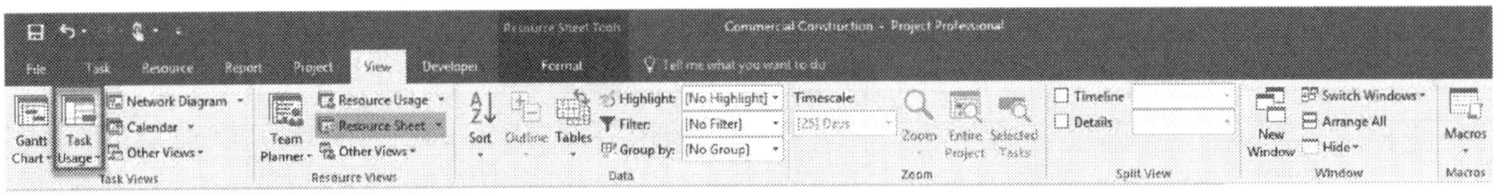

Step 2: Double-click on the resource assigned to the task that requires a contoured work assignment.

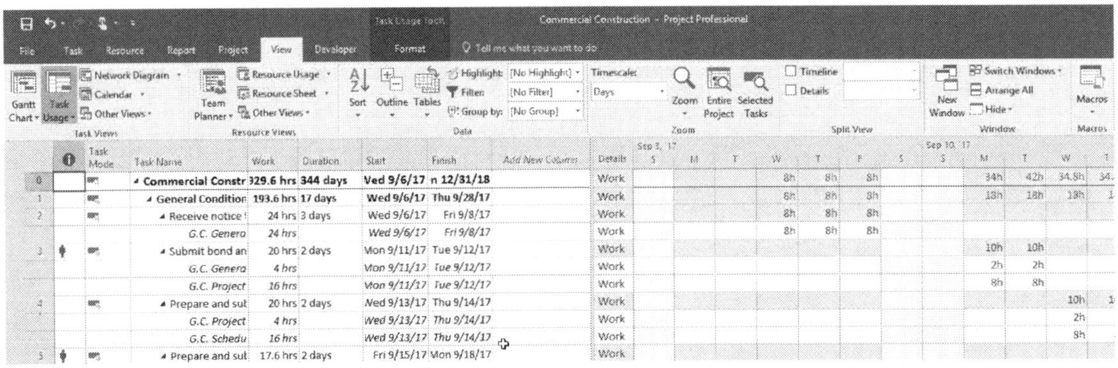

Project displays the Assignment Information dialog box.

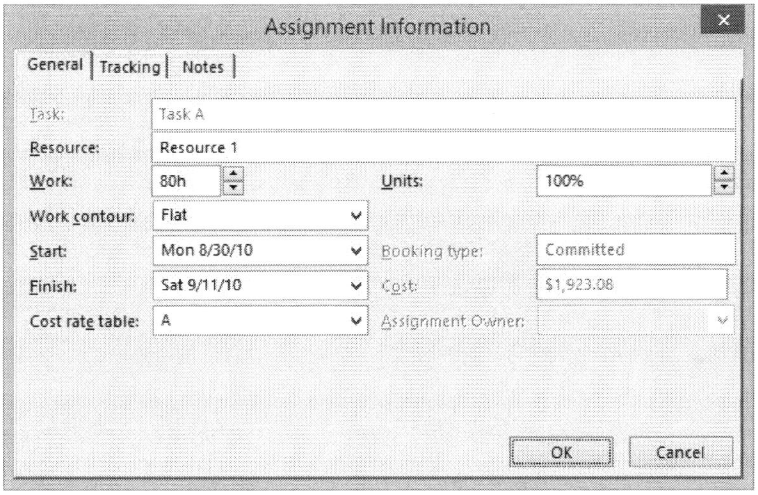

Step 3: In the Work Contour field, select an option from the drop-down list.

Step 4: Select OK.

Project displays the contour indicator in the Indicator Column. It also redistributes the work for the selected task according to the selected contour.

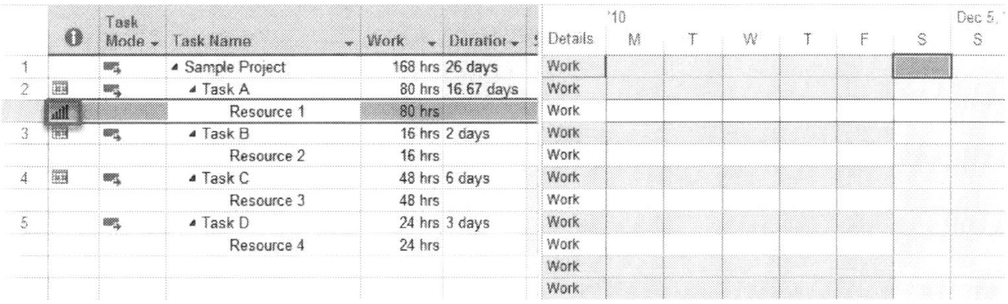

Specifying Resource Availability Dates

How to indicate resource availability.

Step 1: Select Resource Sheet from the View tab.

Step 2: Double-click on the resource to open the Resource Information dialog box.

Step 3: Select the arrow under Available From to select a Start Date for the availability definition from the calendar.

Step 4: Select the arrow under Available To select an End Date for the availability definition from the calendar.

Step 5: Select or enter the Units of availability as a percentage of full time.

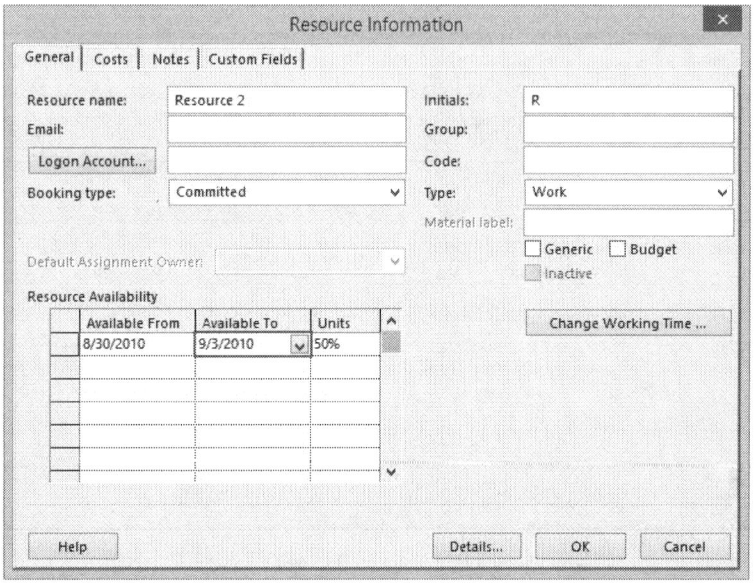

Step 6: Select OK to close the Resource Information dialog box.

Grouping Resources

Use the following procedure to group resources.

Step 1: Select Resource Sheet from the View tab.

Step 2: Enter a group name in the Group column for each resource.

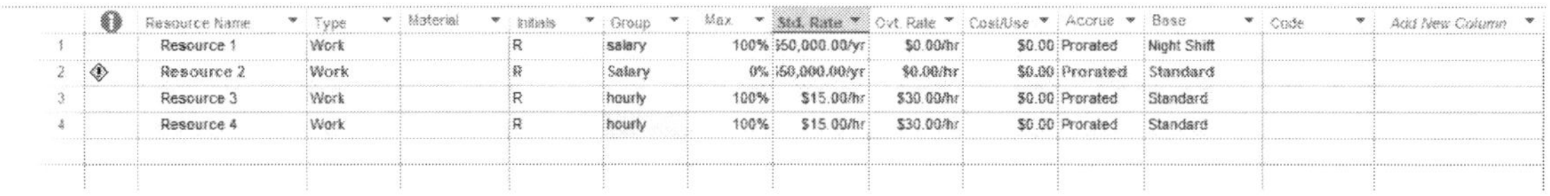

Step 3: Select the Group By drop-down list.

Step 4: Select Resource Group.

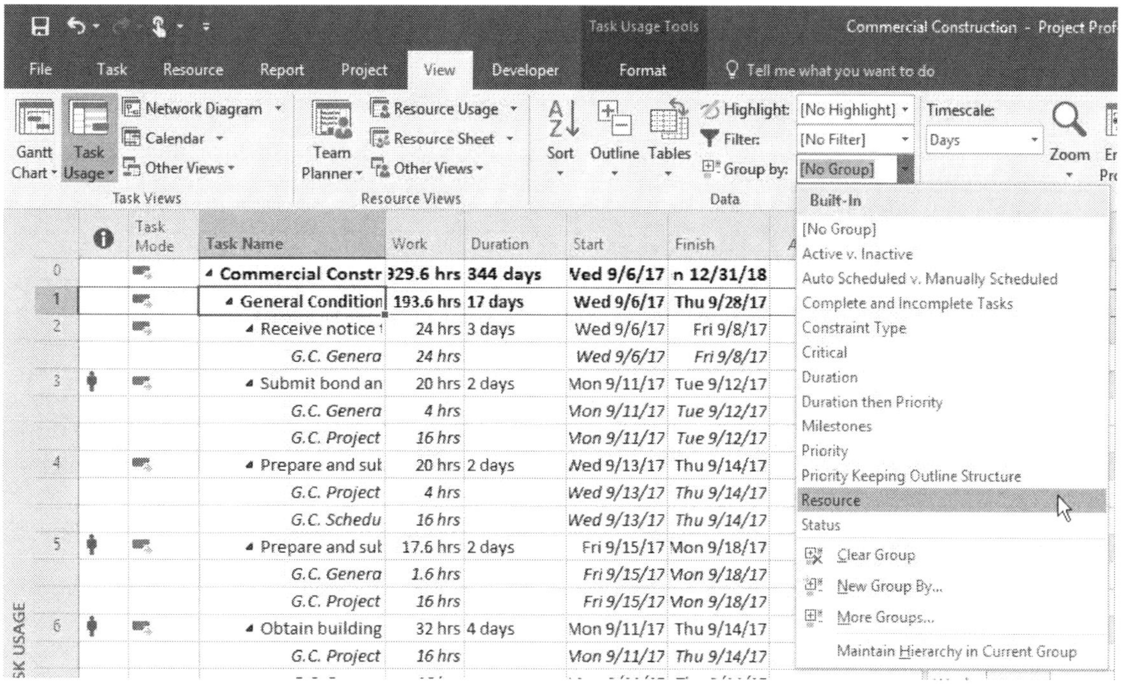

Project displays the resources in the appropriate group.

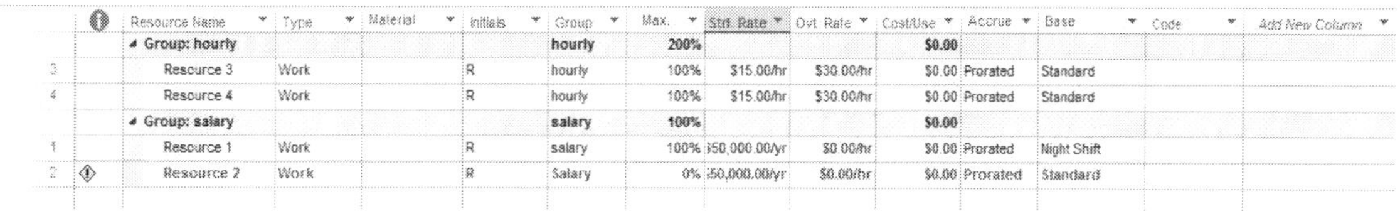

		Resource Name	Type	Material	Initials	Group	Max.	Std. Rate	Ovt. Rate	Cost/Use	Accrue	Base	Code	Add New Column
		Group: hourly				**hourly**	**200%**			**$0.00**				
3		Resource 3	Work		R	hourly	100%	$15.00/hr	$30.00/hr	$0.00	Prorated	Standard		
4		Resource 4	Work		R	hourly	100%	$15.00/hr	$30.00/hr	$0.00	Prorated	Standard		
		Group: salary				**salary**	**100%**			**$0.00**				
1		Resource 1	Work		R	salary	100%	$50,000.00/yr	$0.00/hr	$0.00	Prorated	Night Shift		
2	◈	Resource 2	Work		R	Salary	0%	$50,000.00/yr	$0.00/hr	$0.00	Prorated	Standard		

Removing a Resource Assignment

To remove a resource assignment, use the following procedure.

Step 1: While in Gantt Chart view, select the task for which you want to remove a resource assignment.

Step 2: Select the Resources tab from the Ribbon.

Step 3: Select Assign Resources.

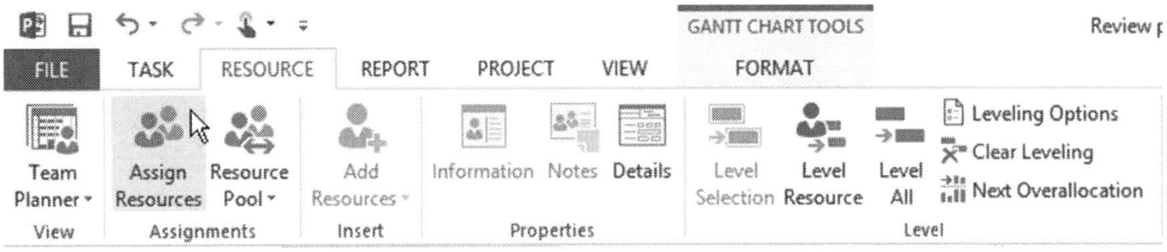

Step 4: In the Assign Resources dialog box, highlight the resource you want to remove. If there is more than one resource assigned, and you want to remove all of them, you can hold the SHIFT or CTRL keys down while selecting the resources.

Step 5: Select Remove.

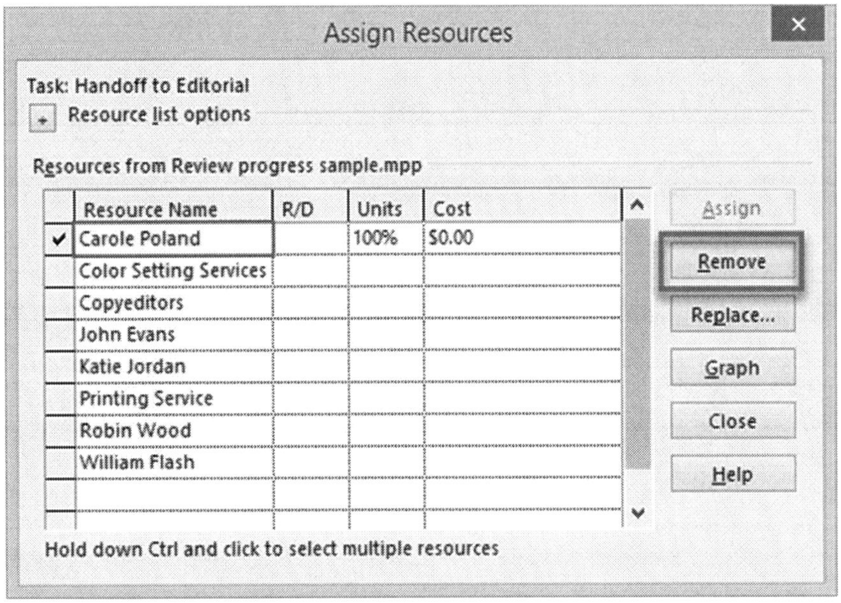

Replacing a Resource Assignment

To replace a resource, use the following procedure.

Step 1: While in Gantt Chart view, select the task for which you want to replace a resource assignment.

Step 2: Right-click the task.

Step 3: Select Assign Resources from the context menu.

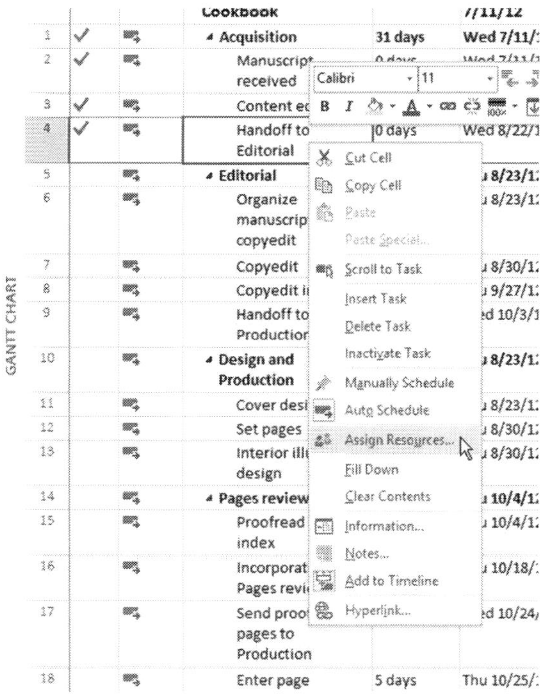

Step 4: In the Assign Resources dialog box, highlight the resource you want to replace.

Step 5: Select Replace.

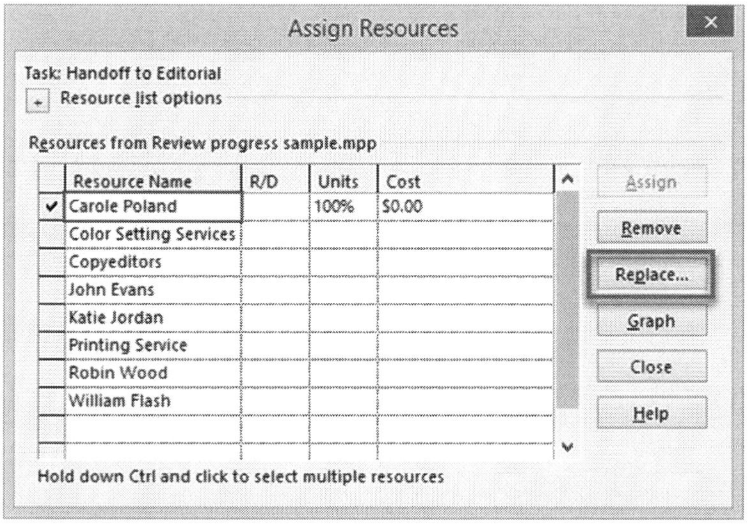

Step 6: In the Replace Resources dialog box, select the new resource to assign to the selected task.

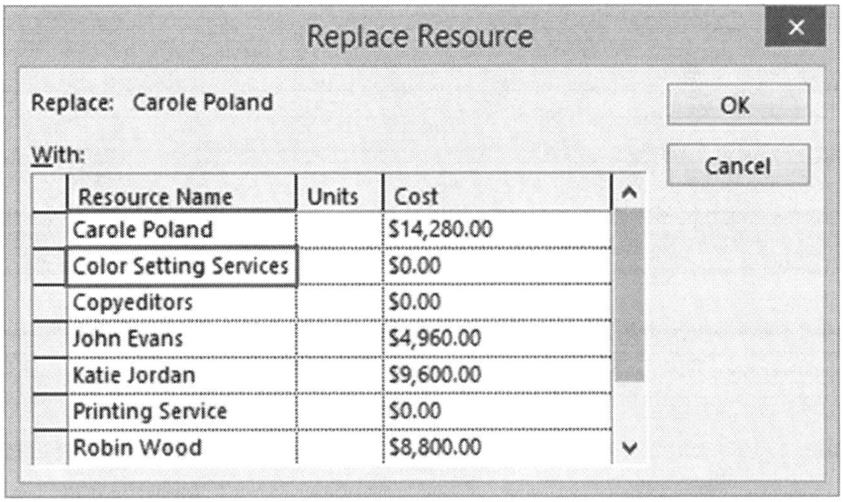

Step 7: Select OK.

Managing Unassigned Tasks

The Resource Usage Sheet, use the following procedure.

Step 1: Select the View tab from the Ribbon.

Step 2: Select Resource Usage.

Step 3: The Resource Usage sheet shows all the unassigned tasks at the top of the sheet.

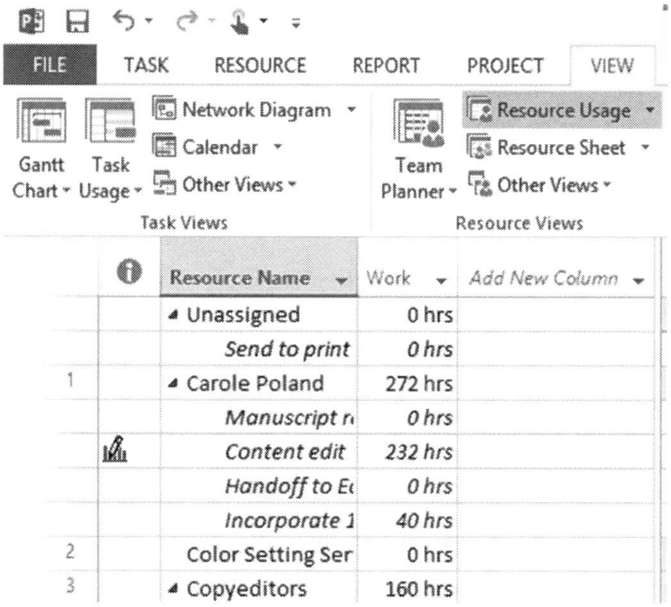

Step 4: You can use the View tools to help filter and manage the information.

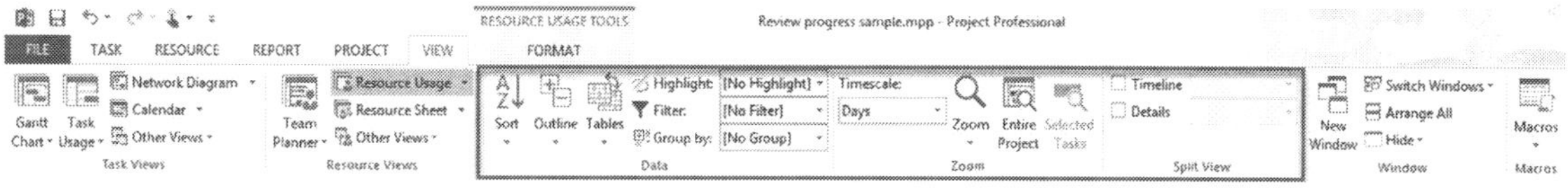

Step 5: You can also use the Resource Usage Tools Format tools to insert columns or custom fields, change the column settings, or choose which information to show on the right side of the screen.

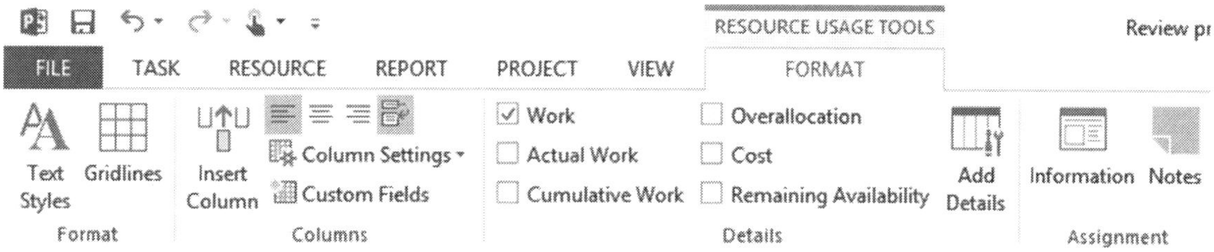

Step 6: You can also select Information to view the Assignment information dialog box.

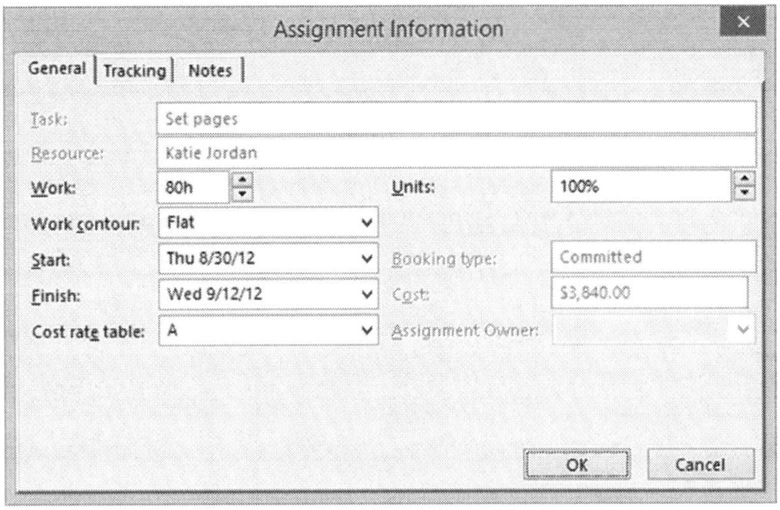

Another way to assign resources is by simply dragging the task assignment from one resource grouping to another. The task assignment appears as a horizontal bar until you release the mouse button.

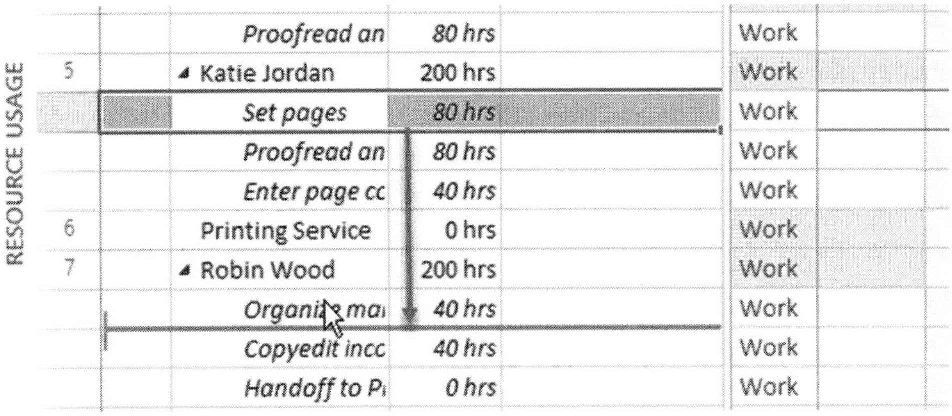

Resolving Resource Conflicts

To reschedule a task to the next available date, use the following procedure.

Step 1: Open the Task Usage sheet. To do so, select the View tab from the Ribbon. Select Task Usage.

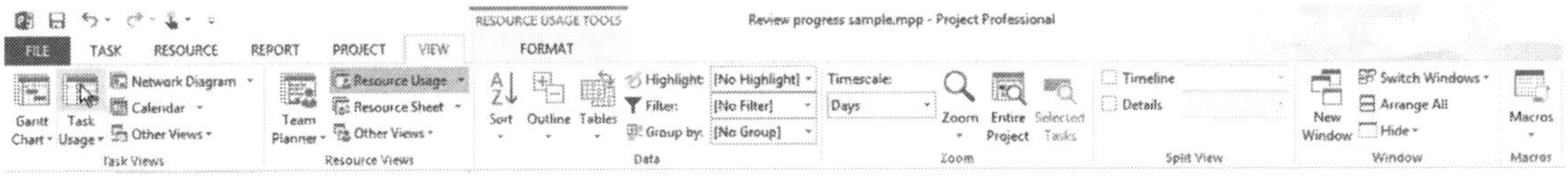

Step 2: Right-click on a task with an over allocation icon.

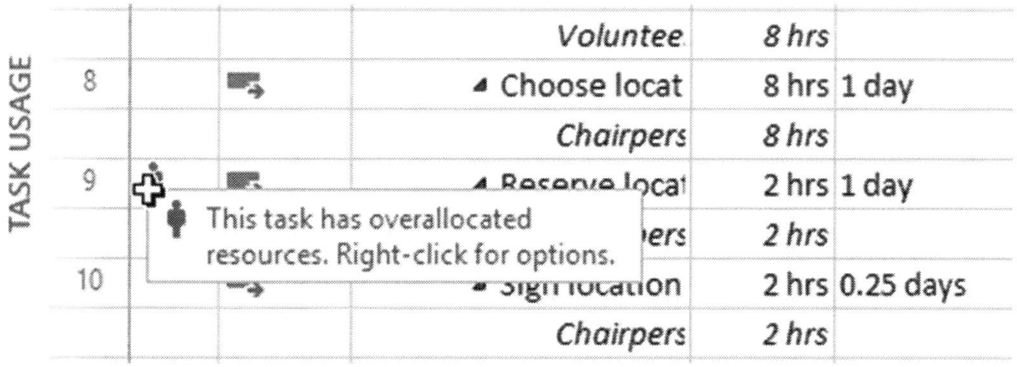

Step 3: Select Reschedule to Available Date from the context menu.

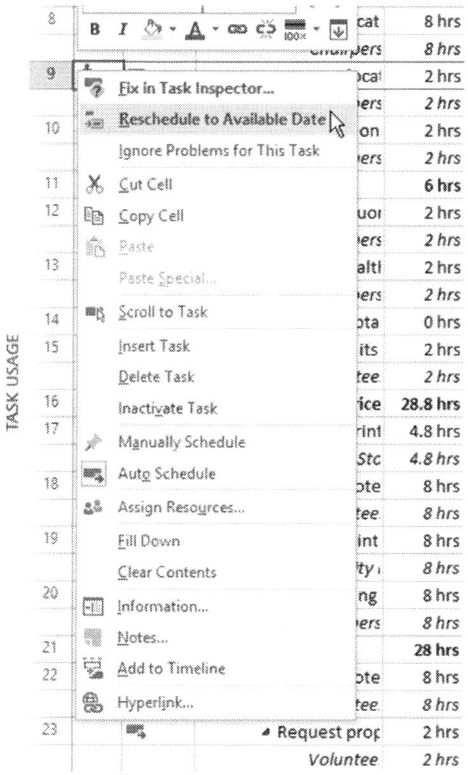

Step 4: The selected task is rescheduled.

Remember that you can use the Clear Leveling Options on the Resource tab if you need to revert to the schedule before you have done any leveling (either by task or for the whole project).

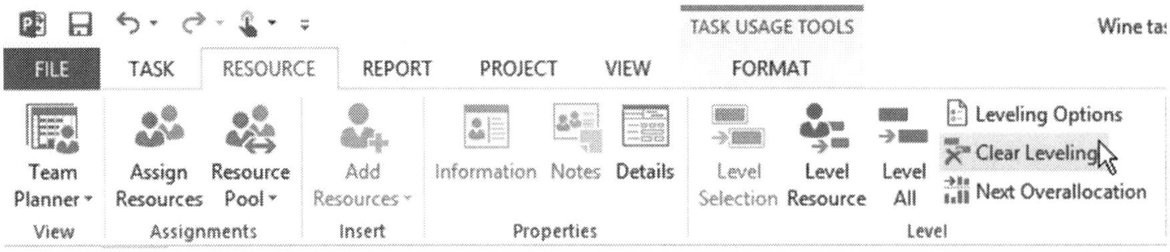

The Task Inspector pane.

Step 1: Task Usage sheet. To do so, select the View tab from the Ribbon. Select Task Usage.

Step 2: Right-click on a task with an over allocation icon.

Step 3: Select Fix in Task Inspector from the context menu.

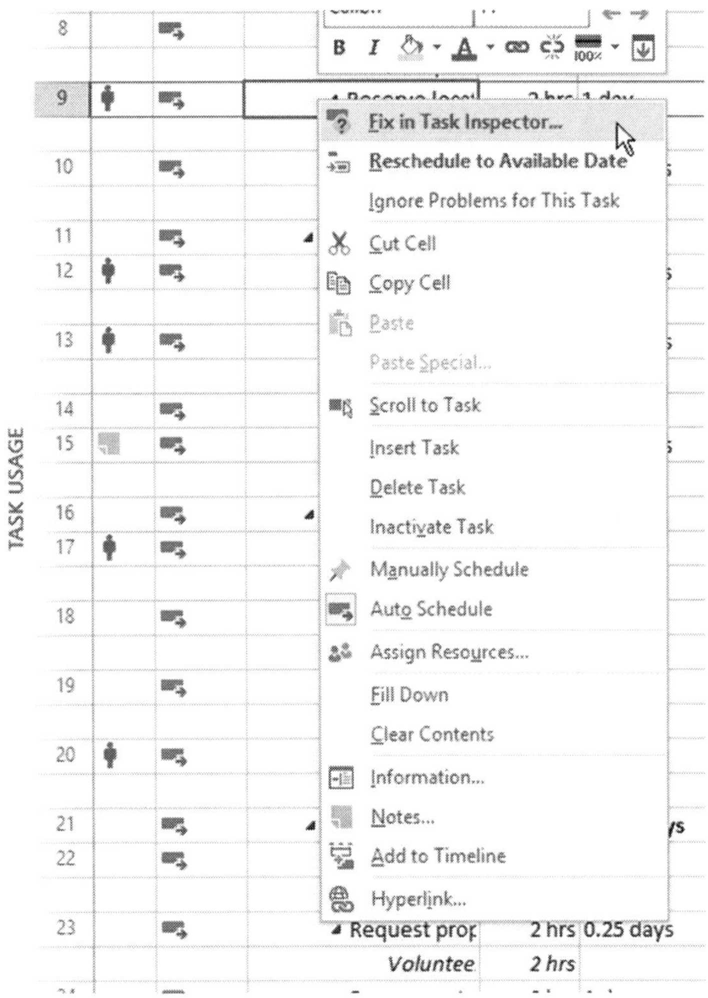

The Task Inspector pane opens on the left side of the screen.

Use the scroll bar to see all the options in the Task Inspector pane.

You can Reschedule the selected task; view the over allocated resources in the Team Planner, or see the factors affecting the task.

Printing a View or Report of Resource Information

To print a resource view, use the following procedure.

Step 1: Open the view that you want to print. For this example, let us use the Resource Usage view.

Step 2: Select the File tab from the Ribbon to open the Backstage view.

Step 3: Select Print.

Step 4: Use the Print Preview area to review your selection. For example, notice that the Resource Usage view is 96 pages when you print all dates. Perhaps you only want to print a week or a month of this view. You can use the Date Range settings to adjust.

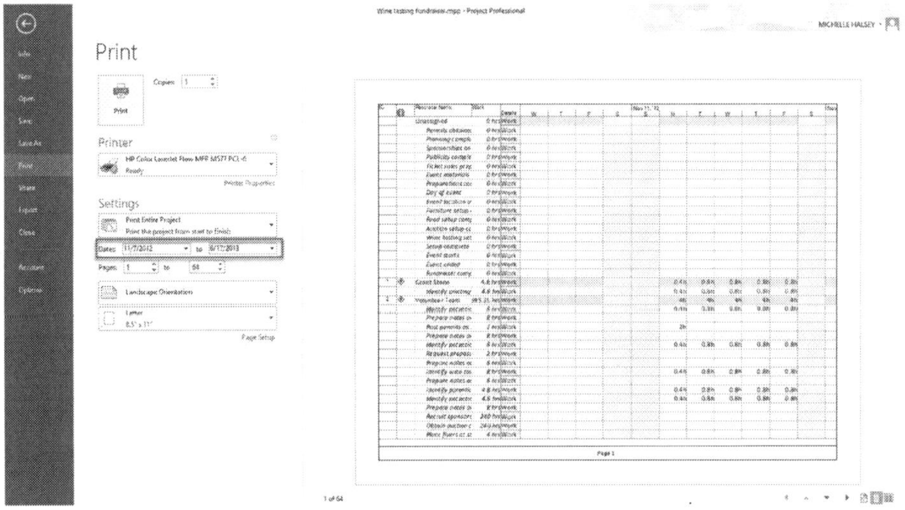

Step 5: Select Print.

To print a resource report, use the following procedure.

Step 1: Select the Project tab from the Ribbon.

Step 2: Select Reports.

Step 3: In the Reports dialog box, select the report category you want to print. Click Select.

Step 4: In the dialog box for the report category you chose, select the report that you want to print and click Select.

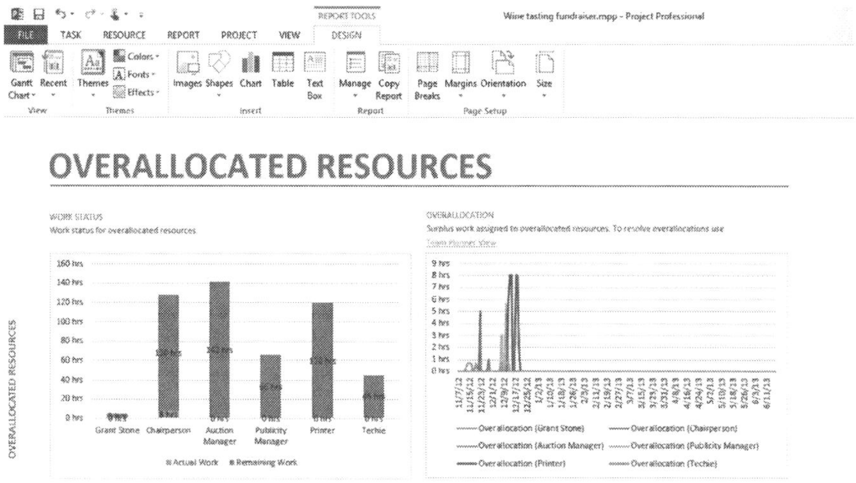

Step 5: Use the Print Preview area to adjust your settings, if necessary.

Step 6: Select Print.

Adding Resources to the Enterprise Resource Pool

To add individual resources to the enterprise resource pool, use the following procedure.

Step 1: Select the Resources tab from the Ribbon.

Step 2: Select Resource Pool. Select Enterprise Resource Pool.

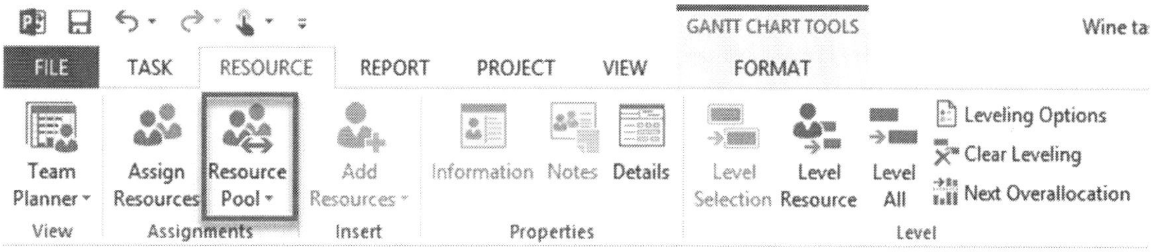

Step 3: Your browser displays Project Web Access. Check to see if it is not the window shown on top.

Step 4: On Project Web Access, select New Resource.

Step 5: Enter the resource information.

- Select work resource, cost resource, or material resource from the Resource Type list.
- Check the Budget or Generic check box to indicate that the resource is a budget resource or a generic (placeholder or skill based) resource.
- Clear the Resource can log on to Project Server check box if you do not want to collect information on tasks that the resource works on, or if you do not want the resource to log on to Microsoft Project Server.
- Enter the Resource Name and other identifying information, including the resource's email address and Resource Breakdown Structure value. You can also enter the hyperlink name and URL if the resource maintains a team website.
- Clear the Leveling box to exclude the resource from leveling in Project.
- In the Timesheet manager box, enter or search for the manager's name if the resource is a timesheet manager.
- In the Default Assignment Owner box, enter or search for the assignment owner's name to indicate an assignment owner for this resource.

Step 6: Indicate the resource's Base Calendar.

Step 7: Indicate whether the resource is committed or proposed by selecting an option from the Default Booking Type.

Step 8: Check the Team Assignment Pool to indicate that the resource is a member of a team and enter the name of the team in the Team Name box. (Or select Browse to find the name of the team.)

Step 9: Use the calendar icon to change the Earliest Available and Latest Available dates for the resource.

Step 10: Select the Group, Code, Cost Center and Cost Type codes for the resource, if applicable.

Step 11: Enter any additional information about the resource that your organization requires in the Resource Custom Fields.

Step 12: Select Save.

Remember that you will need to add the enterprise resource back into your project.

You can also import multiple resources into the enterprise resource pool at once. To import multiple resources, use the following procedure.

Step 1: Select the Resources tab from the Ribbon.

Step 2: Select Add Resources. Select Import Resources to Enterprise.

Project displays the Import Resources wizard on the left with the Resource Sheet view on the right.

Step 1: For each resource that you want to import, select Yes in the corresponding Import column.

Step 2: If you have customized field, select Map Resource Fields to map the custom fields in your project with enterprise custom fields defined by your organization.

Step 3: Select Continue to Step 2 to validate information about the resources and ensure that no errors are created when the resources are imported. If there are errors, they are displayed next to the resource name in the Errors column.

Step 4: Select Save and Finish.

Exporting Resource Data

To export resource information, use the following procedure.

Step 1: Select the File tab from the Ribbon.

Step 2: Select Save As and Browse.

Step 3: In the Save As dialog box, select the file format you want to use for the export from the Save As Type drop-down list.

Step 4: Enter the File Name.

Step 5: Select Save.

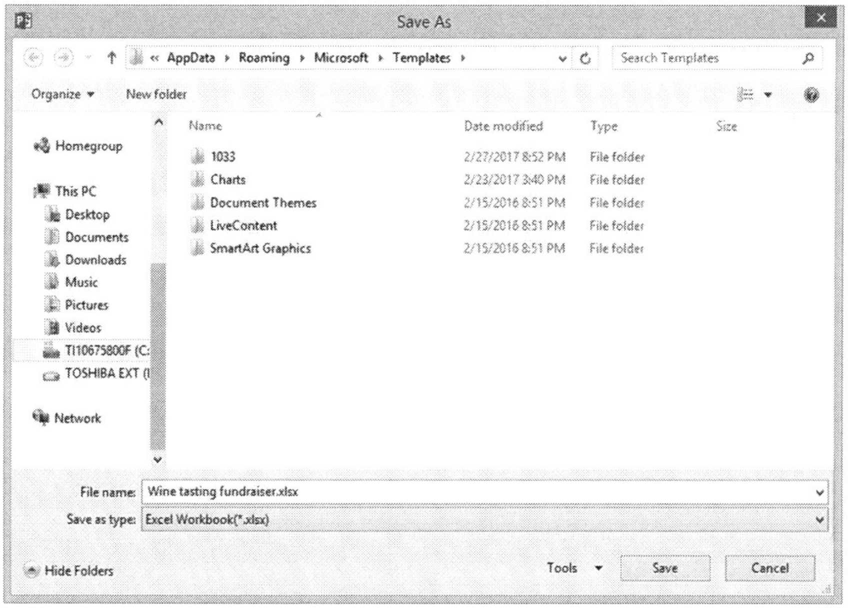

Project opens the Export Wizard to export the data you want into the proper fields for the destination file.

Step 6: Select Selected Data. Select Next.

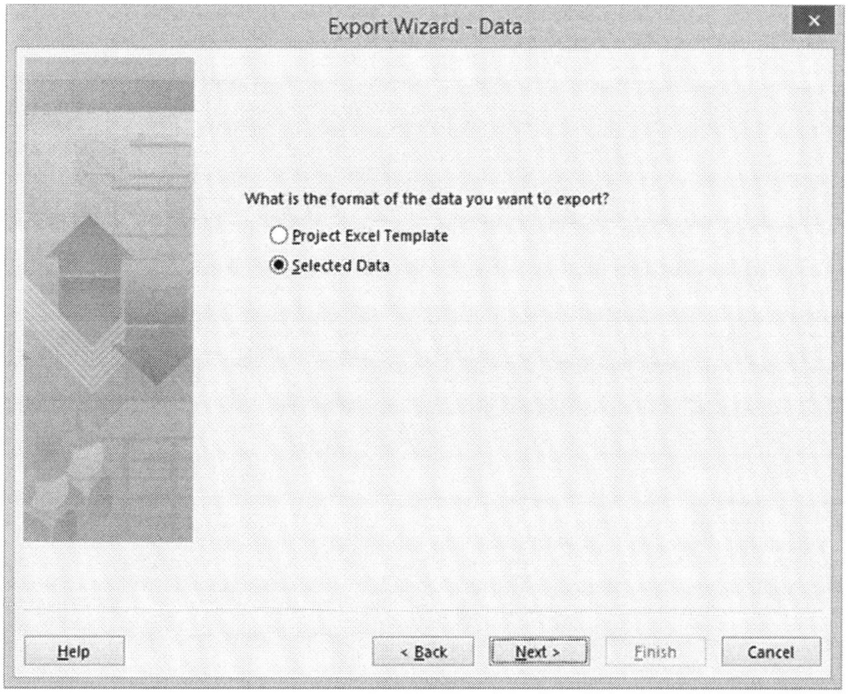

Step 7: Select New map unless you have previously saved a map file. Select Next.

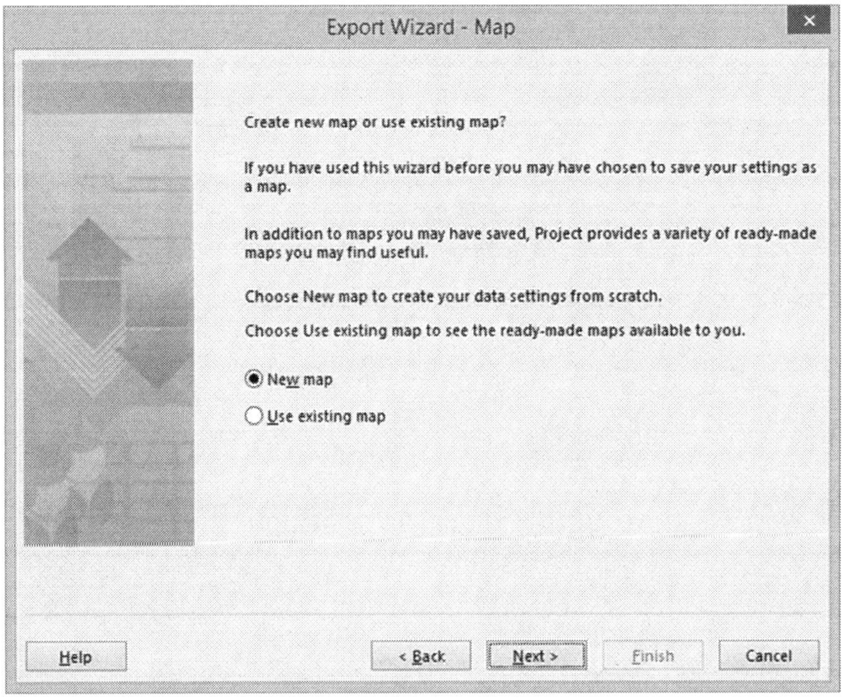

Step 8: Check the Resources box. If desired, select Include assignment rows in output. Select Next.

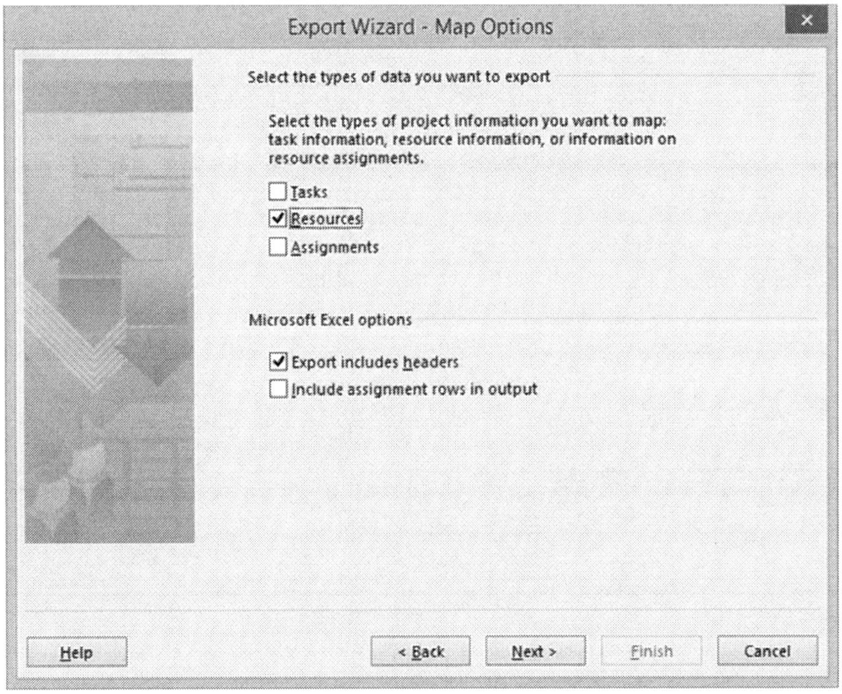

Step 9: On the Resource Mapping screen, complete the following steps:

Step 10: Enter the Destination Worksheet name for the destination worksheet, if you want to change it.

Step 11: Select the Export Filter from the drop-down list, if you want to change it.

Step 12: For each Microsoft Project Field, select a column name from the drop-down list in the first column of the Resource mapping dialog box. Then enter a heading in the Excel Field column.

Step 13: You can use the up and down arrows to rearrange the order of the items.

Step 14: You can use Add All, Clear All, Insert Row, and Delete Row to manage the mapping information.

Step 15: Continue until you have mapped all the RESOURCE information from Microsoft Project to Excel workbook to fields.

Step 16: Select Next.

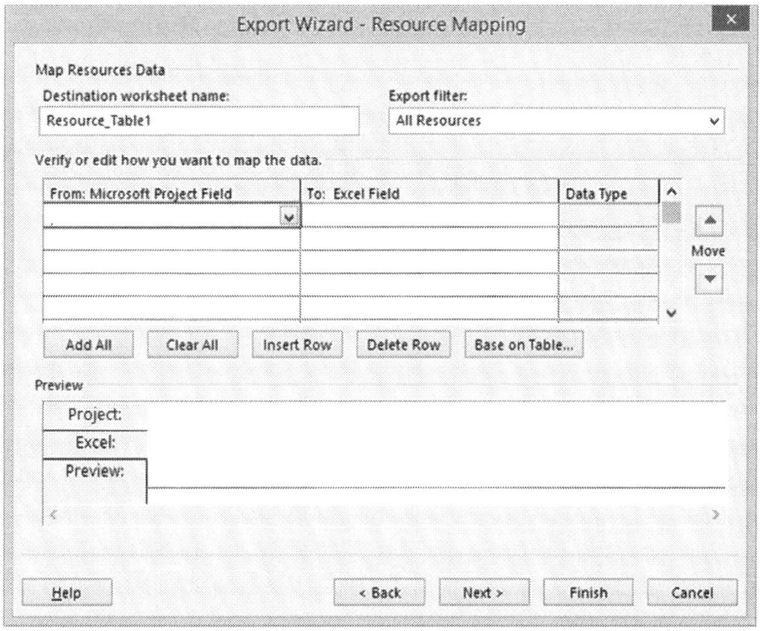

Step 17: On the last screen, you have the opportunity to save a map file. Select Finish.

Importing Resource Data

To import resource information, use the following procedure.

Step 1: Select the File tab to open the Backstage View.

Step 2: Select Open.

Step 3: In the Open dialog box, select the type of file you want to import from the drop-down list next to File Name. In this example, we will choose Excel Workbook.

Step 4: Select the File you want to import.

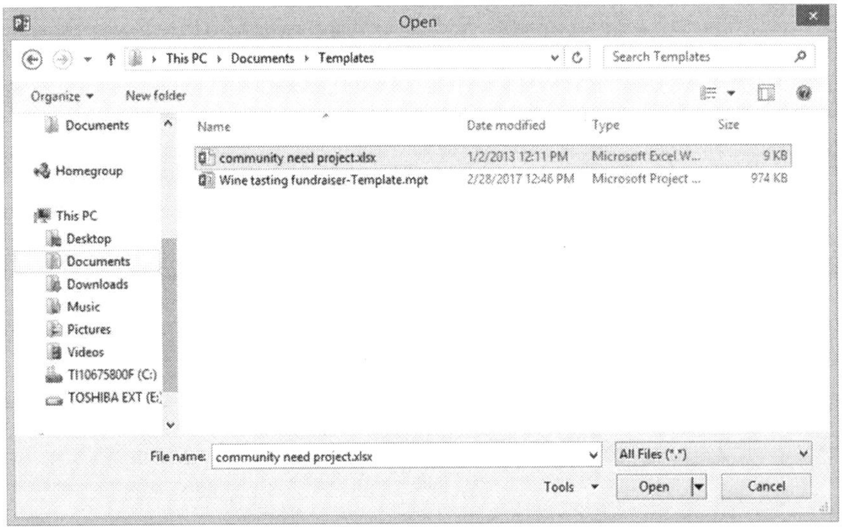

Step 5: Select Open.

Step 6: In the Import Wizard dialog box, select Next.

Step 7: Select whether to use an existing Map or a New Map. For the purposes of this example, we will select New Map. Select Next.

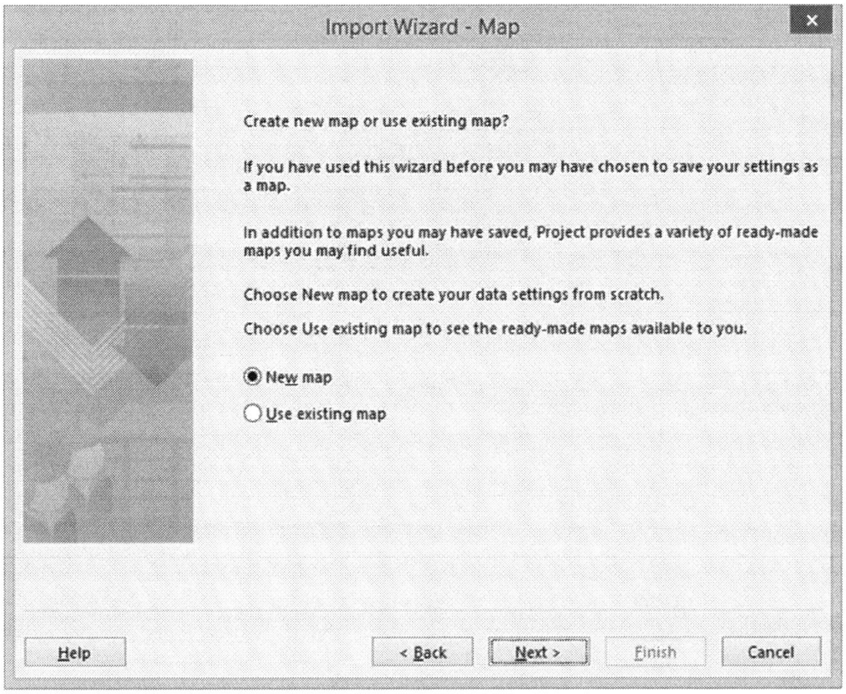

Step 8: Select the option to import the data either to Append the Data to the Active Project, or if have some of the resource information already in your project and you want to add missing details or additional resources, select Merge the Data to the Active project.

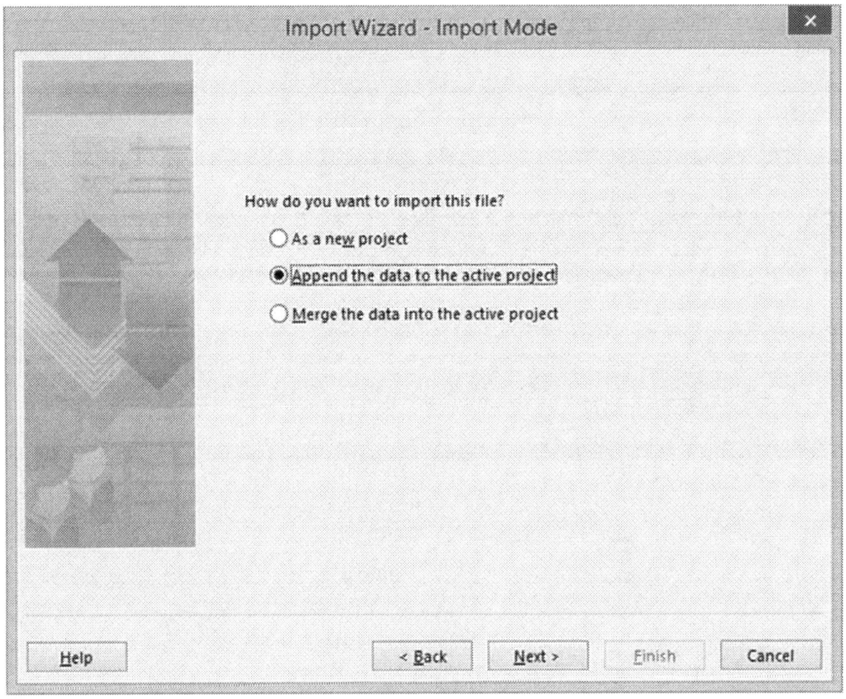

Step 9: The next screen on the Import Wizard allows you to indicate what type of information the Excel file includes. Check the Resources box as well as the Import includes headers. Select Next.

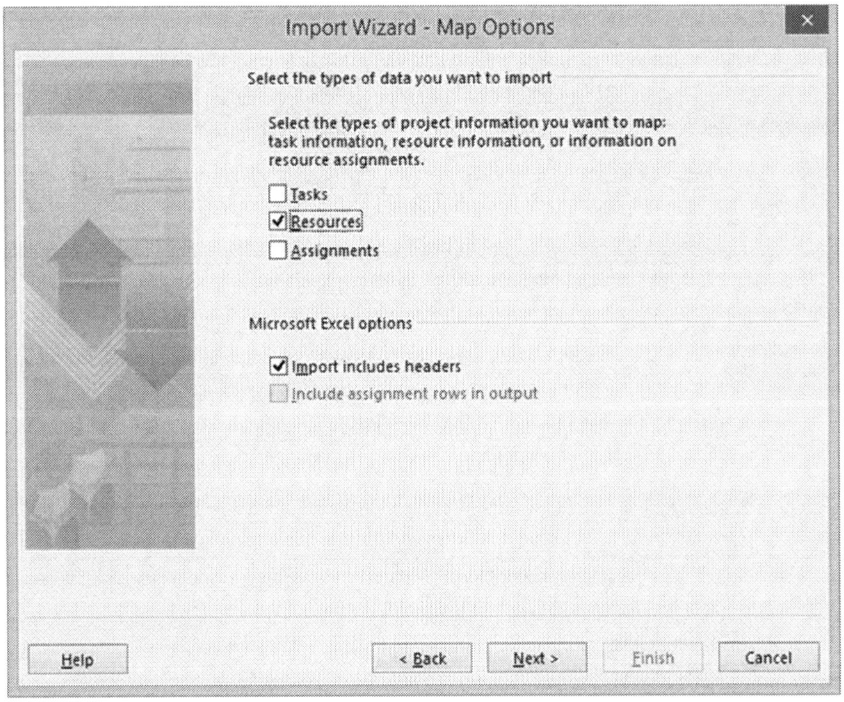

Step 10: The next screen on the Import Wizard maps the actual Excel workbook columns to Project fields for the RESOURCES.

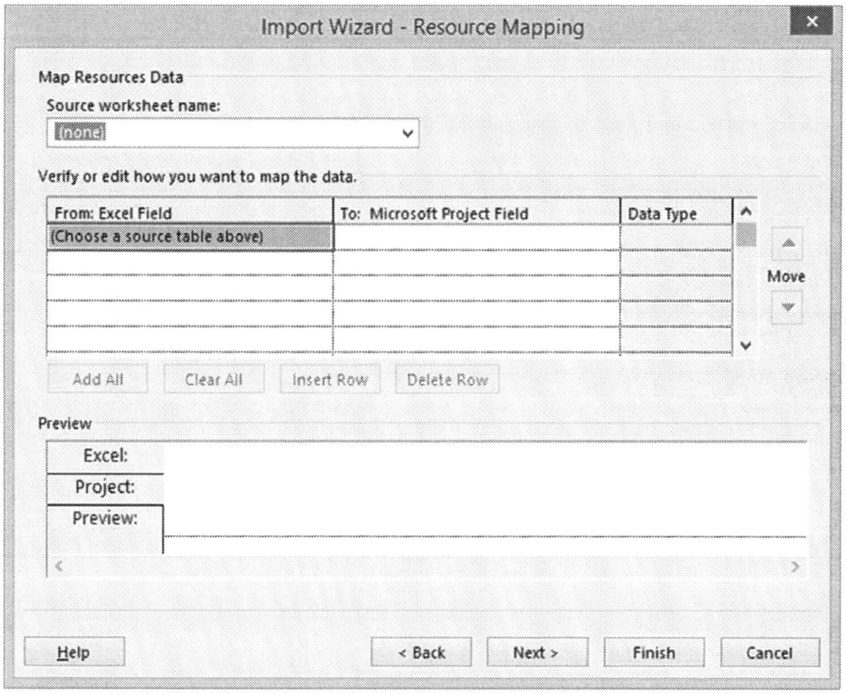

- Select the Source Worksheet name from the drop-down list.
- For each column in the Excel worksheet, select a column name from the drop-down list in the first column of the Resource mapping dialog box. Then select a Microsoft Project Field from the drop-down list.
- You can use the up and down arrows to rearrange the order of the items.
- You can use Add All, Clear All, Insert Row, and Delete Row to manage the mapping information.
- Continue until you have mapped all the RESOURCE information from the Excel workbook to Microsoft Project fields.
- Select Next.

Step 11: On the final screen of the Import wizard, you can save your map for future use, or simply select Finish.

Modifying Resource Information in the Resource Center

To change enterprise resource information, use the following procedure.

Step 1: Select the Resource tab from the Ribbon.

Step 2: Select Resource Pool

Step 3: Select Enterprise Resource Pool.

The Resource Center opens in Internet Explorer, displaying a list of the enterprise resources.

Step 4: Check the box next to the resource you want to modify and select Edit Resource. You can also select multiple resources and select Bulk Edit to make the same changes to more than one resource.

Step 5: Change the resource information as needed.

Step 6: Select Save.

If you need to change the resource calendar, availability, contours, or cost, you must make the changes in the enterprise resource pool

To make changes in the enterprise resource pool for these types of changes, use the following procedure.

Step 1: Select the Resource tab from the Ribbon.

Step 2: Select Resource Pool

Step 3: Select Enterprise Resource Pool.

The Resource Center opens in Internet Explorer, displaying a list of the enterprise resources.

Step 4: Check the box next to the resource(s) you want to modify and select Open.

A new project opens with the selected resources displayed in the Resource Sheet Tools view.

Step 5: Update the resource information as needed.

Step 6: Select the File tab from the Ribbon.

Step 7: Select Save.

Viewing Availability Across Multiple Projects

To view resource usage for a resource who has been assigned tasks for multiple projects, use the following procedure.

Step 1: Open the project.

Step 2: Open the Resource Usage view.

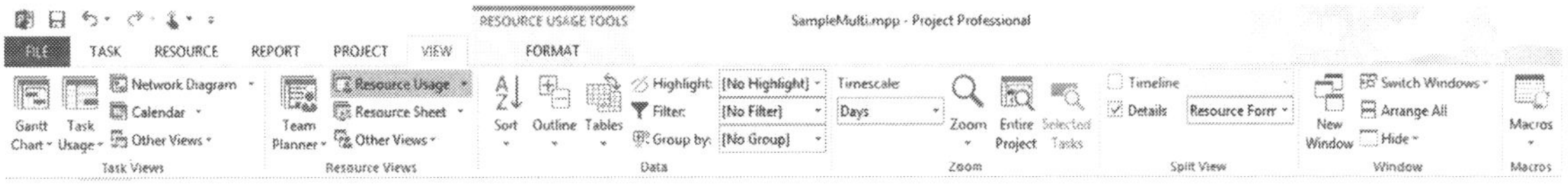

Step 3: Scroll down to the resource named Grant Stone.

All tasks from projects that share resources are shown with the resource’s name.

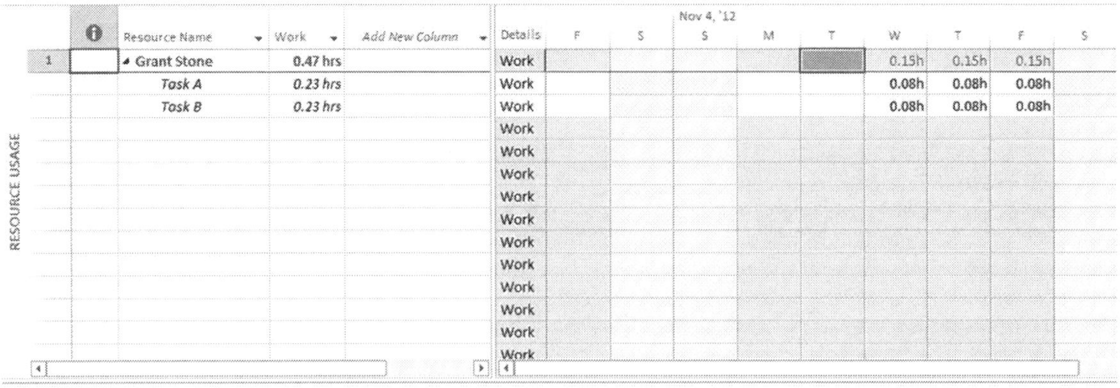

You can also view the details for a resource, including all assignments from multiple projects.

Step 1: Select the Resource tab from the Ribbon.

Step 2: Select Details.

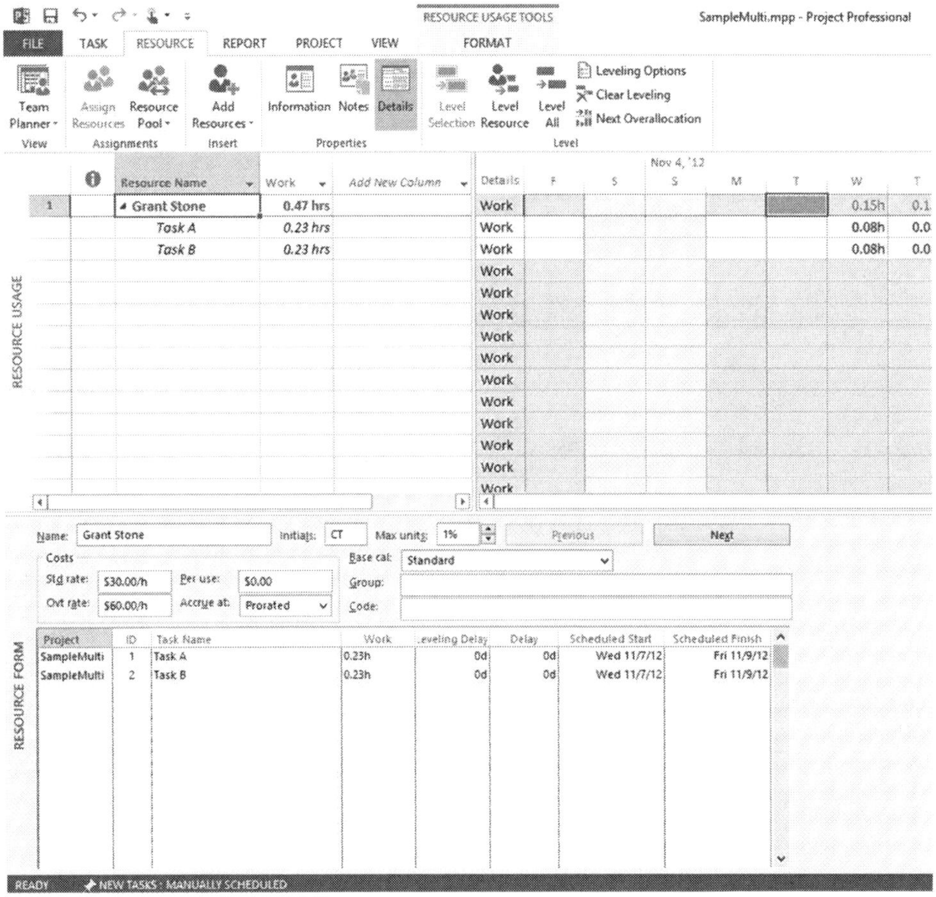

Notice in the details area at the bottom, the project name for each task is listed on the left.

Chapter 14 – Working with Costs

For Project to successfully calculate costs for a project, you must enter accurate information, including varying pay rates. A work resource has a standard pay rate and an overtime rate. You can also enter specific pay rates for a certain day or for an assignment. In this chapter, we will look at resource pay rates, including material resource consumption rates. Finally, we will look at entering task fixed costs.

Adding Pay Rates for a Resource

Use the following procedure to enter pay rates.

Step 1: Select Resource Sheet from the View tab on the Ribbon.

Step 2: Enter the standard and overtime rate for each resource.

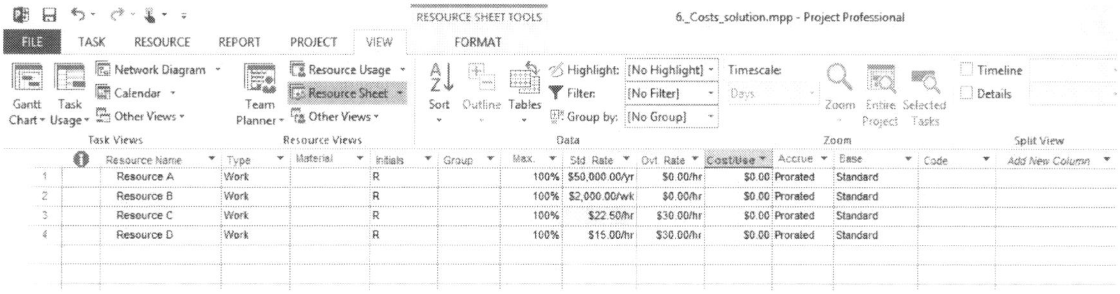

Specifying Pay Rates for Different Dates

To enter a pay rate for a specific date, use the following procedure.

Step 1: Select Resource Sheet from the View tab on the Ribbon.

Step 2: Double-click the resource whose pay rate you want to define to open the Resource Information dialog box.

Step 3: Select the Costs tab.

Step 4: In the second row, enter the new standard, overtime, or a per use cost rate.

Step 5: Select the down arrow in the Effective Date column to select the effective date from the calendar.

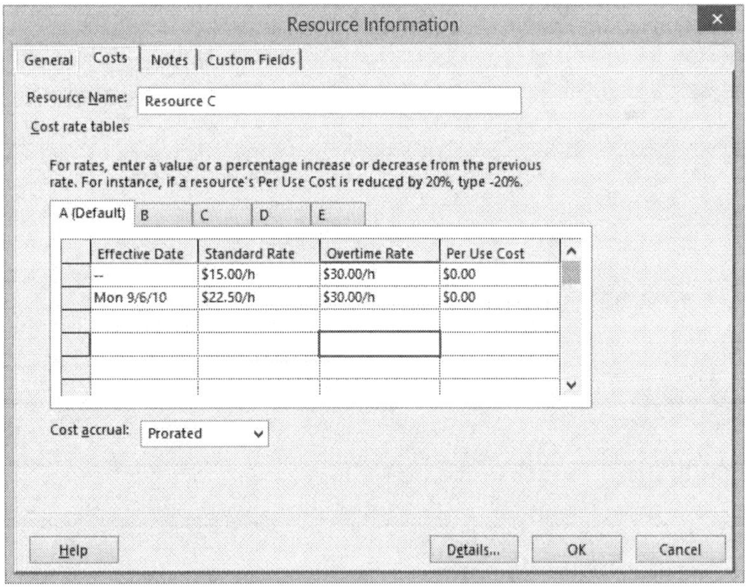

Step 6: Click on a lettered tab to enter additional cost rate tables. A is the default table. Repeat steps 4 and 5 for each table.

Step 7: Select OK to closet the Resource Information dialog box.

Applying a Different Pay Rate to an Assignment

Use the following procedure to apply a different pay rate to an assignment

Step 1: Select Task Usage from the View tab to open the Task Usage sheet.

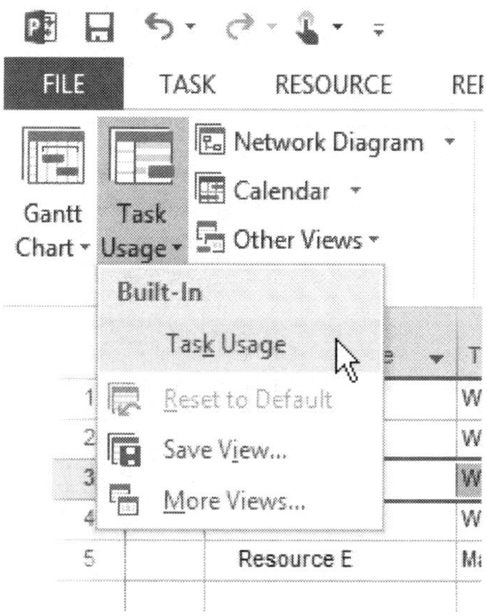

Step 2: Double-click on the resource assigned to the task that requires a different pay rate.

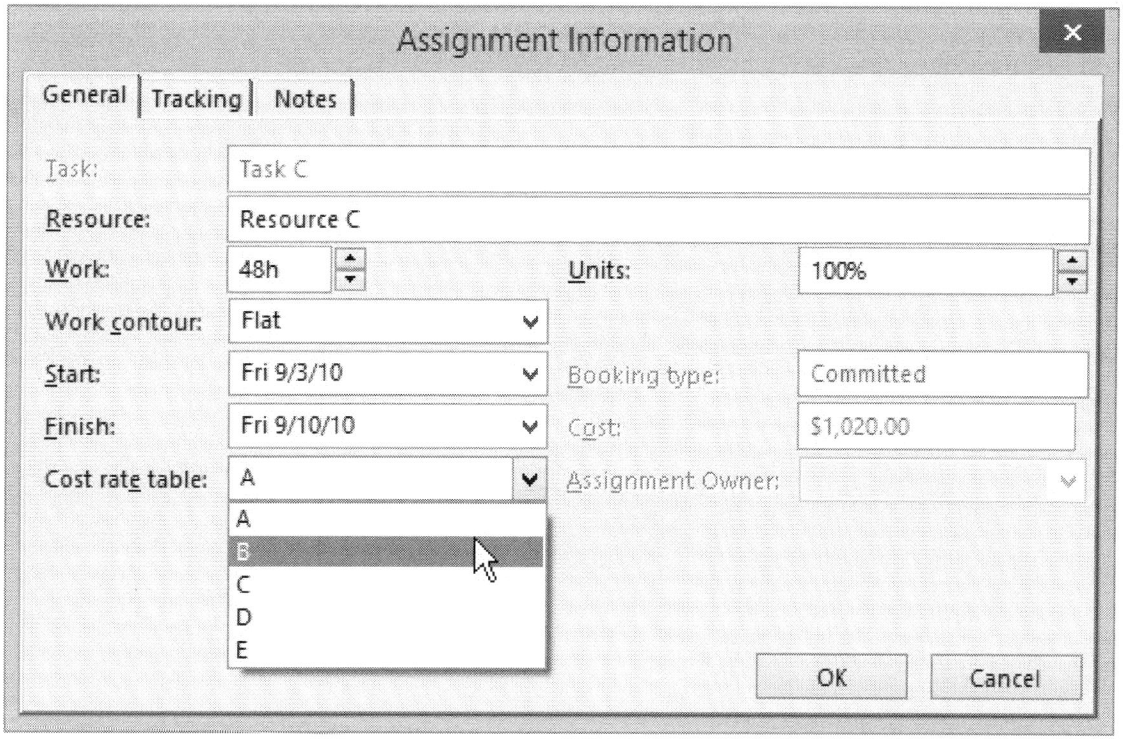

Project displays the Assignment Information dialog box.

Step 3: Select the Cost Rate Table that you want to use from the drop-down list.

Step 4: Select OK.

If desired, return to the Gantt Chart view and show the participants the changed costs based on the new table. Project highlights the changes in the selected Changed Cell highlight color. (Note: the default highlight color is very light. You may want to change it using the Text Styles tool on the Format tab.)

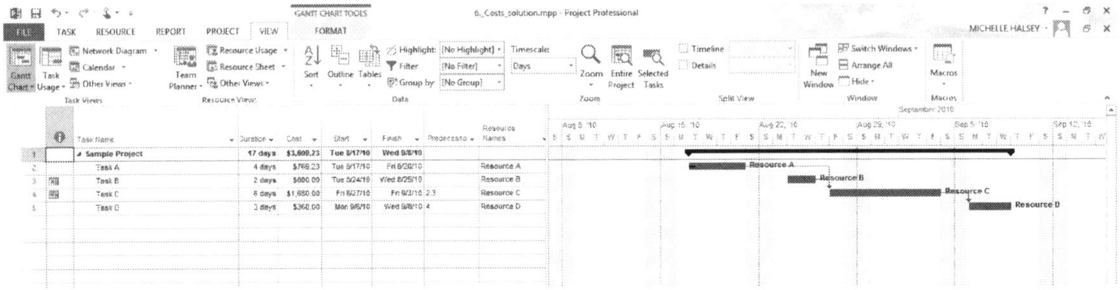

Using Material Resource Consumption Rates

To assign a material resource to a task, use the following procedure.

Step 1: Double-click on the task to open the Task Information dialog box. In this example, select Task E.

Step 2: Select Resource E for the task. Enter 10 as the units for the material resource.

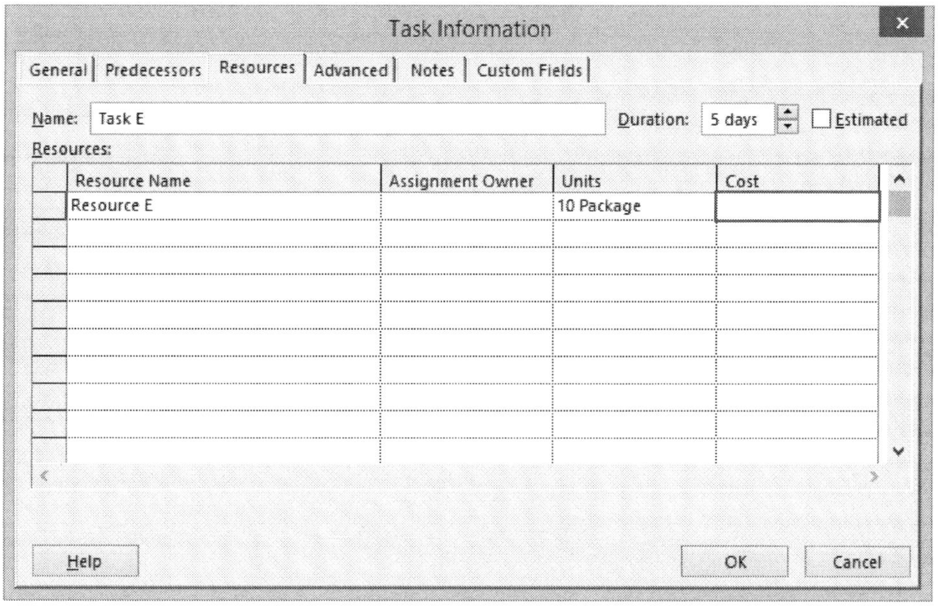

Step 3: Select OK to close the Task Information dialog box.

Step 4: Select Task Usage from the View tab.

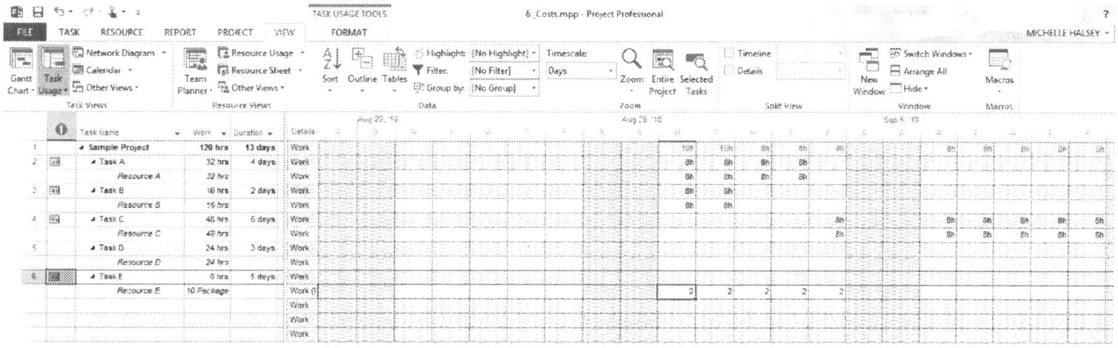

You can also manually override the calculated consumption rates by entering a new value in the columns for that resource for that task.

Entering Task Fixed Costs

Use the following procedure to enter task fixed costs.

Step 1: Select Table from the View menu. Select Cost.

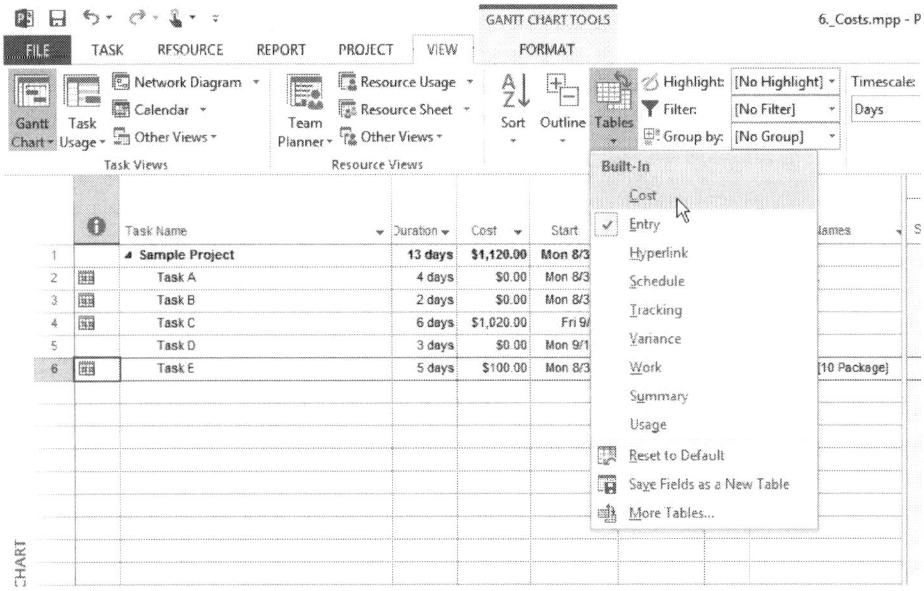

Step 2: Select the task for which you want to enter a fixed cost.

Step 3: Enter the amount in the Fixed cost column.

	Task Name	Fixed	Fixed Cost	Total Cost	Baseline	Variance	Actual	Remaining
1	**Sample Project**	**$0.00**	**Prorated**	**$1,145.00**	**$0.00**	**$1,145.00**	**$0.00**	**$1,145.00**
2	Task A	$0.00	Prorated	$0.00	$0.00	$0.00	$0.00	$0.00
3	Task B	$0.00	Prorated	$0.00	$0.00	$0.00	$0.00	$0.00
4	Task C	$0.00	Prorated	$1,020.00	$0.00	$1,020.00	$0.00	$1,020.00
5	Task D	$25.00	Prorated	$25.00	$0.00	$25.00	$0.00	$25.00
6	Task E	$0.00	Prorated	$100.00	$0.00	$100.00	$0.00	$100.00

Chapter 15 – Balancing the Project

Project management involves constantly balancing opposing needs of costs, resources, and tasks to be done. In this chapter, we will look at some of the ways to balance a project, including both scheduling overtime for a resource and identifying and leveling resources that are over allocated. This chapter will explain how to level resources manually and automatically.

Scheduling Resource Overtime

To enter overtime work, use the following steps.

Step 1: Select Resource Usage from the View tab of the Ribbon.

Step 2: Right-click the Work column and select Insert Column from the context menu.

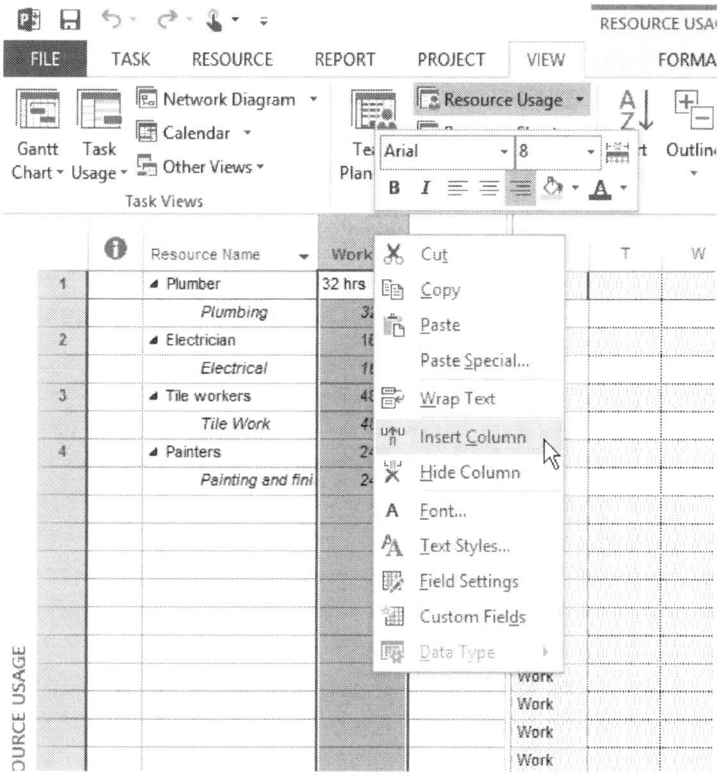

Step 3: Select Overtime Work from the Field Name-drop-down list.

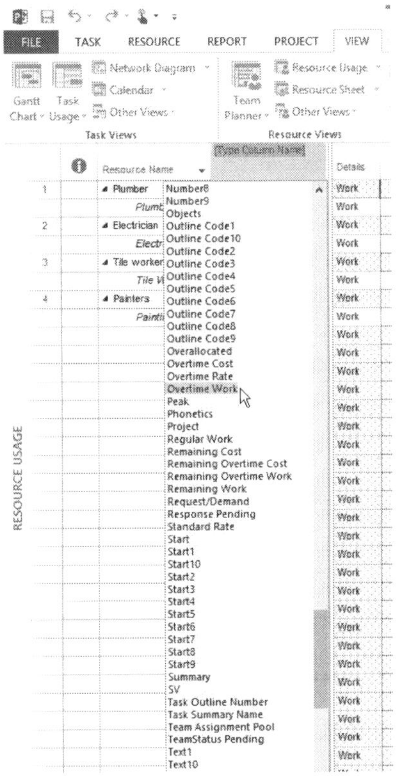

Step 4: In the Overtime Work column of the Resource Usage sheet, use the up and down arrows to adjust the number of overtime hours for each task.

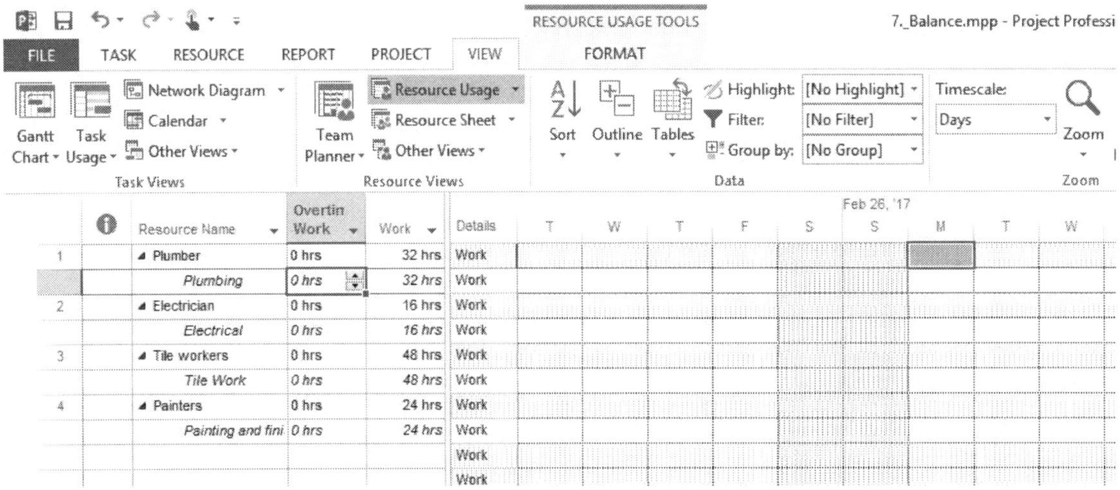

Project redistributes the work for the task to account for the overtime hours. The total duration for the task does not change.

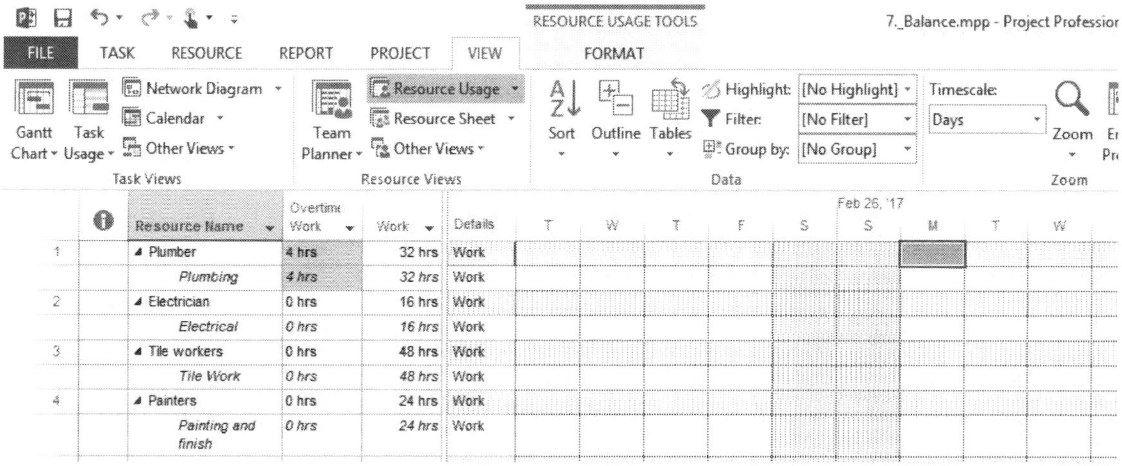

Identifying Resource Over Allocation

Use the following procedure to review the resource graph.

Step 1: Select the View tab from the Ribbon.

Step 2: Select Other Views from the Resource area.

Step 3: Select Resource Graph.

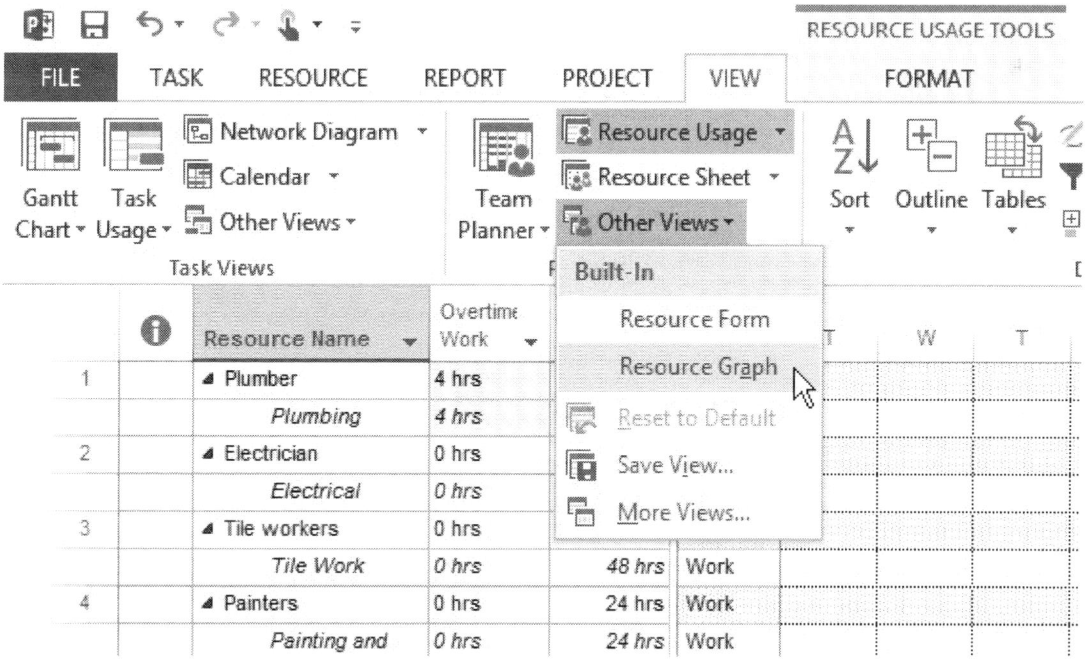

Step 4: Select the Zoom Entire Project tool from the View tab to show the right dates on the chart.

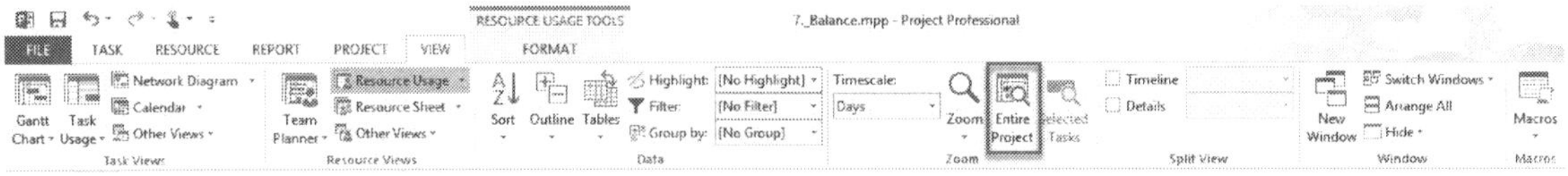

Step 5: Use the scroll bar for the left side of the chart to scroll through the different resources. Or use the wheel on your mouse to scroll through the resources.

The graph shows any resources that are over allocated as a percentage.

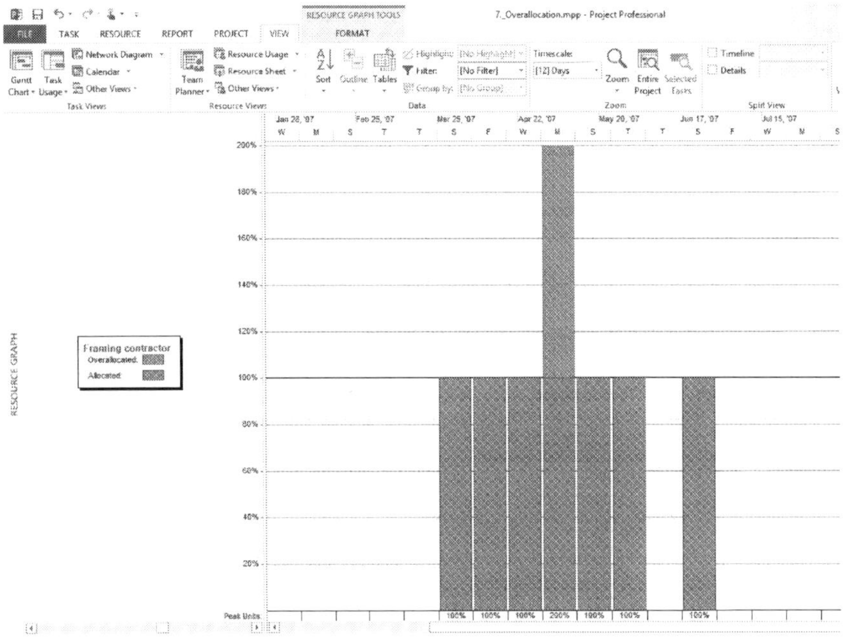

The Resource Sheet highlights over allocated resources in red.

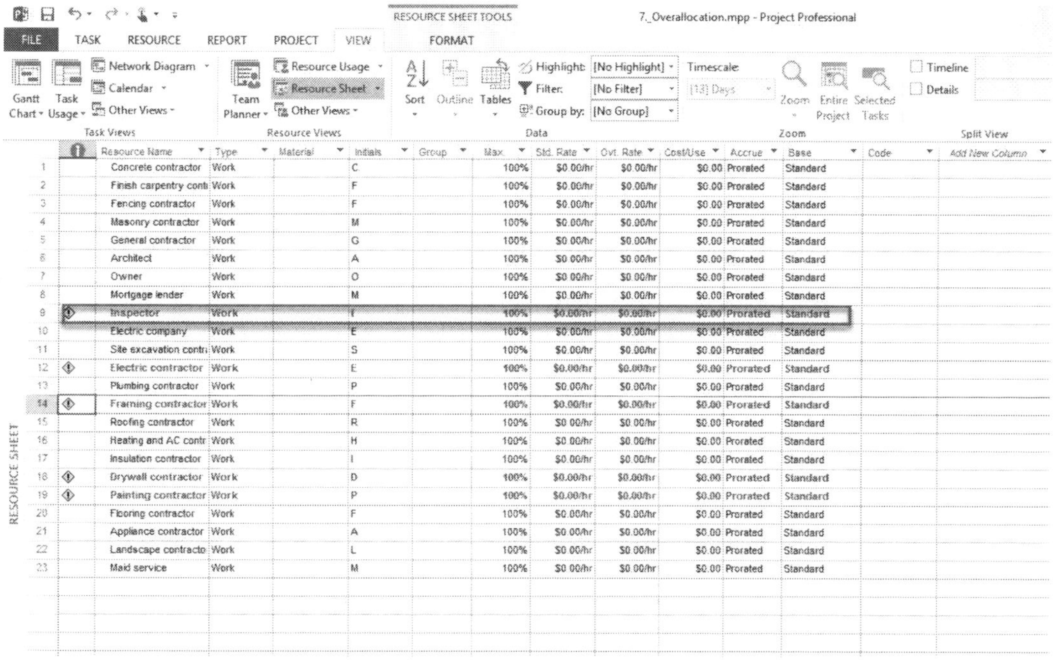

The Resource Usage sheet also highlights over allocated resources in red. You can expand the tasks for each resource to see which overlapping tasks result in an over allocation.

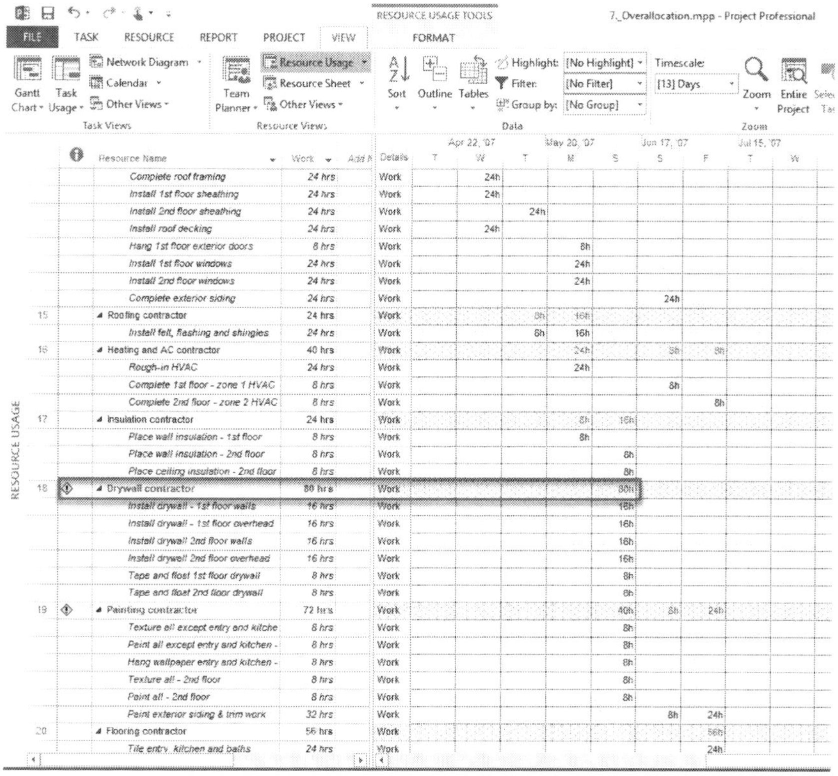

Setting Leveling Options

Use the following procedure to open the Resource Leveling dialog box.

Select Resource Usage from the View tab. You can start from another view, but this view allows you to see the number of hours for each task. It also highlights the over allocated resources in red.

Step 1: Select the Resource tab from the Ribbon.

Step 2: Select Leveling Options.

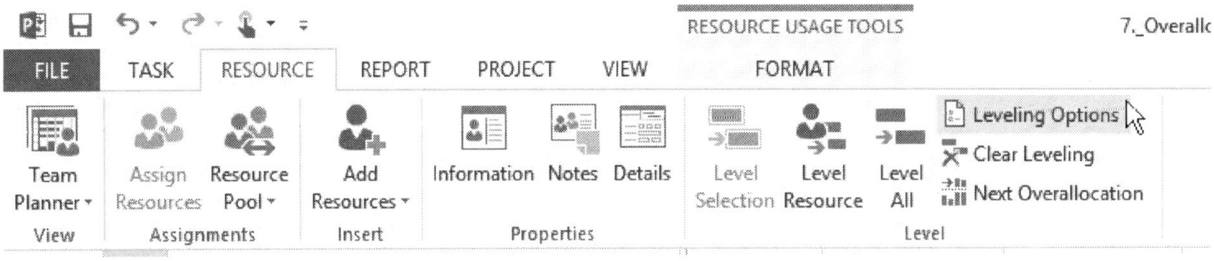

Project displays the Resource Leveling dialog box.

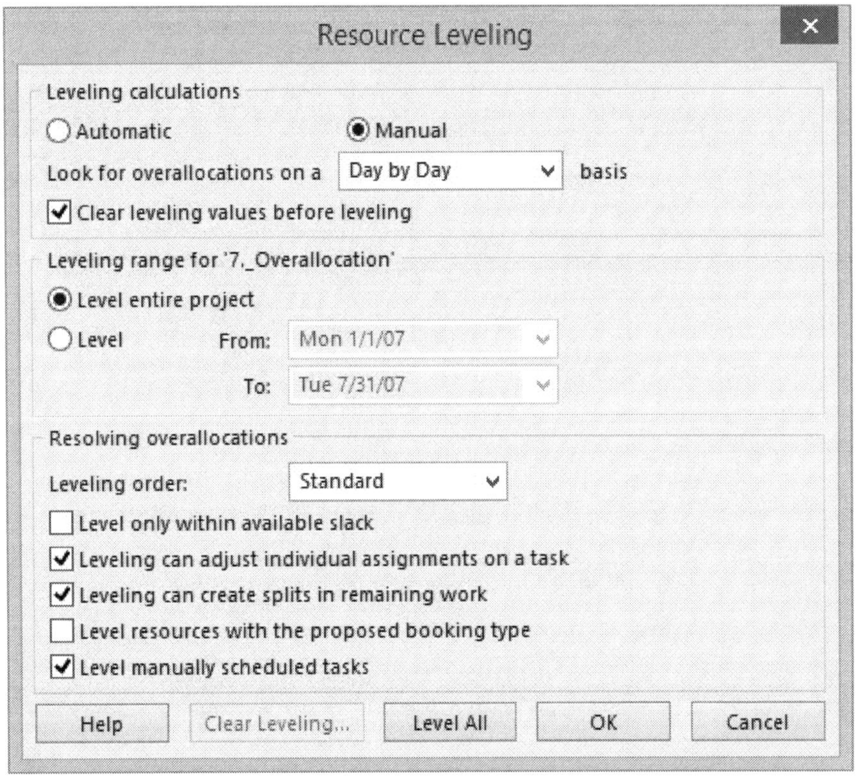

Step 3: Choose either Automatic or Manual. If you choose Automatic, make sure to clear the Clear leveling values before leveling checkbox. This way, Project will not clear previous leveling changes before making new changes to future over allocated resources.

Step 4: Select an option to indicate how you want Project to Look for Over allocations from the drop-down list. Day by Day is the default selection. For this example, choose Hour by Hour.

Step 5: Select whether to level the entire project or a selected date range. If you choose the second option, select the From and To dates from the drop-down calendars.

Step 6: Select a Leveling order from the drop-down list. The Standard method examines predecessor dependencies, slack, dates, priorities, and constraints. ID Only looks at tasks in ascending order of ID number to look for places to level. Priority, Standard looks at priority level of the tasks first before considering the other criteria.

Step 7: Check the Level only within available slack box to prevent a delay of your project finish date. If this box is checked, you may get an error message if Project cannot find enough slack time to level the schedule.

Step 8: Check the Leveling can adjust individual assignments on a task box to allow leveling to adjust when a resource works on a task independently of when other resources working on the same task.

Step 9: Check the Leveling can create splits in remaining work box to allow for interruptions in tasks.

Step 10: Check the Level resources with the proposed booking type box to include proposed resources in the leveling.

Step 11: Check the Level manually scheduled tasks box to allow leveling to change manually scheduled tasks.

Step 12: Select Level All or OK to save your changes without leveling all resources now.

Balancing Resource Over Allocations Automatically

Use the following procedure to level resources automatically.

Step 1: Select the resource(s) you want to include in the leveling from the Resource Usage sheet (or another resource view). Press the Ctrl key while selecting nonconsecutive resources.

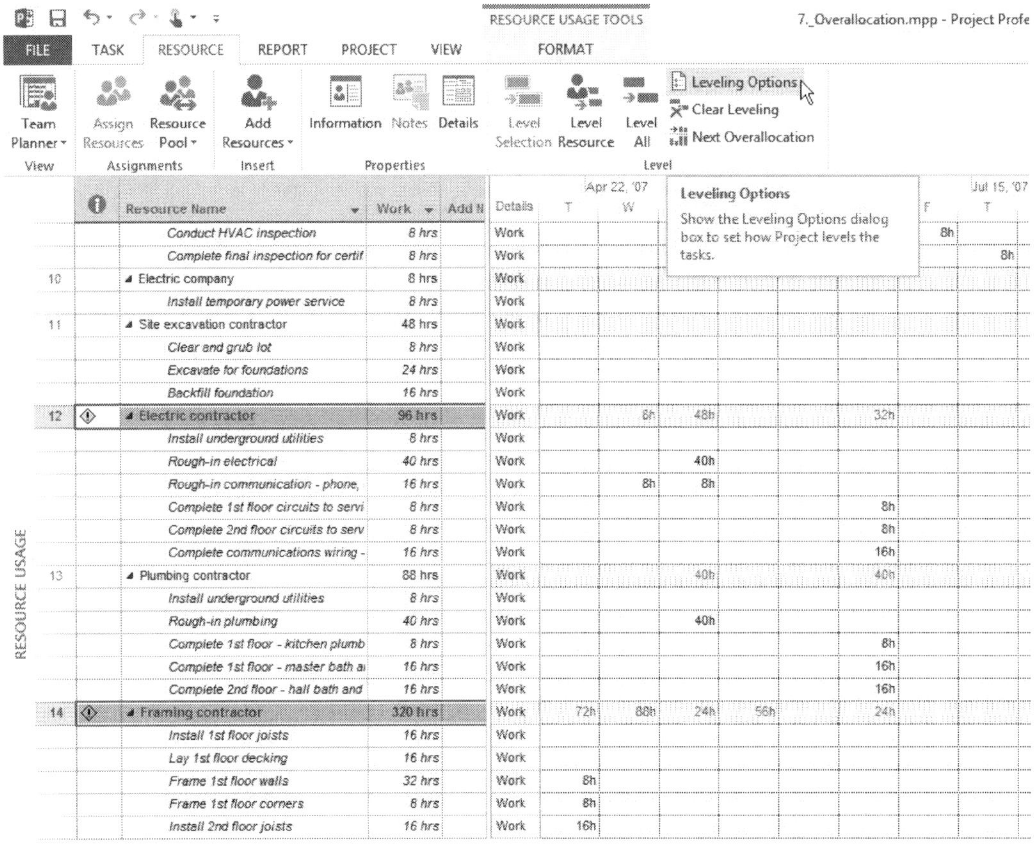

Step 2: Open the Resource Leveling dialog box (from the previous topic).

Step 3: Select Automatic.

Step 4: Set the other leveling options (see the previous topic for more information).

Step 5: Select Level All.

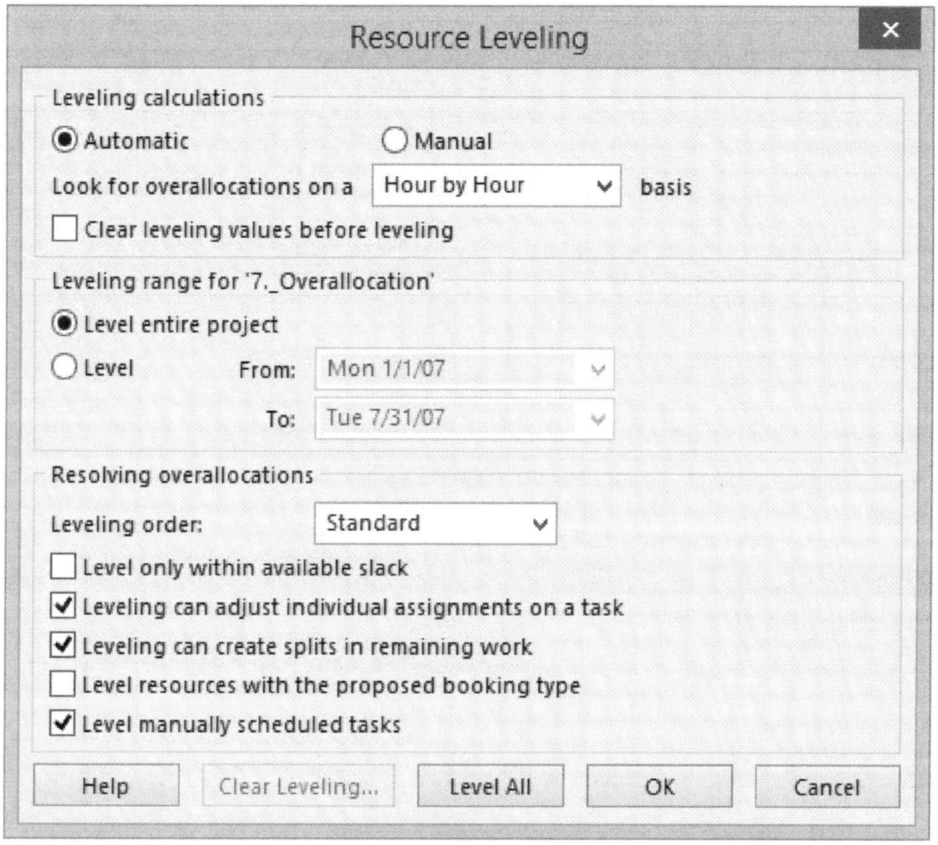

Project levels the selected resources now, but it will also level any future changes or additions that cause the selected resource(s) to be over allocated.

Balancing Resource Over Allocations Manually

Use the following procedure to clear previous leveling.

Step 1: Select the Resource tab from the Ribbon.

Step 2: Select Clear Leveling.

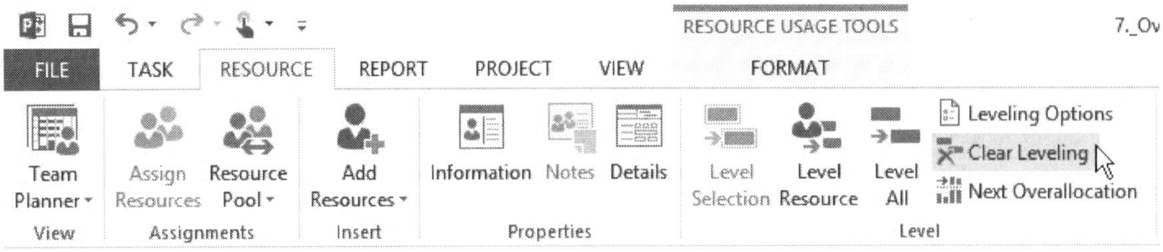

To level resources manually one resource at a time.

Step 1. Open the Resource Leveling dialog box (from the previous topic).

Step 2: Select Manual.

Step 3: Set the other leveling options (see the previous topics for more information).

Step 4: Select OK.

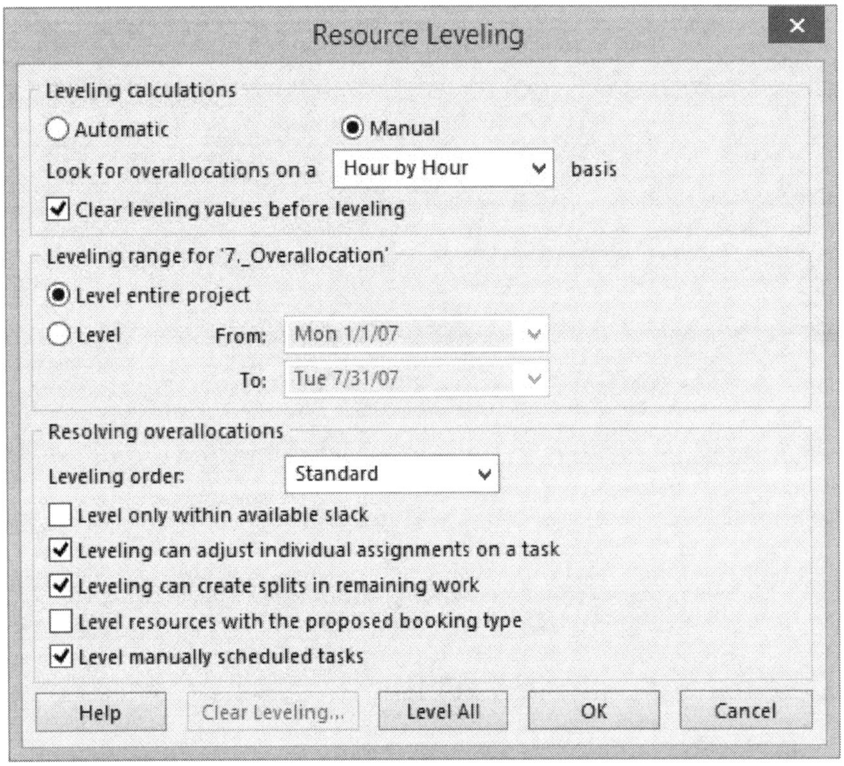

Step 5: Select the resource you want to level.

Step 6: Select Level Resource from the Resource tab on the Ribbon.

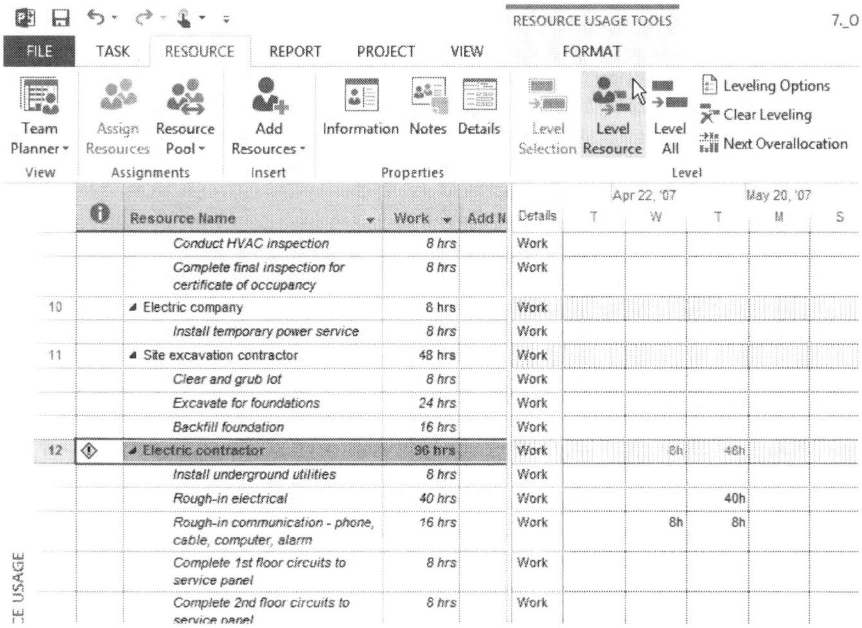

Step 7: Project displays the Level Resource dialog box.

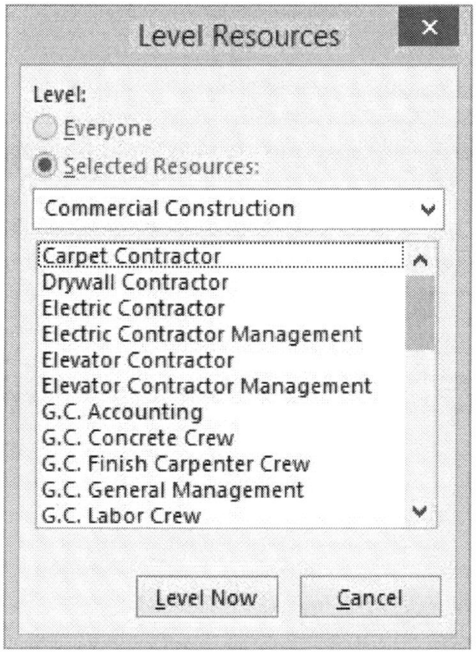

Step 8: Make sure the resource you want to level is highlighted and select Level Now. Project reschedules the tasks related to the over allocation of that resource.

Step 9: To move to the next over allocation, select Next Overallocation.

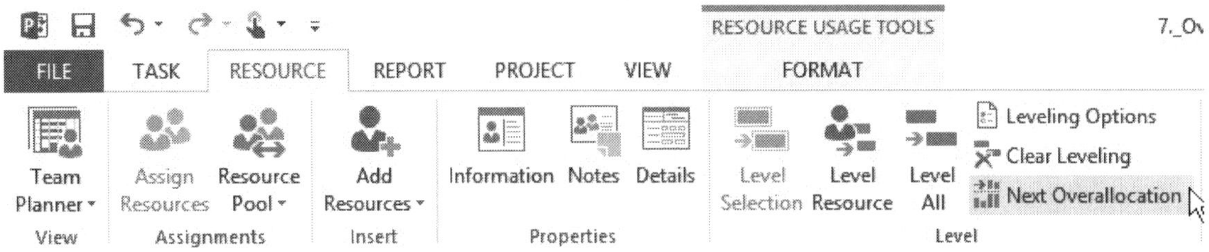

Remember that you can view the results of the leveling both by reviewing the Resource Usage sheet and by reviewing the Leveling Gantt view.

To level all resources manually.

Step 1: Open the Resource Leveling dialog box (from the previous topic).

Step 2: Select Manual.

Step 3: Set the other leveling options (see the previous topics for more information).

Step 4: Select OK.

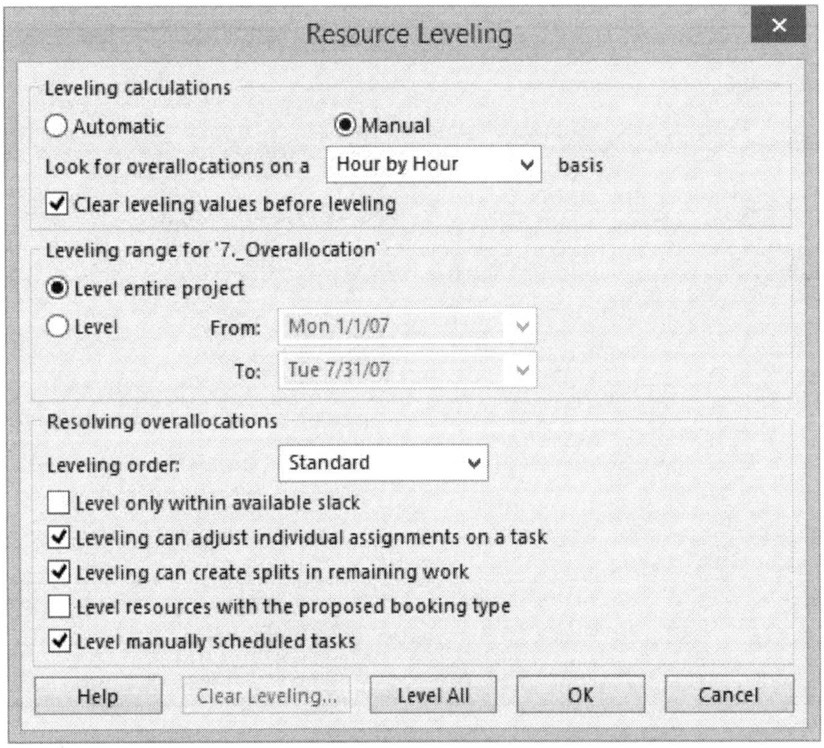

Step 5: Select Level All from the Resource tab on the Ribbon.

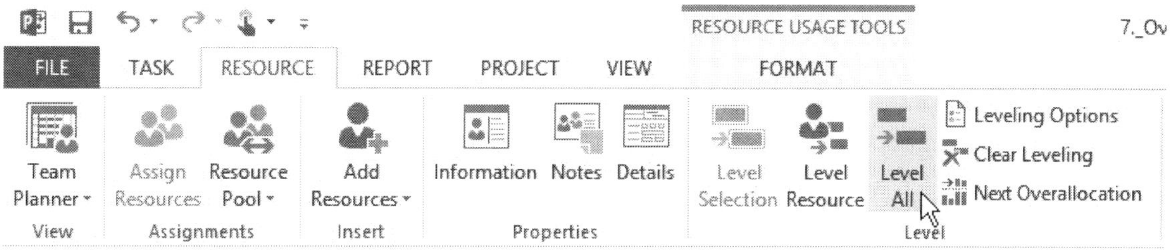

Project reschedules the tasks related to the over allocation of all resources.

Remember that you can view the results of the leveling both by reviewing the Resource Usage sheet and by reviewing the Leveling Gantt view.

Chapter 16 – Updating Project Progress

This chapter covers updating the project progress. You will learn how to save a baseline plan to use as comparison as you review the progress of your project. This chapter also explains how to update the entire project. You will also learn how to update the specifics of a project, including task actual values, completion percentages, actual work, and actual costs.

Saving a Baseline Plan

Use the following procedure to set a baseline.

Step 1: Enter the tasks, durations, and other details of your base project before setting a baseline.

Step 2: To set a baseline for selected tasks, select the tasks you want to track from the Gantt chart view.

Step 3: Select the Project tab from the Ribbon.

Step 4: Select Set Baseline. Select the Set Baseline option.

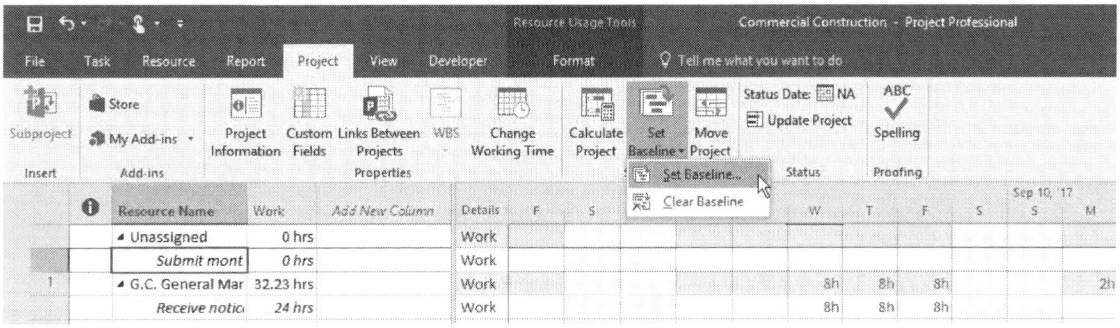

Project displays the Set Baseline dialog box.

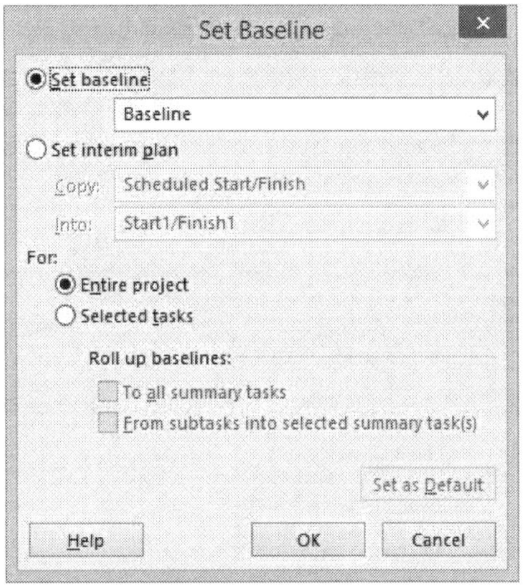

Step 5: From the Set baseline drop-down list, select which baseline you want to set. You have the choice of the default baseline, or up to 10 other baselines, which are numbered 1 – 10.

Step 6: Select Entire Project or Selected tasks.

Step 7: If you select Selected Tasks, indicate how to roll up the baselines by checking the To all summary tasks box and/or the From subtasks into selected summary task(s) box.

Step 8: Select OK.

The Baseline column stores the baseline information. To add the Baseline column to see the baseline information. Use the following procedure.

Step 1: Right-click a column in the Gantt Chart view and select Insert Column from the context menu.

Step 2: Select a field to enter from the Field name drop-down list. For each baseline you can set, you can enter the following columns to your table:

- Baseline Budget Costs
- Baseline Budget Work
- Baseline Cost
- Baseline Duration
- Baseline Finish
- Baseline Fixed Cost
- Baseline Fixed Cost Accrual
- Baseline Start
- Baseline Work

Step 3: Select OK. Repeat to add more baseline information to your current view.

The selected baseline column(s) display the baseline values for the selected task(s).

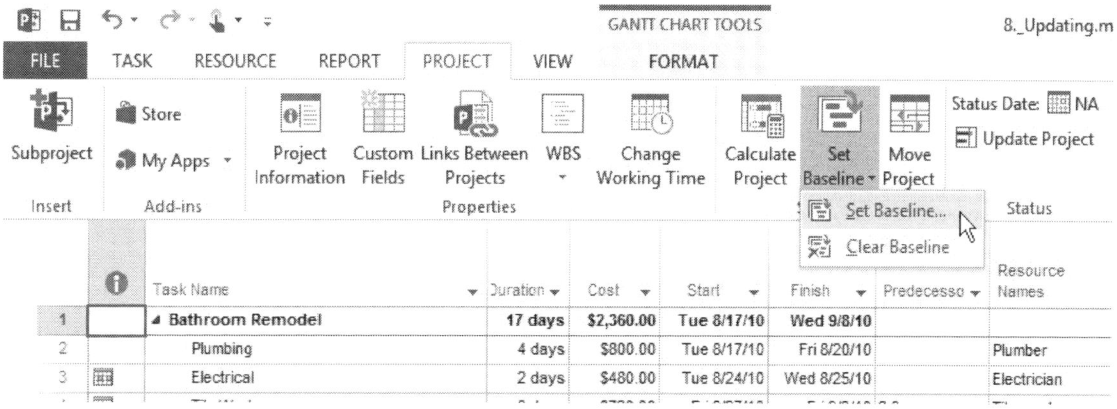

Updating the Entire Project

Use the following procedure to update the entire project.

Step 1: If you want to update only certain tasks, select them from the Gantt chart view first.

Step 2: On the Project Tab select Update Project.

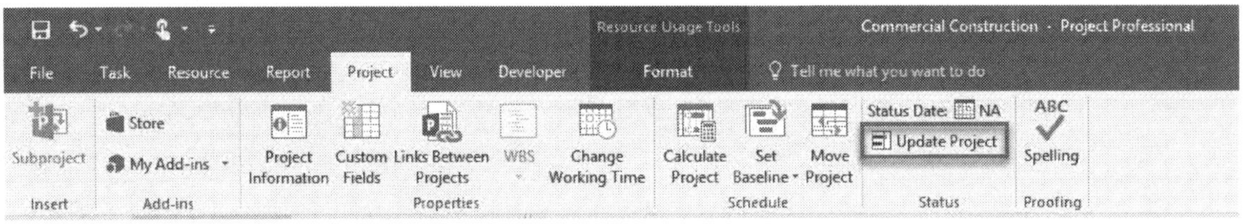

Project displays the Update Project dialog box.

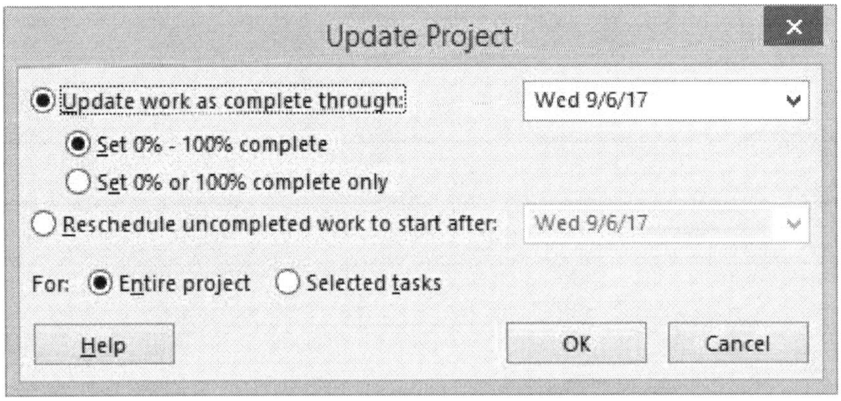

Step 3: Leave the default setting of Update work as complete through. Select the date (for this example, use 8/28/10) from the drop-down calendar.

Step 4: Select Set 0% - 100% complete to have project indicate the percentage complete based on the duration and the schedule. To have Project set any tasks that are complete to 100%, and all others to 0%, select Set 0% or 100% complete only.

Step 5: Select Entire project or Selected tasks.

Step 6: Select OK.

Project updates the schedule and displays the Percent Complete on the Gantt Chart view. Any completed tasks now have a task completed icon in the Indicators column.

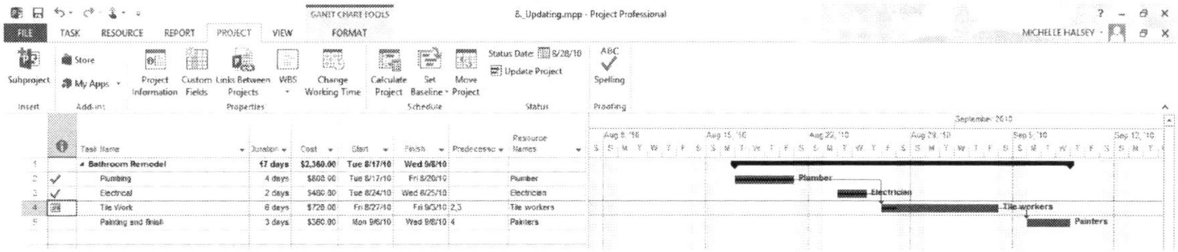

Updating Task Actual Values

Use the following procedure to update the task actual values.

Step 1: Select the task you want to update. If multiple tasks have the same values, you can hold the CTRL key while you select multiple tasks.

Step 2: Select the Task tab from the Ribbon.

Step 3: Select the arrow next to Mark on Track. Select Update Tasks.

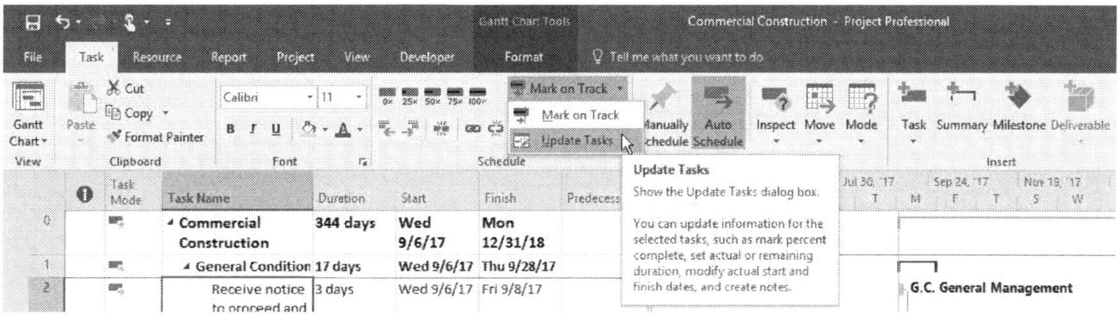

Project displays the Update Tasks dialog box.

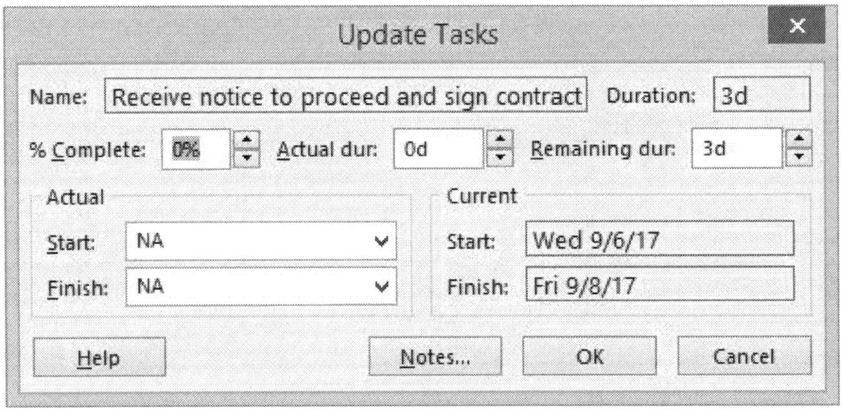

Step 4: Set the % Complete by entering the percentage or using the up and down arrows.

Step 5: Set the Actual Duration by entering the number and the duration abbreviation or by using the up and down arrows.

Step 6: Set the Remaining Duration by entering the number and duration abbreviation or by using the up and down arrows.

Step 7: Select the Actual Start date from the drop-down calendar.

Step 8: Select the Actual Finish date from the drop-down calendar.

Step 9: Select the Notes button to enter any notes about the task.

Step 10: Select OK.

Updating Actual Work

To enter actual work.

Step 1: Select the View tab.

Step 2: Select Task Usage to switch to the Task Usage sheet.

Step 3: Right click on the right side of the sheet and select Actual Work from the context menu.

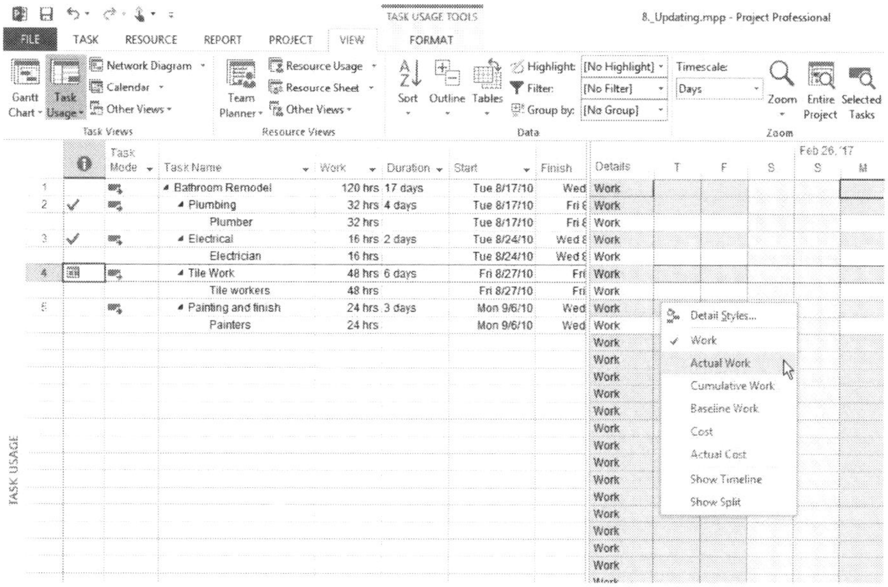

Step 4: Change the table for the Task Usage sheet by selecting Table from the View Tab. Select Work.

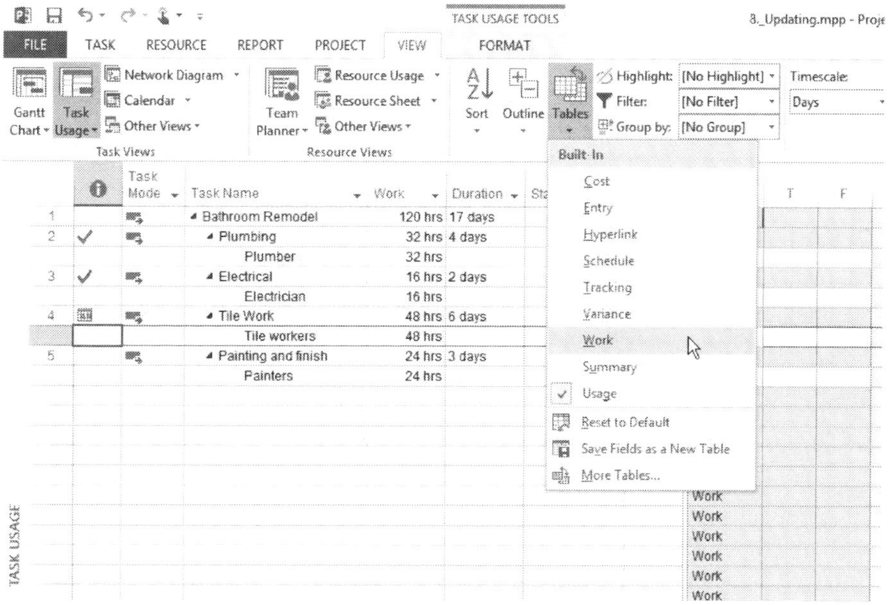

Step 5: Enter the actual work values in the appropriate dates on the right side of the sheet. You can also adjust the total actual work.

Updating Actual Costs

Use the following procedure to turn off the automatic calculation of actual costs

Step 1: Select the File tab from the Ribbon.

Step 2: Select Options.

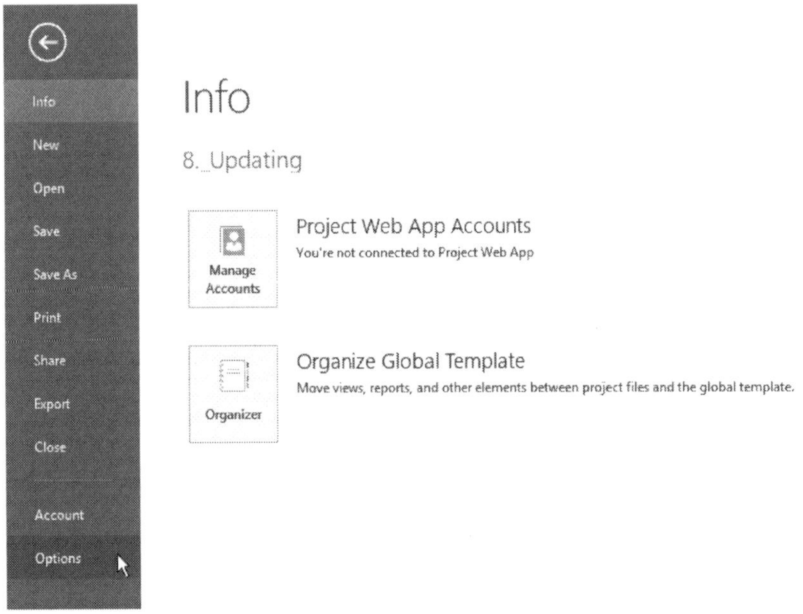

Project opens the Project Options dialog box.

Step 3: Select the Schedule tab.

Step 4: Clear the Actual costs are always calculated by Project box.

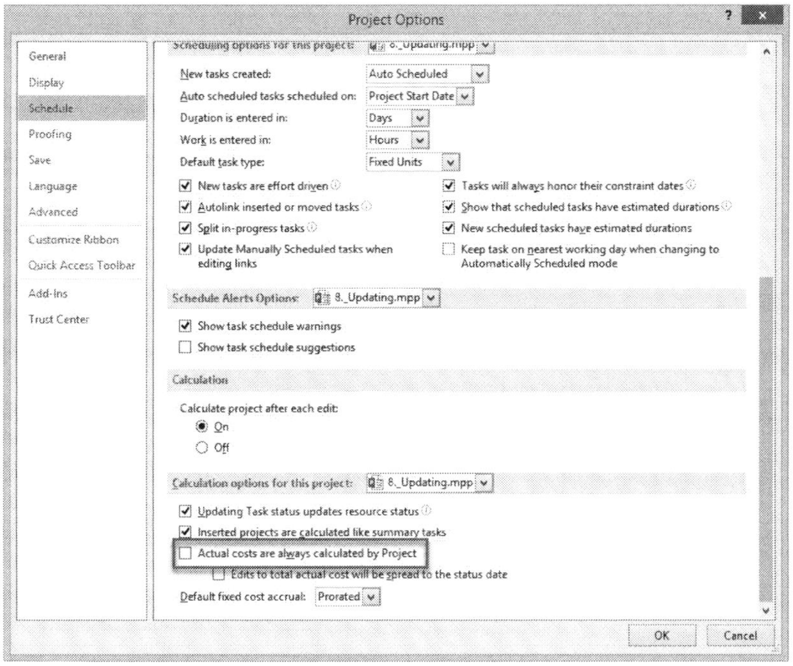

Entering actual costs.

Step 1: Select Task Usage from the View menu.

Step 2: Right click the right side of the sheet and select Actual Cost from the context menu.

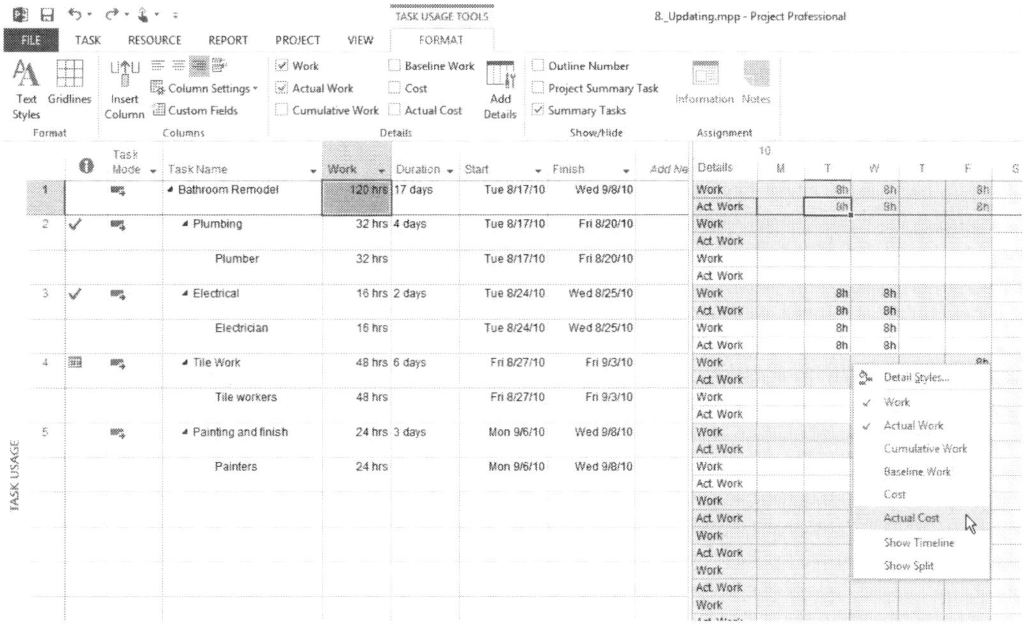

Step 3: Change the table for the Task Usage sheet by selecting Table from the View menu. Select Tracking.

Step 4: Enter the actual cost values in the appropriate dates on the right side of the sheet. You can also adjust the total actual cost. You may have to press TAB or scroll to see the Actual Cost column on the left side of the sheet.

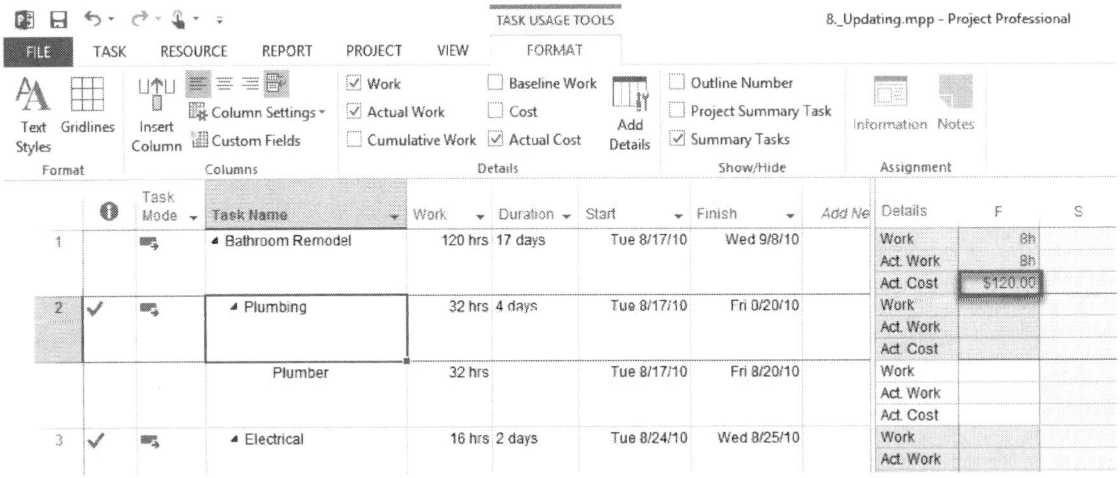

Chapter 17 – Checking Project Progress

An important aspect of project management is checking on the project's progress and adjusting in your plan where necessary. This chapter will explain how to view project statistics and costs. It will also show you how to check variances in work or cost. You will also learn how to identify slipped tasks and save an interim plan.

Viewing Project Statistics

Use the following procedure to review the project statistics.

Step 1: Select the Project tab.

Step 2: Select Project Information.

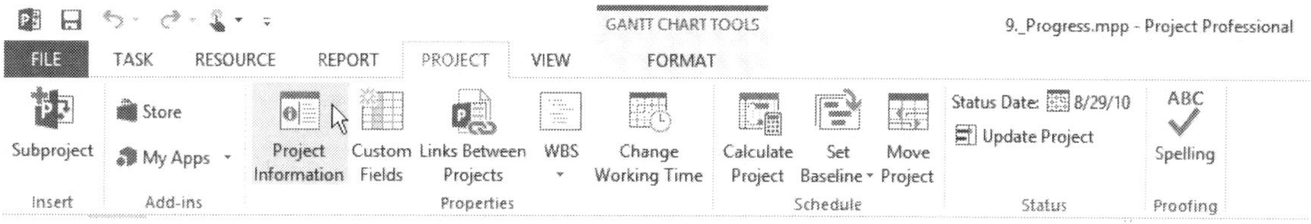

Project opens the Project Information dialog box.

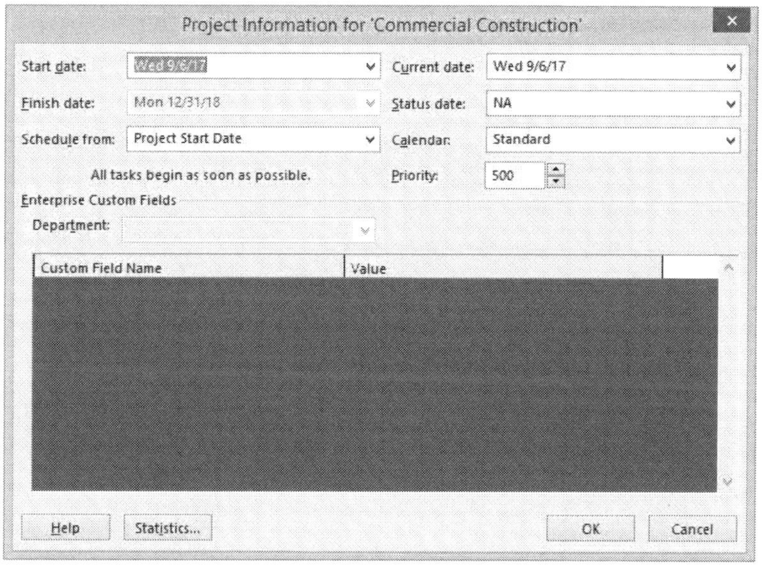

Step 3: Select the Statistics button.

Project displays the Project Statistics dialog box.

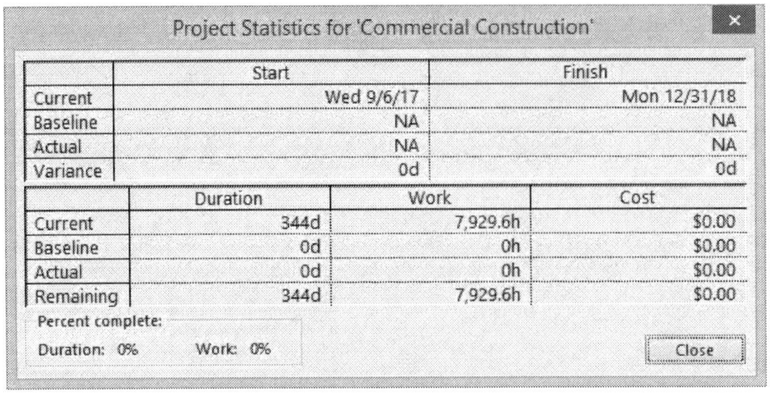

Viewing Project Costs

Use the following procedure to switch to and view the Cost table.

Step 1: Starting from the Gantt Chart view, select Tables from the View Tab. Select Cost.

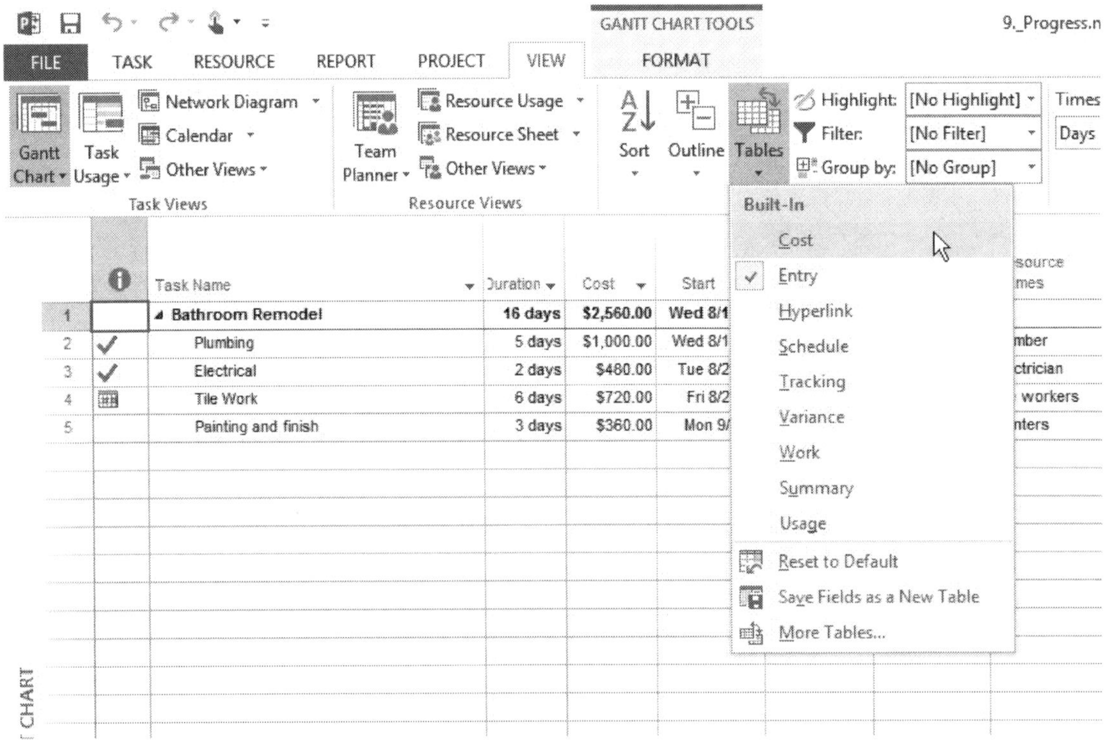

Step 2: Drag the divider to the right to see more of the Cost columns at one time.

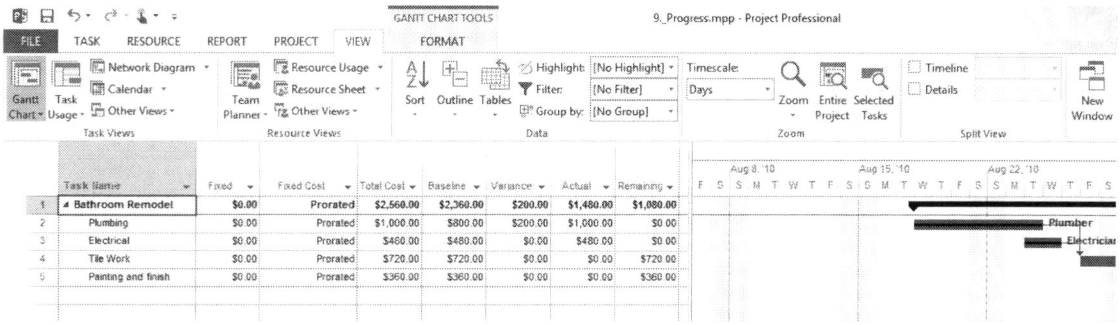

You can add columns to the current view. Discuss the different cost-related columns available beyond what is shown in the Cost table.

Checking Duration Variance

Use the following procedure to check the duration variance.

Step 1: From the Gantt Chart view, right-click the Duration column and select Insert Column from the context menu.

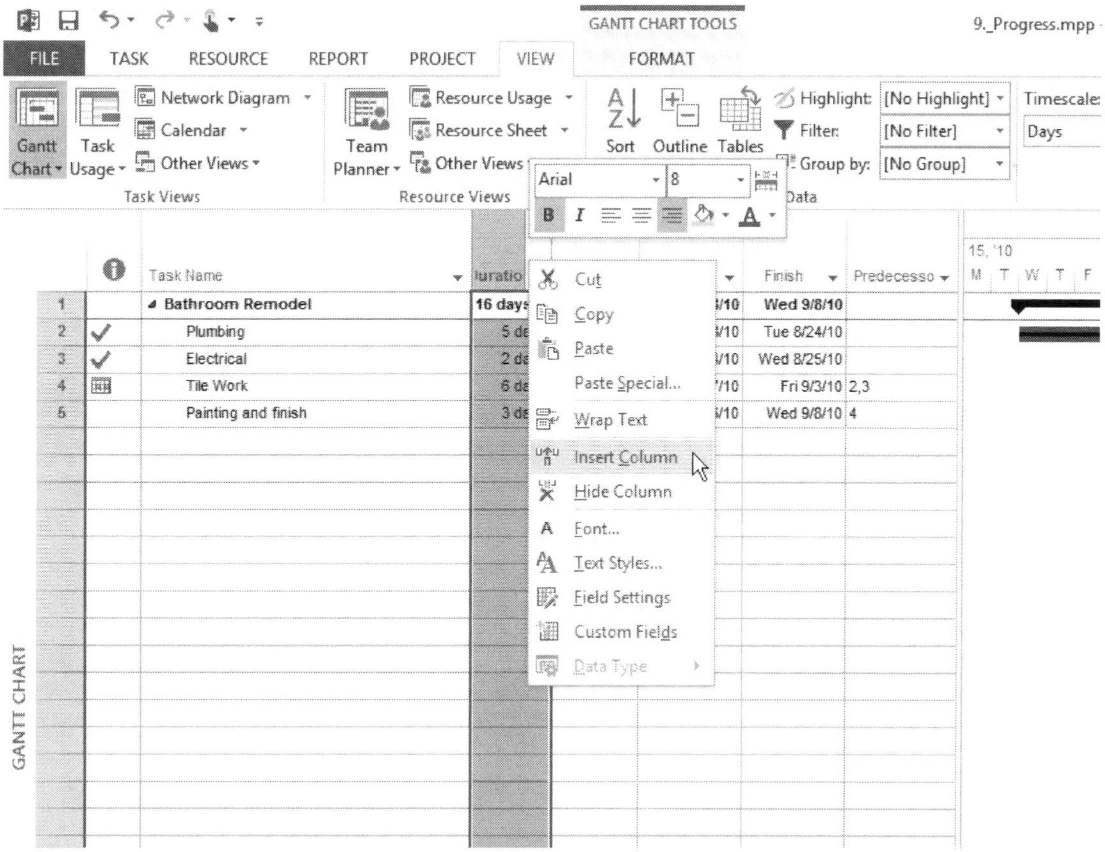

Step 2: Select Duration Variance as the Field Name.

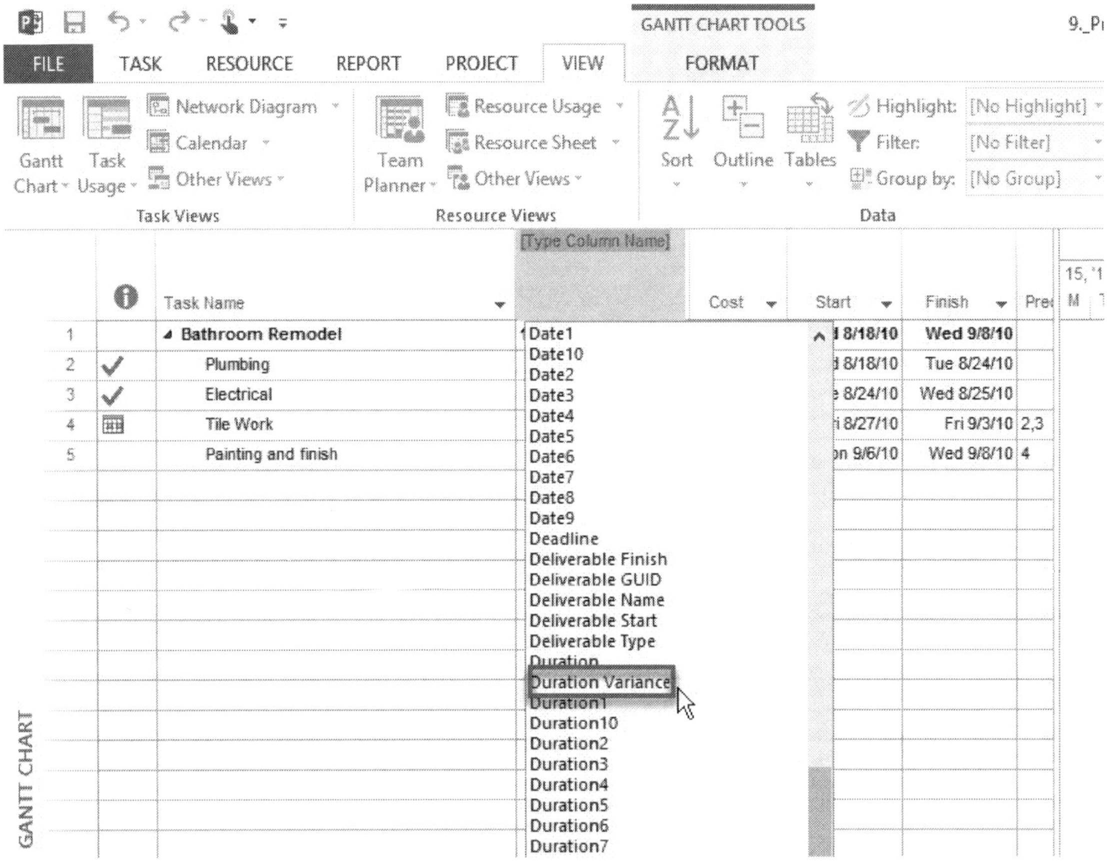

View the duration variance for the Summary Tasks and the detail tasks.

Checking Work Variance

Use the following procedure to check the work variance.

Step 1: Select Task Usage from the View tab.

Step 2: Right-click one of the column headings and select Insert Column from the context menu.

Step 3: Select Work Variance as the Field Name.

View the work variance for the Summary Tasks and the detail tasks.

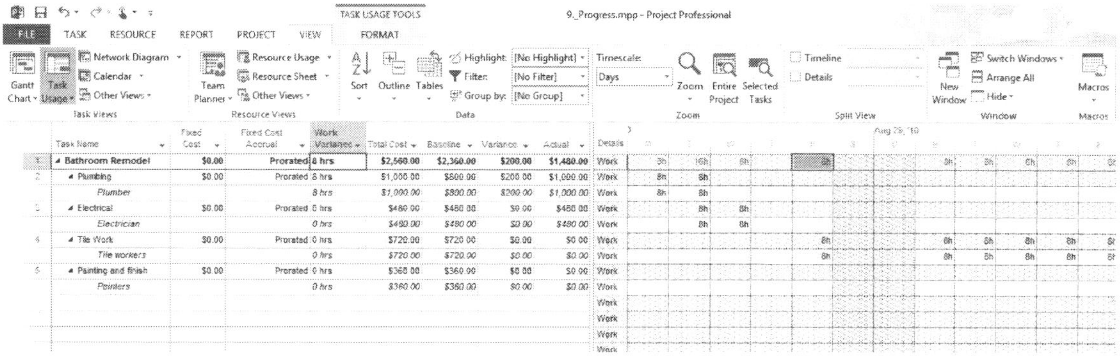

Checking Cost Variance

Use the following procedure to check the cost variance.

Step 1: Select Tables from the View Tab. Select Cost.

View the cost variance for the Summary Tasks and the detail tasks.

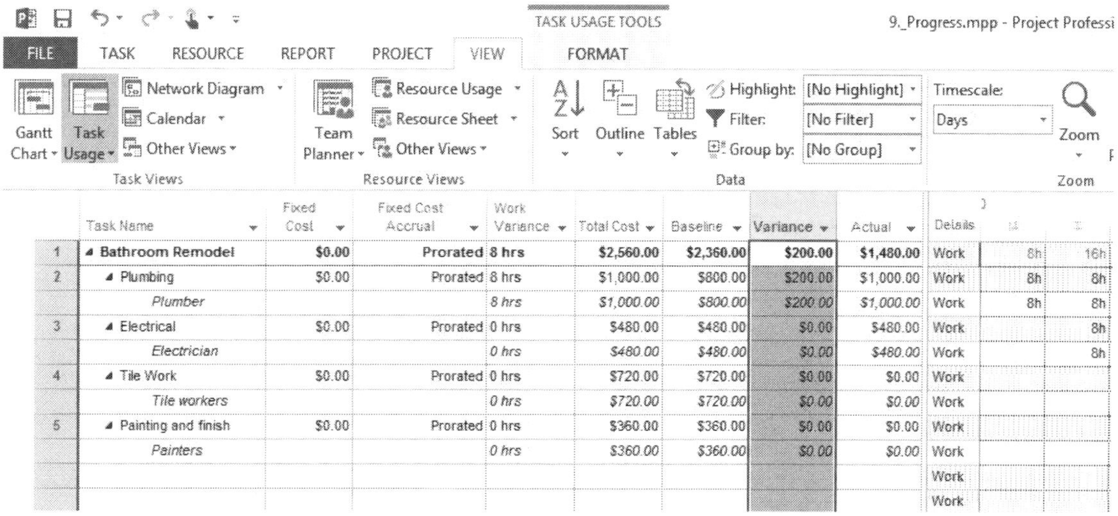

Identifying Slipped Tasks

To check the work variance.

Step 1: Select the arrow next to Gantt Chart on the View Tab. Select Tracking Gantt.

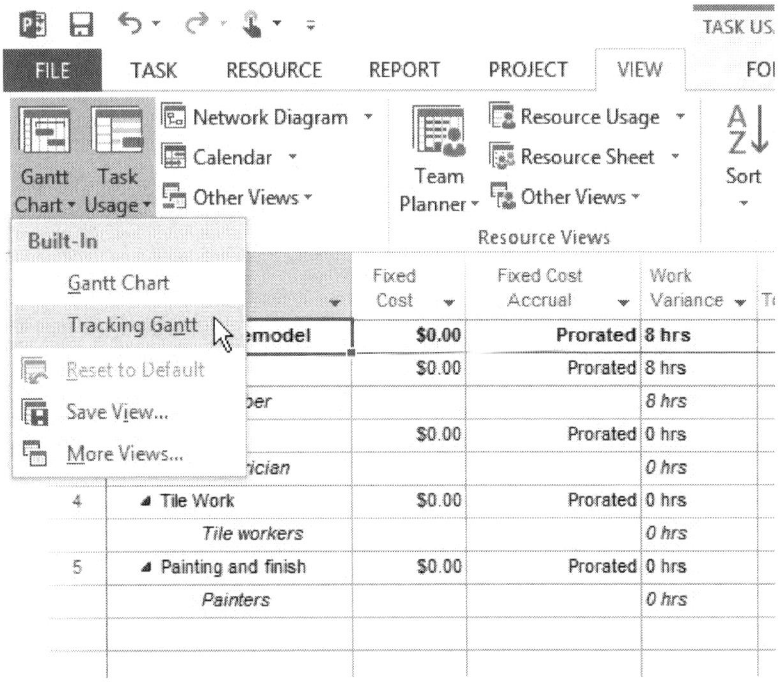

Step 2: Select Tables from the View Tab. Select Variance.

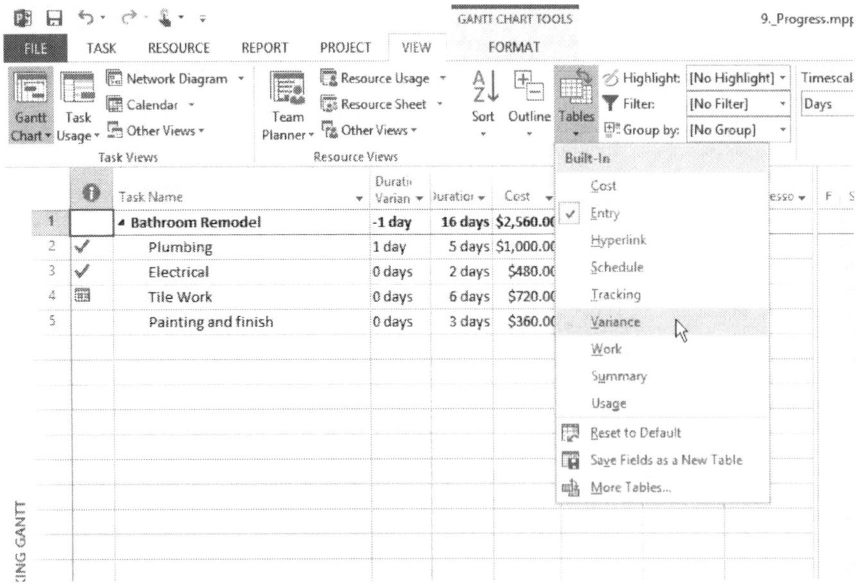

Step 3: Select Filter from the View Tab. Select More Filters.

Step 4: Select Slipping Tasks and select Apply.

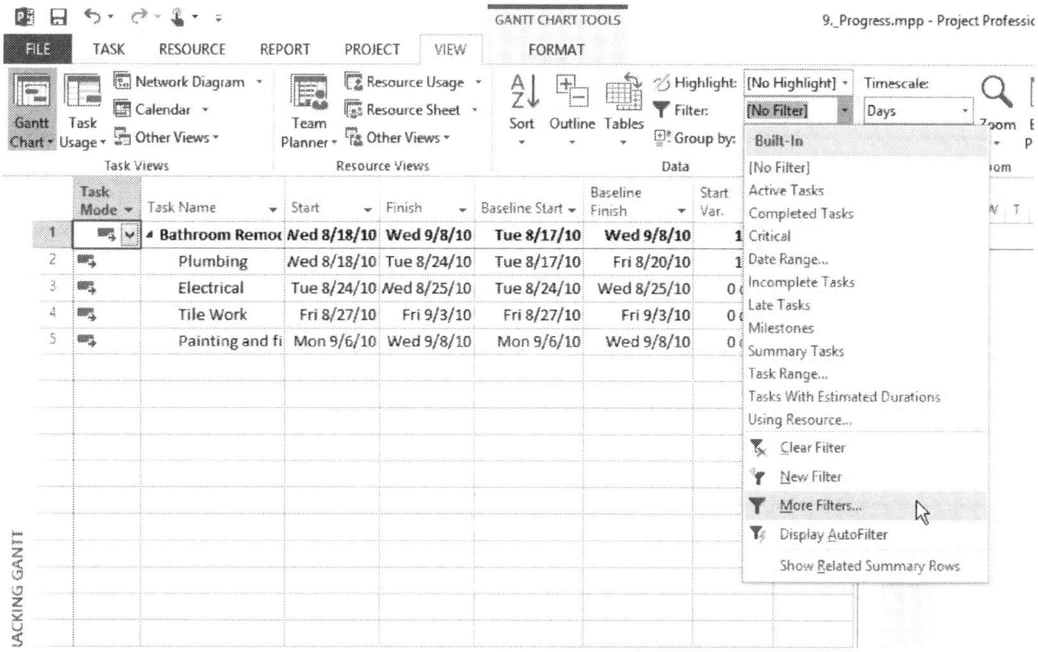

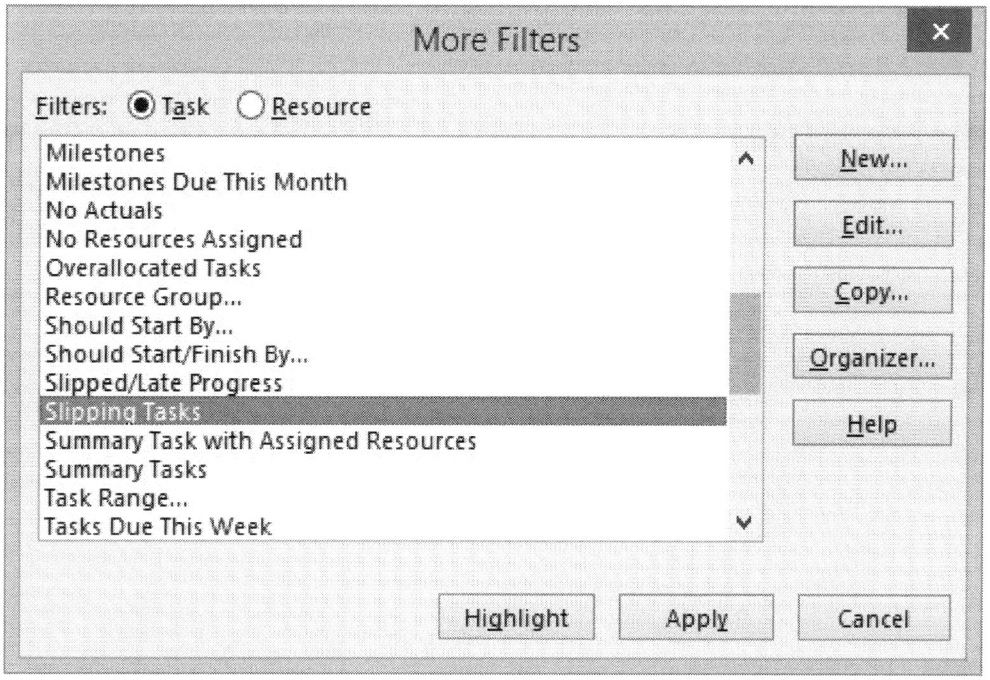

Saving an Interim Plan

Setting an interim plan.

Step 1: Select Set Baseline from the Project tab.

Step 2: Select Set Interim Plan.

Step 3: Select a value that you want to save from the Copy drop-down list. The current start date, finish date, and baseline values are not numbered.

Step 4: Select a value into which you want to copy values from the Into drop-down list. Interim plans are stored in the Start and Finish fields.

Note: If you select a baseline in both the Copy and the Into fields, you will save a baseline, rather than an interim plan. If you select a baseline in the Copy box, and a start and finish interim plan in the Into box, only the start date and finish date from the baseline will be copied to the interim plan.

Step 5: Select Entire project or Selected Tasks.

Step 6: Select OK.

Chapter 18 – Working with Reports

In this chapter, you will learn how to work with reports. You will learn how to customize basic and visual reports. You will also learn how to create a custom report. Finally, you will learn how to sort a report.

Customizing a Basic Report

Use the following procedure to define the report contents.

Step 1: Select the Report tab from the Ribbon.

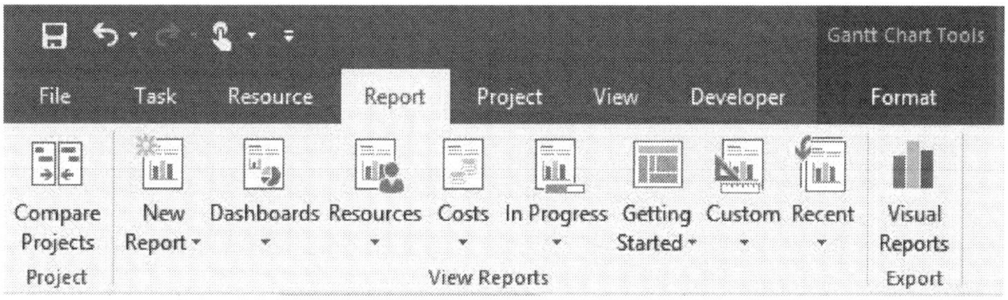

Step 2: Select the arrow under Dashboard and select Upcoming tasks.

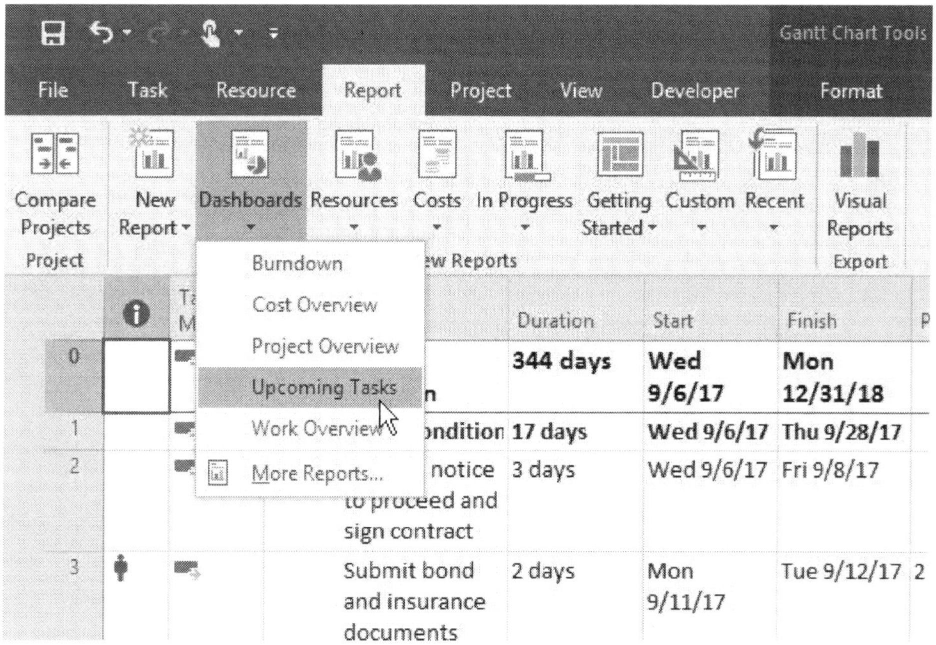

Step 3: Click the Table to open the field list.

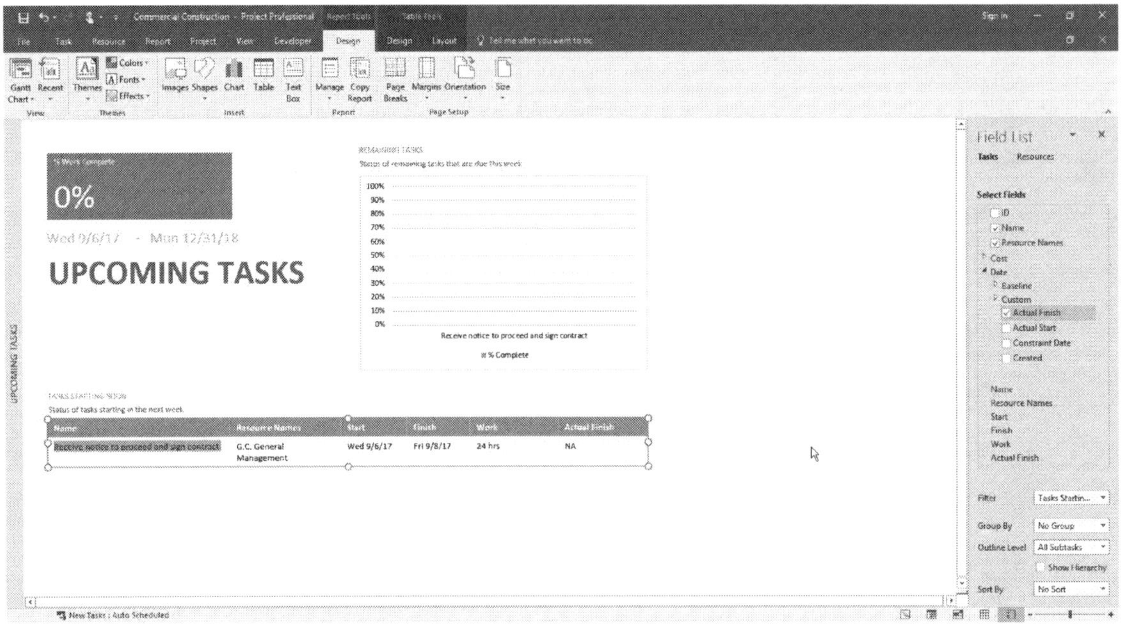

Step 4: Select the Desired fields.

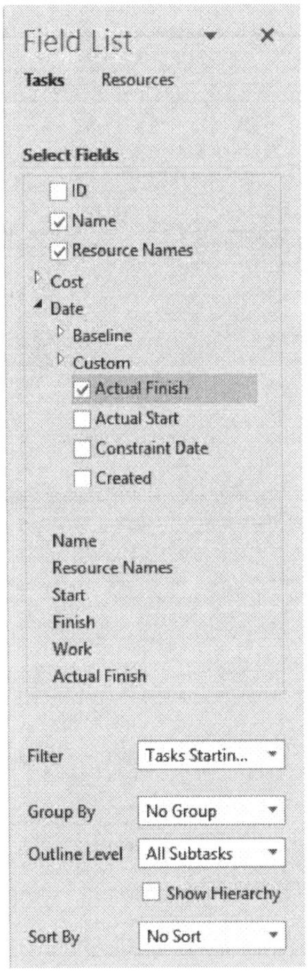

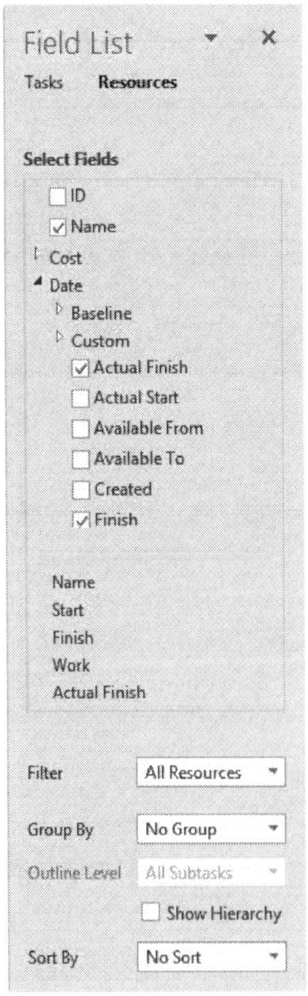

Step 5: On the Report, click Chart to open the design ribbon in Report Tools.

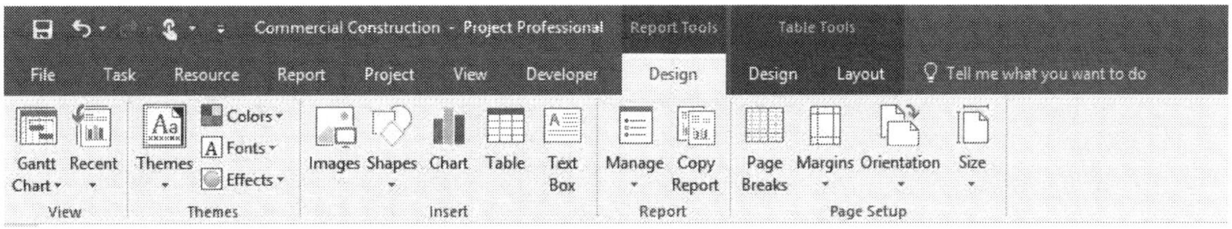

Step 6: Make the desired formatting changes for the Design and Layout.

Design Ribbon of Table Tools

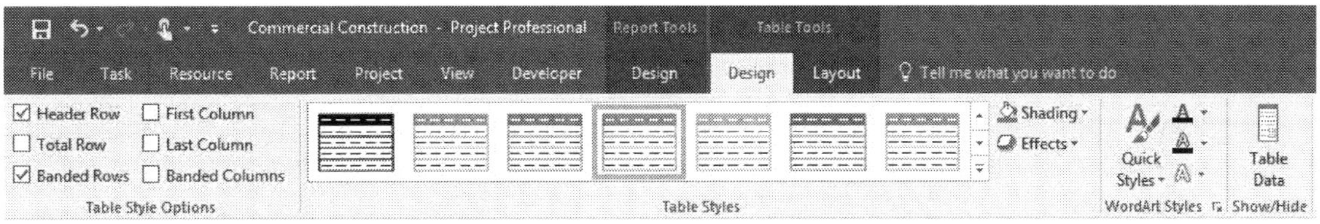

Layout Ribbon of Table Tools

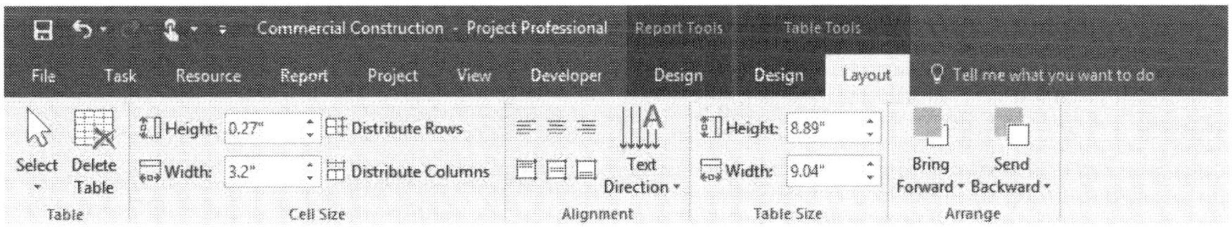

Creating a New Report

Step 1: Select the Report tab from the Ribbon.

Step 2: Select arrow under New Report.

Step 3: Select Blank, Chart Table or Comparison.

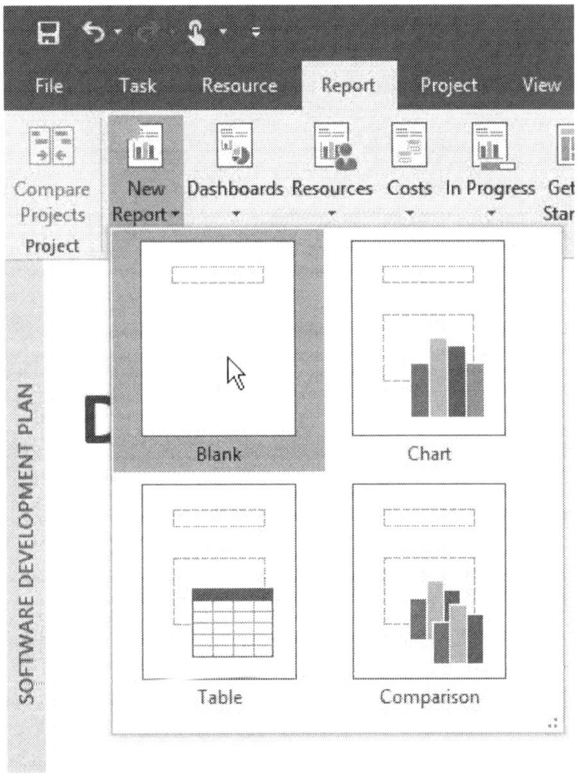

Step 4: Select an option from the New Reports list.

Step 5: Give the report a name.

Project displays the Design ribbon of the Report Tools options. Use the functions on the Design ribbon to develop the report layout.

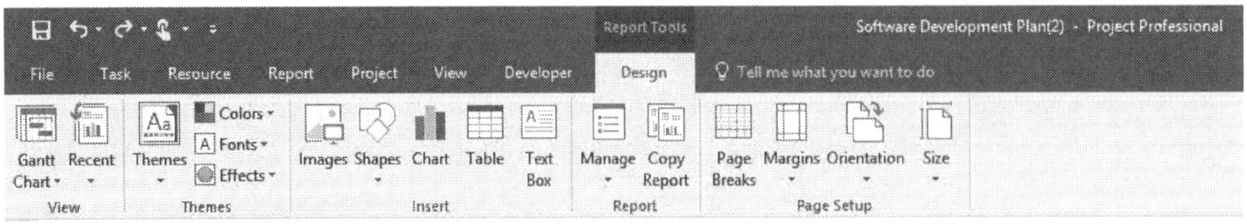

The Reports Organizer.

Step 1: Select the File tab from the Ribbon.

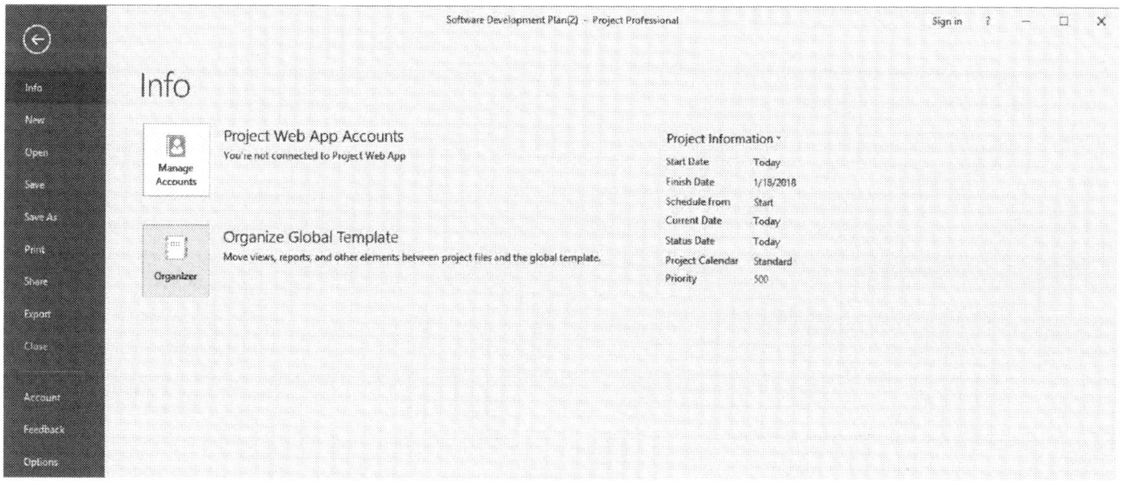

Step 2: Select Info.

Step 3: Select Organize Global Template.

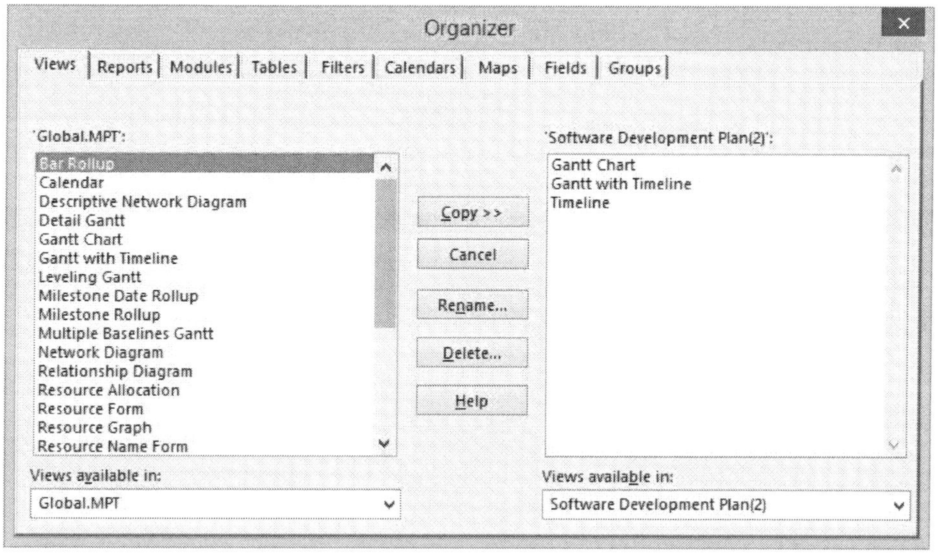

The left side of the dialog box lists available reports. You can select a different Project file from the drop-down list (it must be open). The right side of the dialog box lists reports that are in your current project. To copy a report, highlight it on the left side and select Copy.

Customizing a Visual Report

Use the following procedure to customize a visual report.

Step 1: Select the Report tab from the Ribbon.

Step 2: Select Visual Reports.

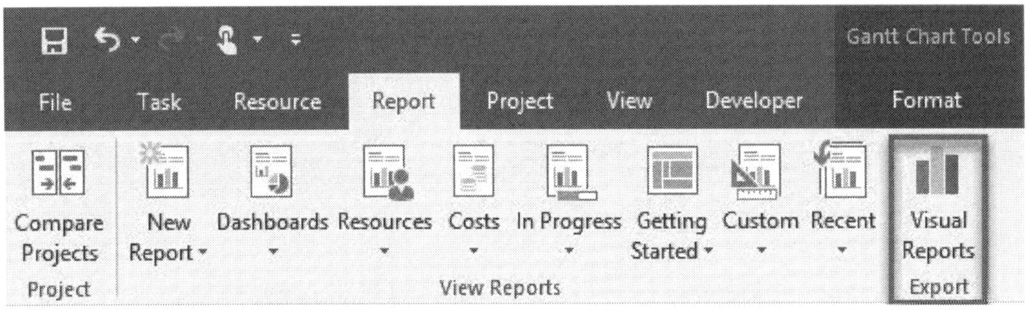

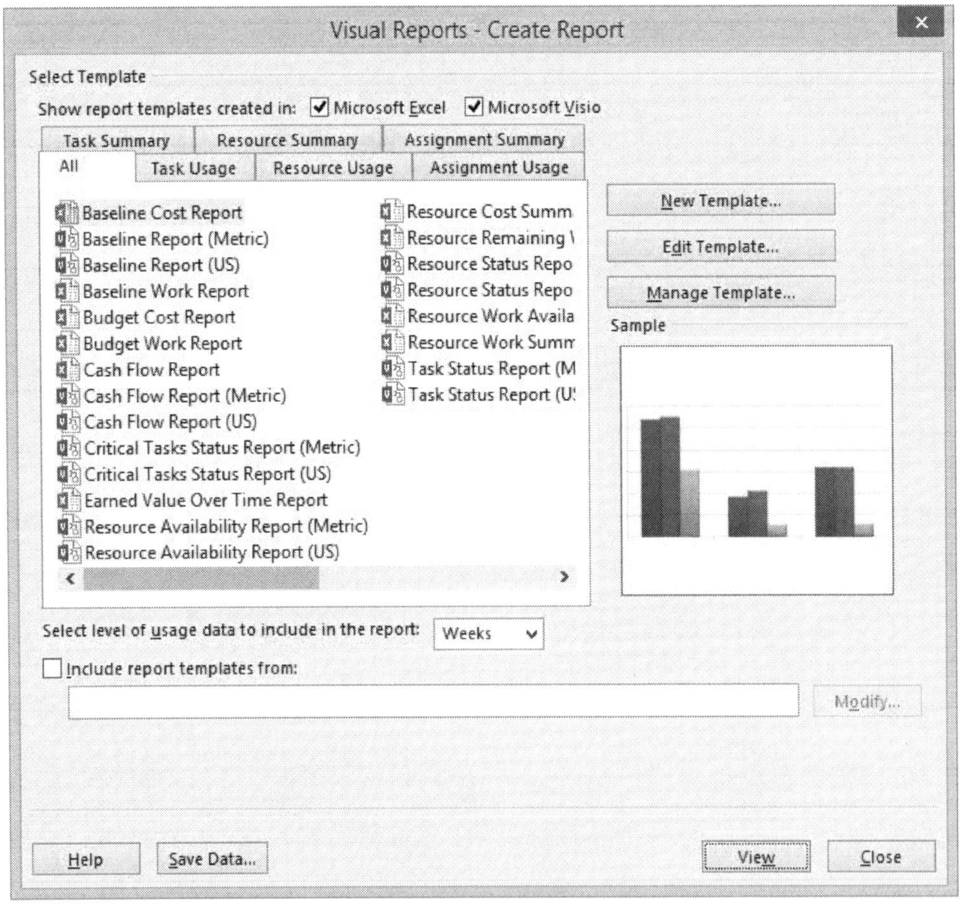

Step 3: Select a report from the list to use as a starting point. You must have either Excel or Visio to be able to run visual reports. The list only shows reports you have available if you do not have both programs installed.

Step 4: Select Edit Template.

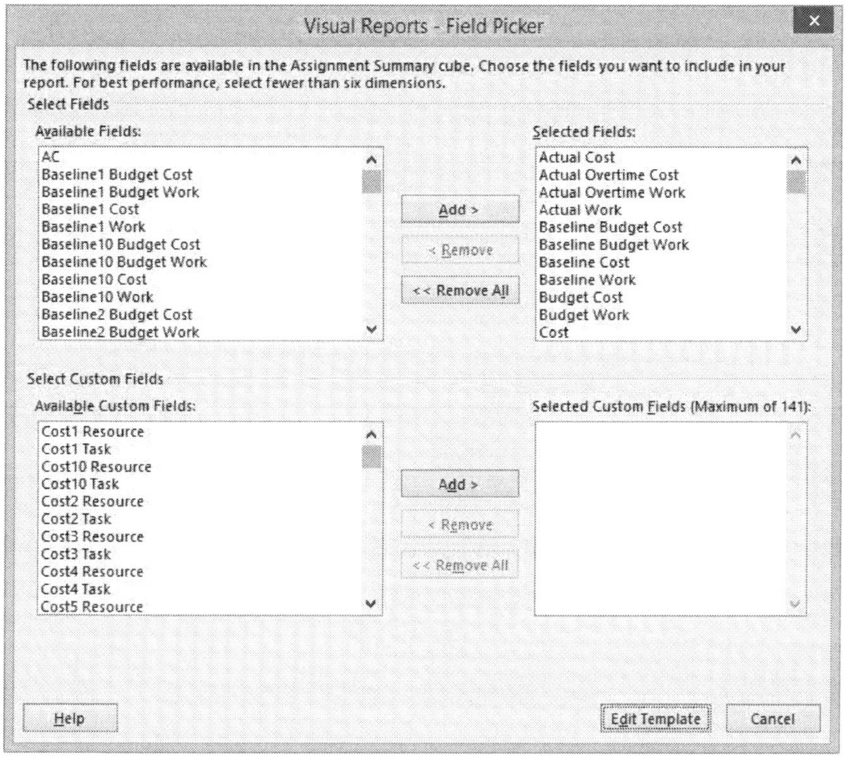

Step 5: In the Selected Fields column on the right, highlight fields you do not want included and select Remove. In the Available Fields column on the left, highlight the fields you do want to include on the report and select Add. For best report, include six or fewer fields in the report.

Step 6: Select Edit Template.

Step 7: The selected report opens in either Excel or Visio, depending on your selections.

Sorting a Report

Use the following procedure to sort a report.

Step 1: Select the Report tab from the Ribbon.

Step 2: Select the arrow under Dashboard.

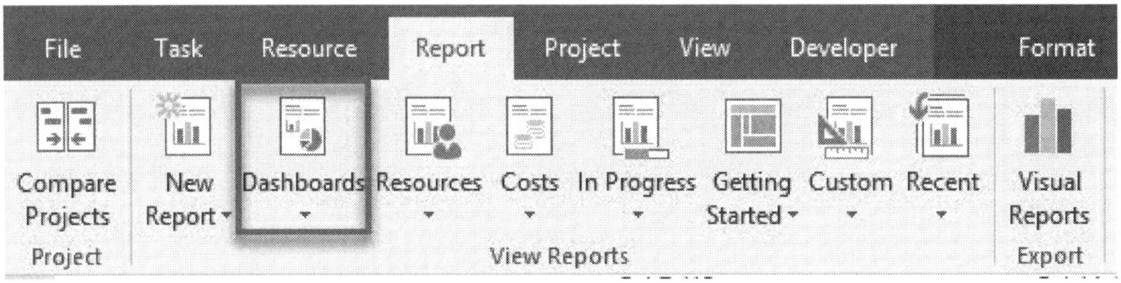

Step 3: Select Upcoming Tasks.

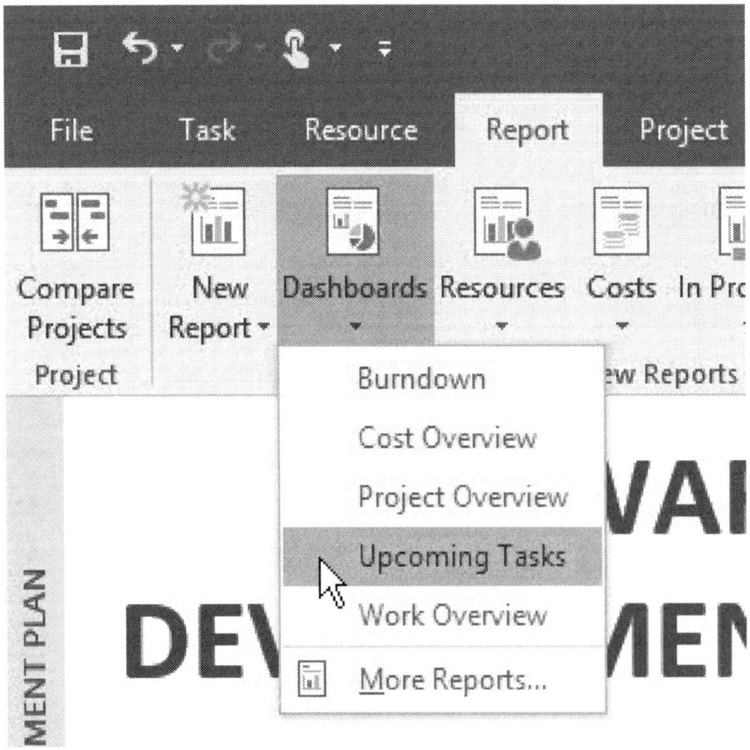

Step 4: Select a component of the Upcoming Tasks report to open a field list.

Step 5: Select the Sort option from the Sort By: drop-down menu.

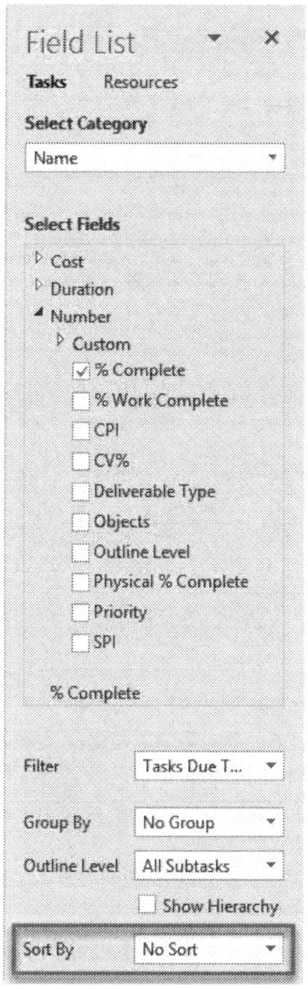

Step 6: For up to three sorting options, select the Custom Sort option from the drop-down lists. Select Ascending or Descending for each field.

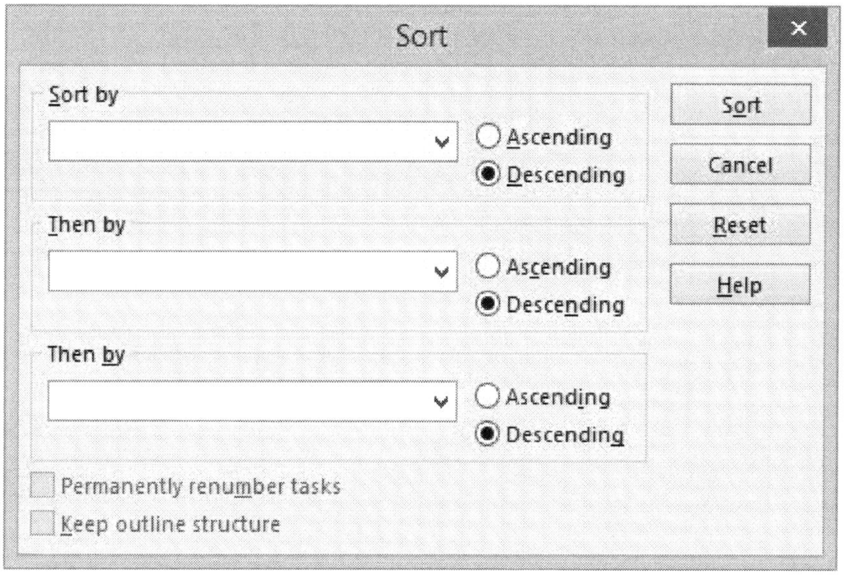

Step 7: Check the Keep outline structure box to keep the outline structure for the sorted tasks.

Step 8: Select the Sort button.

Step 9: Now, in the Report options window, choose the Select button.

Chapter 19 – Working with Multiple Projects

This chapter explains how to handle multiple projects. You will learn how to create links between projects. Since working with a single file is always faster if you can help it, this chapter will also explain how to consolidate projects. You will learn how to view multiple project critical paths and consolidated project statistics. Finally, this chapter explains how to create a resource pool.

Inserting a Subproject

Use the following procedure to link a project in a master project

Step 1: In the blank project, from the Gantt Chart view, highlight the row where you want to insert the project.

Step 2: Select the Project tab.

Step 3: Select Subproject.

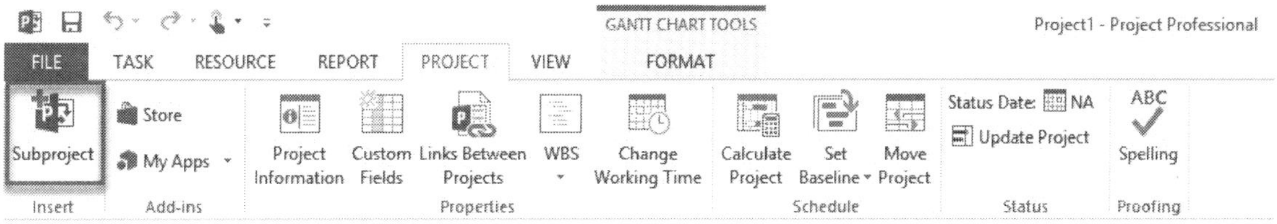

Step 4: Highlight the project you want to insert.

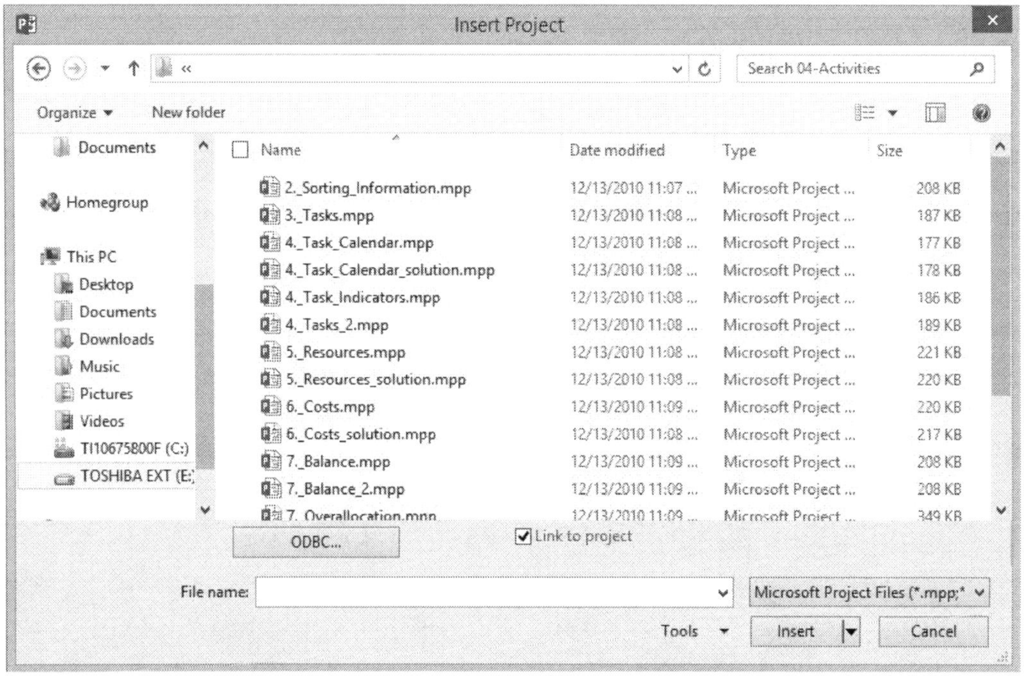

Note: To insert multiple projects, hold down CTRL and click the projects in the order that you want to insert them.

Step 5: Make sure the Link to project box is checked. However, if you do not want to update the subprojects with any changes made to the master project, clear the box.

Step 6: Select Insert.

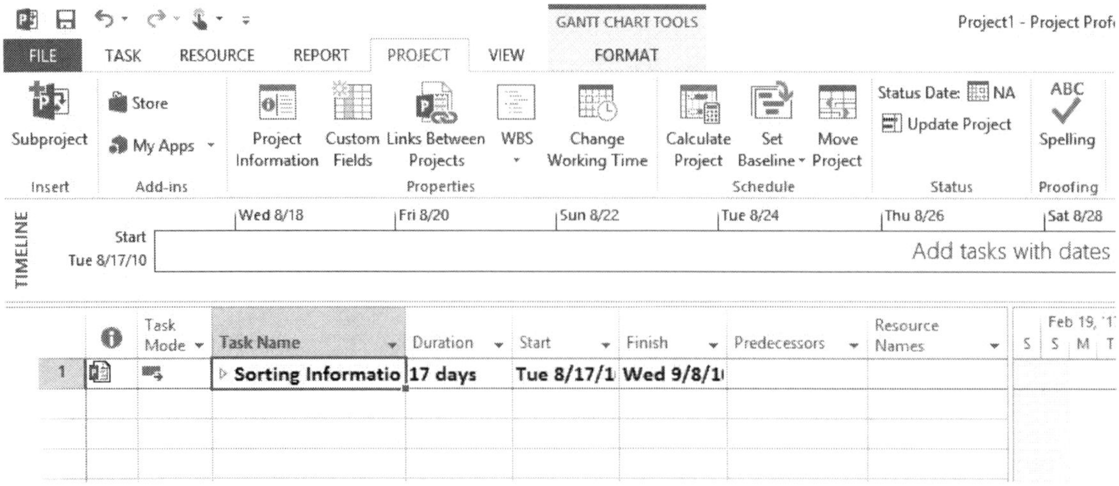

Consolidating Projects

To consolidate projects for printing, use the following procedure.

Step 1: Open the project files you want to combine.

Step 2: Select New Window from the View tab on the Ribbon.

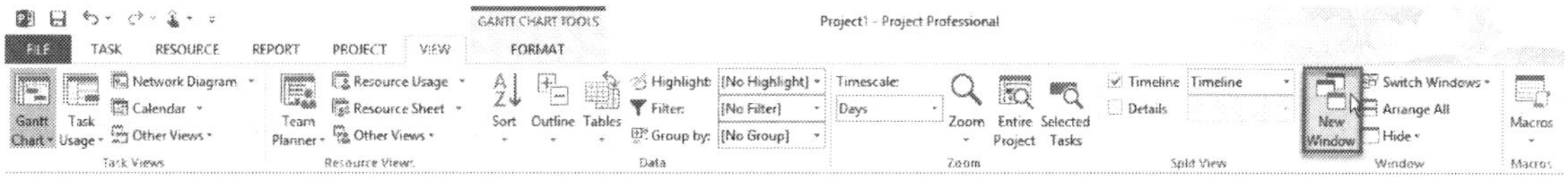

Step 3: Highlight the first file you want to appear in the consolidated window. Hold the CTRL key to select subsequent projects in the order you want them to appear.

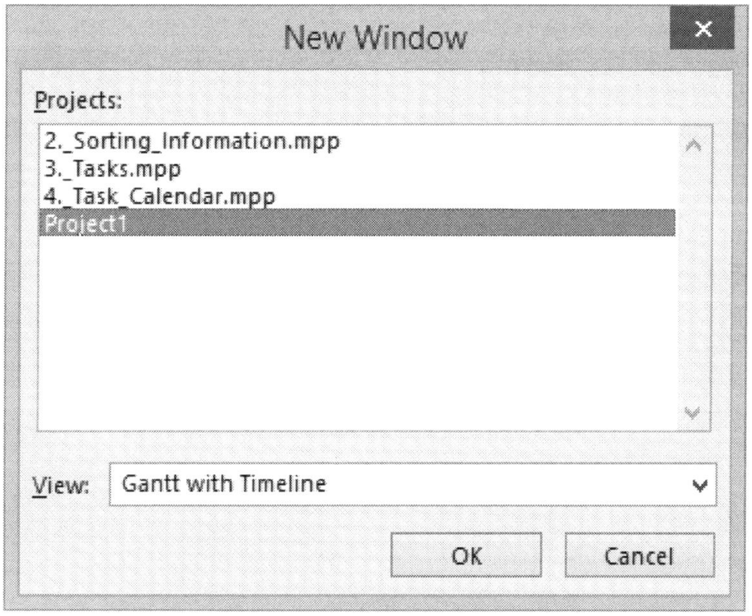

Step 4: Select a View option from the drop-down list.

Step 5: Select OK.

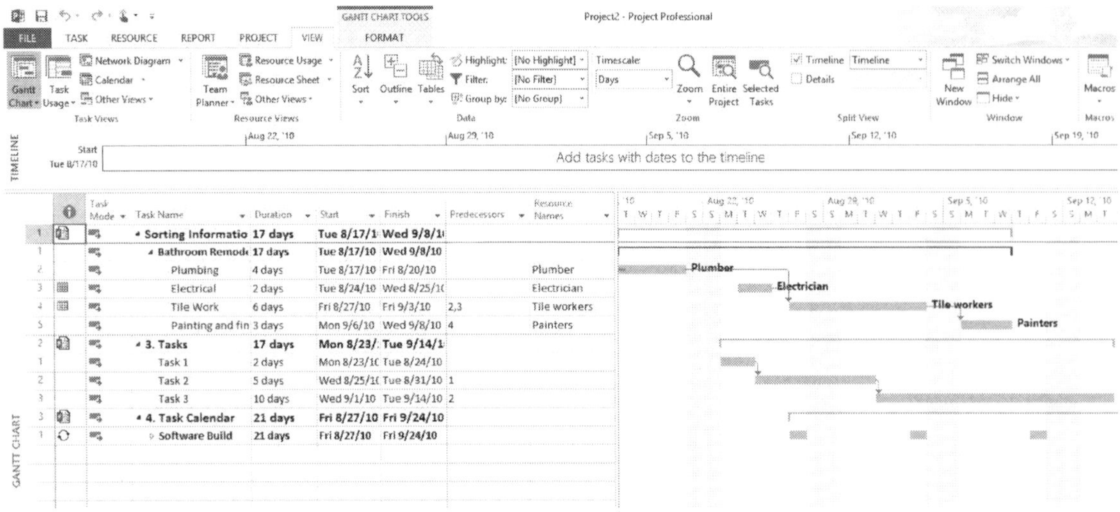

Viewing Multiple Project Critical Paths

Use the following procedure to display multiple critical paths.

Step 1: Select the File tab from the Ribbon.

Step 2: Select Options.

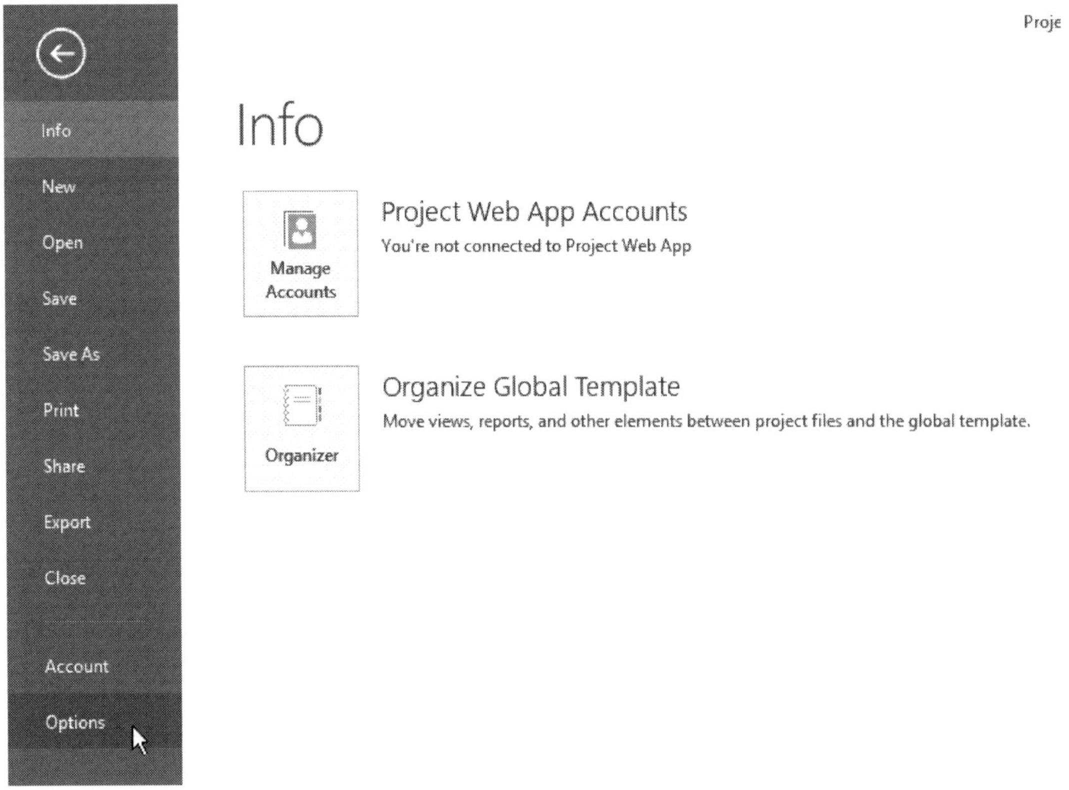

Project opens the Project Options dialog box.

Step 1: Select the Advanced tab.

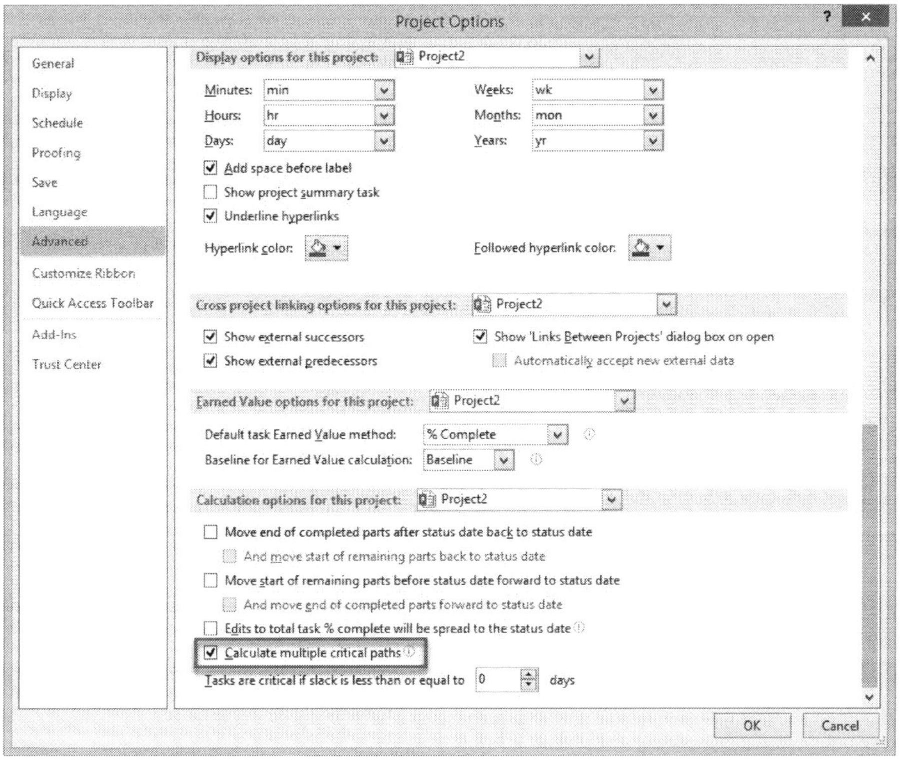

Step 2: Check the Calculate Multiple Critical paths box.

Step 3: Select OK.

Step 4: Select More Views from the View menu.

Step 5: Select Detail Gantt.

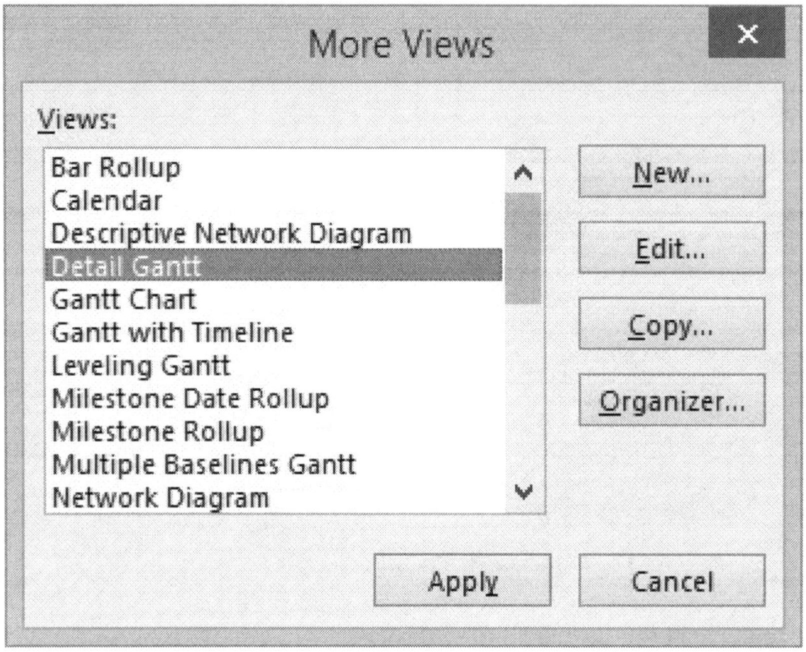

Step 6: Select Apply.

Viewing Consolidated Project Statistics

Use the following procedure to review the project statistics.

Step 1: Select Project Information from the Project tab.

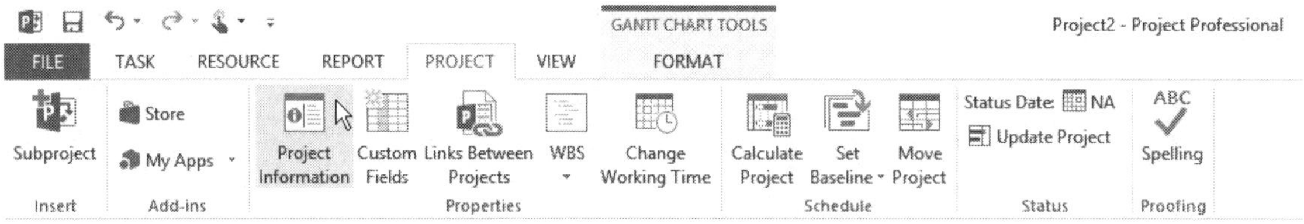

Project opens the Project Information dialog box.

Step 2: Select the Statistics button.

Project displays the Project Statistics dialog box.

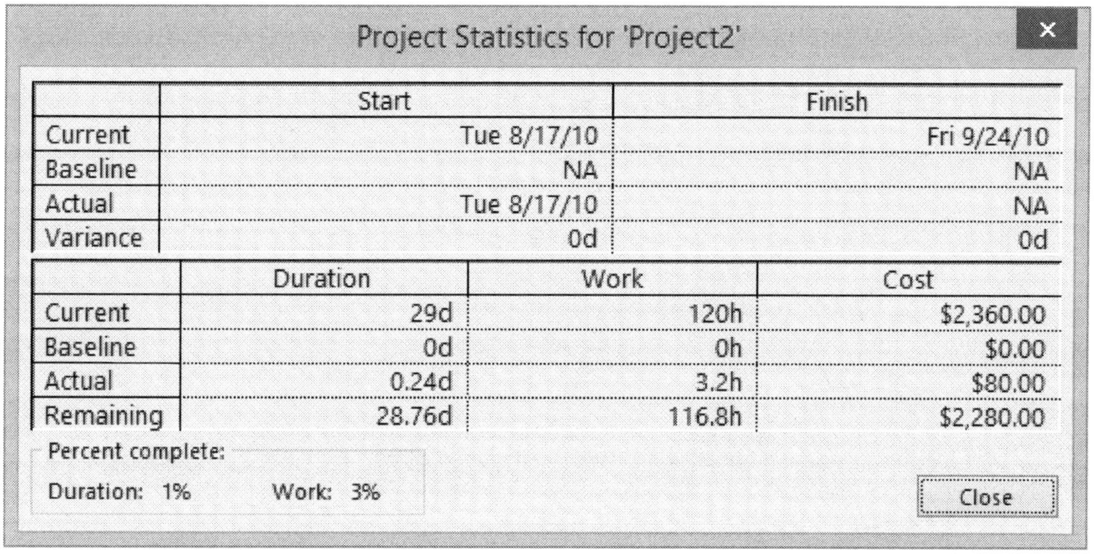
Project Statistics for 'Project2'

	Start	Finish
Current	Tue 8/17/10	Fri 9/24/10
Baseline	NA	NA
Actual	Tue 8/17/10	NA
Variance	0d	0d

	Duration	Work	Cost
Current	29d	120h	$2,360.00
Baseline	0d	0h	$0.00
Actual	0.24d	3.2h	$80.00
Remaining	28.76d	116.8h	$2,280.00

Percent complete:

Duration: 1% Work: 3%

Close

Creating a Resource Pool

Step 1: The project that includes the resources becomes the resource pool. Open that file, plus the file that will share resources, which is the sharer project.

Step 2: Select the sharer project from the Window area of the View tab.

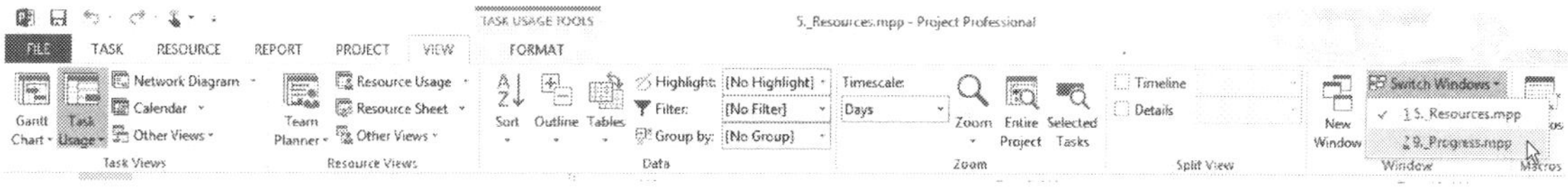

Step 3: Select the Resource tab.

Step 4: Select Resource Pool.

Step 5: Select Share Resources.

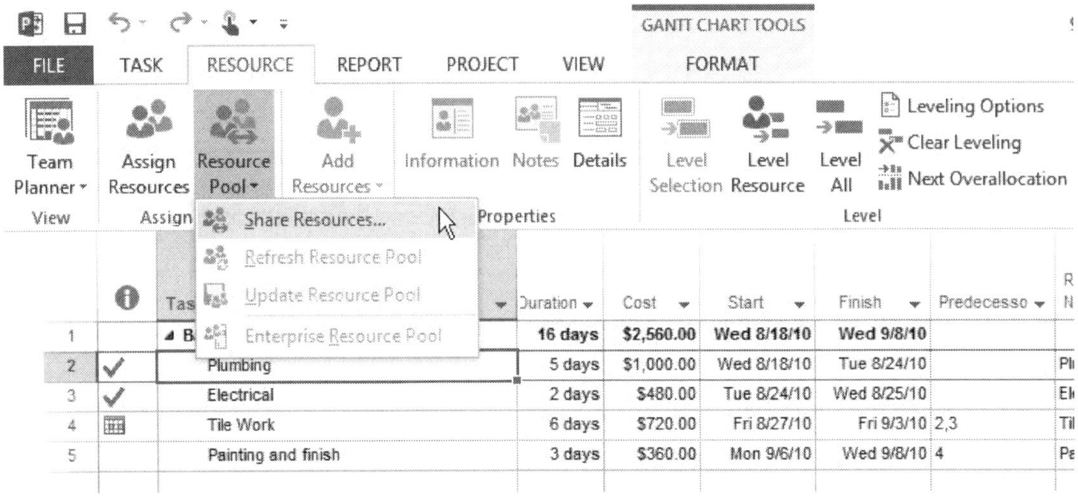

Project displays the Share Resources dialog box.

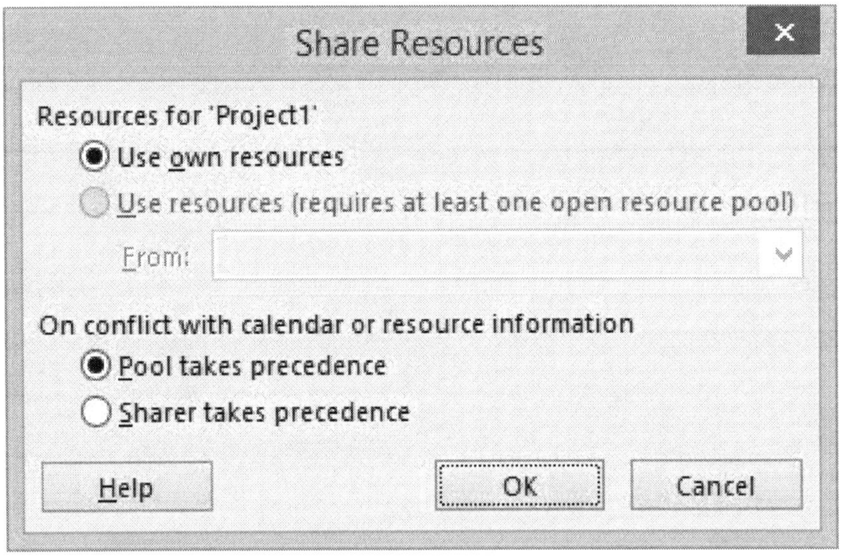

Step 6: Indicate whether to use the project's own resources or the resources from another project. If you select Use Resources From, select the project from the drop-down list.

Step 7: Select how to handle resource conflicts. This indicates whether to overwrite any duplicate resource information, such as rates.

Step 8: Select OK.

View the Resource Sheet to see that the resources are now available to use in the active project. If your project already had resources entered, the resources from both projects are combined.

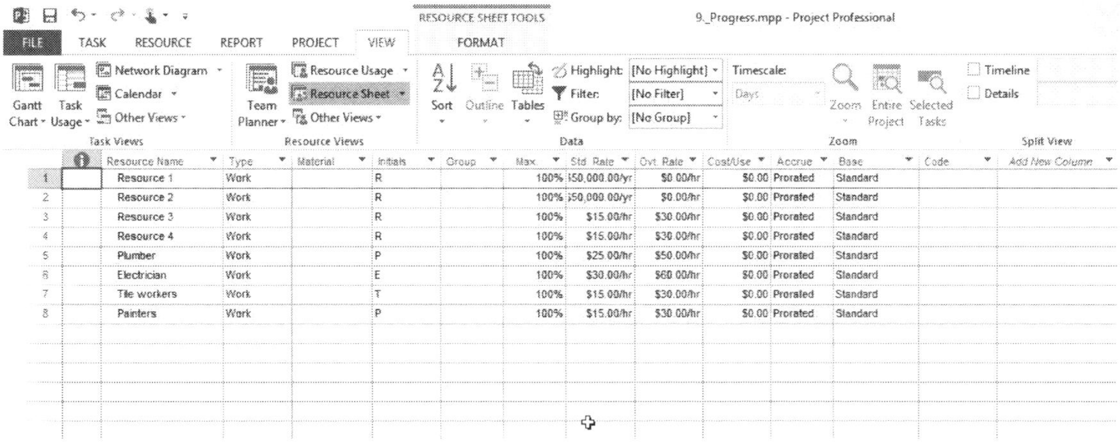

Chapter 20 – Working with the Project Environment

This chapter introduces you to the Project Options dialog box and working with the Project environment. You will learn about customizing the user interface and display options first. On the Scheduling tab, you will learn about the calendar and scheduling customization. You can control proofing, saving, and language options as well. We will take a brief look at the many options on the Advanced tab, where you can control several settings for how Project works. You can customize the Ribbon or the Quick Access Toolbar. You will also practice setting some common settings by setting the default task type and the default assignment unit format.

Setting General and Display Options

To open the Project Options dialog box, use the following procedure.

Step 1: Select the File tab from the Ribbon to open the Backstage View.

Step 2: Select Options.

The Project Options dialog box opens to the General tab.

- Select an option from the drop-down box to select a new Color scheme.
- Select an option from the drop-down box to indicate how you want to use ScreenTips.
- Select a Default view from the drop-down list.
- Select a Date format from the drop-down list.
- Enter your User name and Initials to personalize your copy of Project.

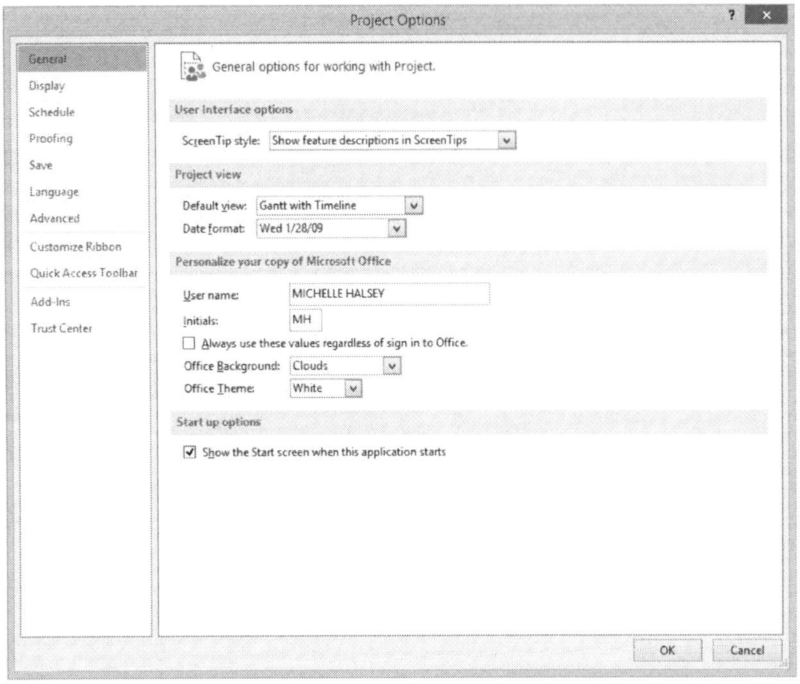

The Display tab of the Project Options dialog box.

Step 1: Select the Display tab from the left.

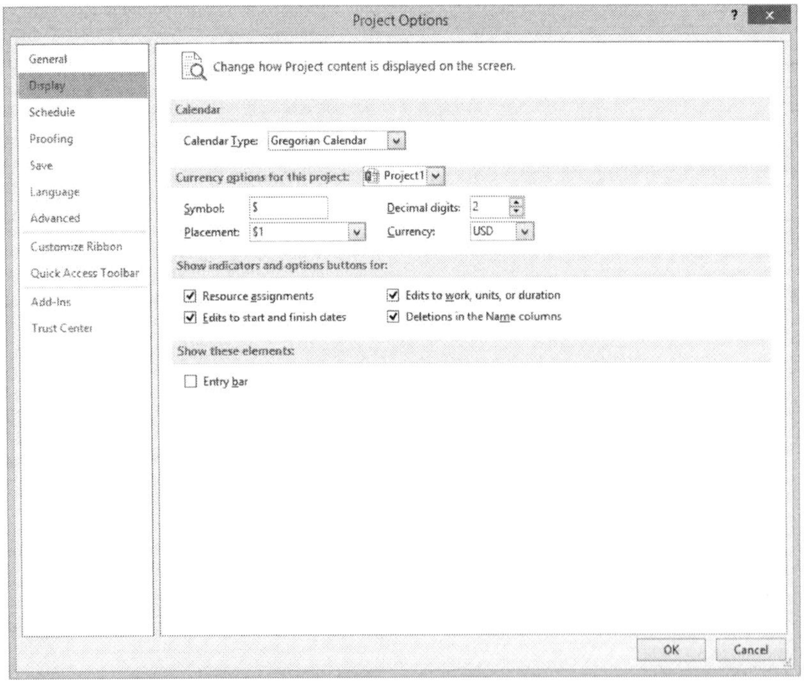

Step 2: Select the Calendar type from the drop-down list.

Step 3: If you have more than one project open, select the Project from the drop-down list to set currency options for that project. You can have different currency options for each project.

Step 4: Enter the Symbol you want to use.

Step 5: Enter or select the number of Decimal places you want to use.

Step 6: Select a Placement option from the drop-down list.

Step 7: Select a Currency option from the drop-down list.

Step 8: Check the boxes to indicate which indicators, options buttons, and elements you want to show.

Setting Calendar and Schedule Options

The Schedule tab on the Project Options dialog box.

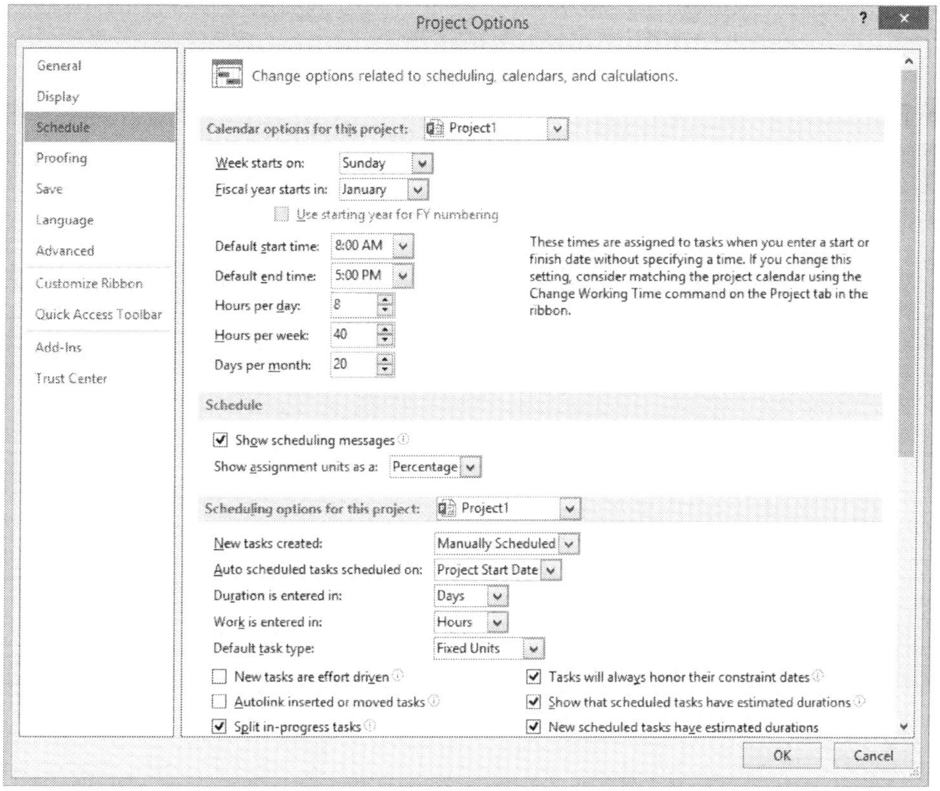

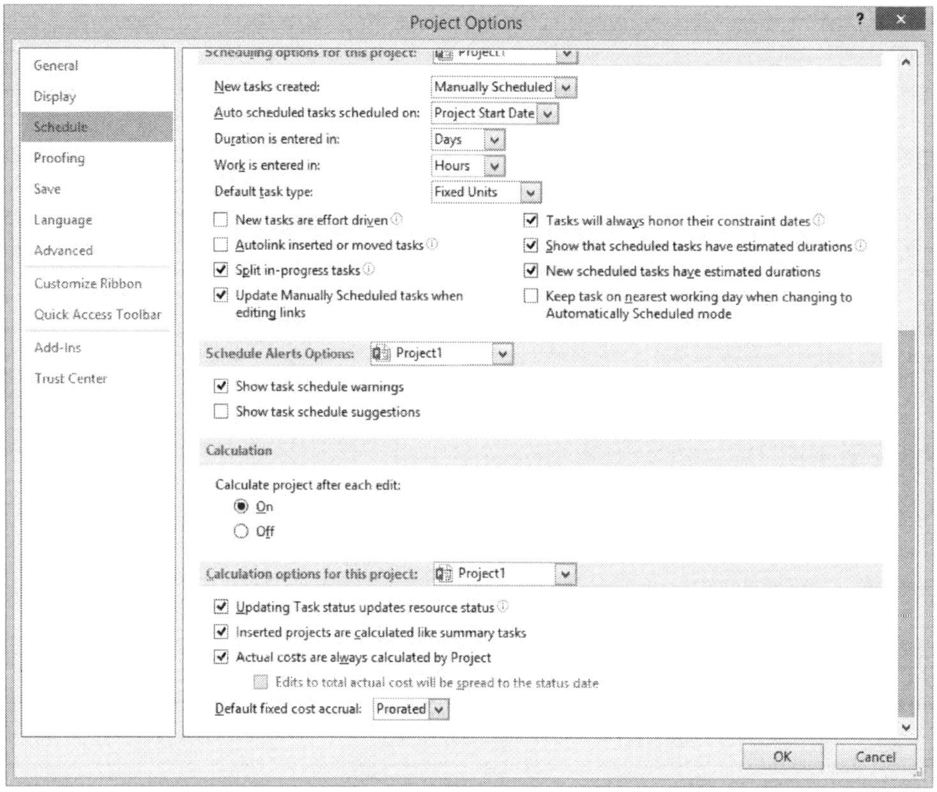

Step 1: If you have more than one project open, select the Project from the drop-down list to set calendar options for that project. You can have different calendar options for each project.

- Select a Week Starting Day from the drop-down list.
- Select a Starting Month for the fiscal year form the drop-down list
- Select a Default Start and End time from the drop-down lists.
- Enter the Hours per Day, the Hours per Week, and the Days per Month.
- Check the box to indicate whether to Show scheduling messages.
- Select an option from the Show Assignment Units drop-down list.

Step 2: If you have more than one project open, select the Project from the drop-down list to set scheduling options for that project. You can have different scheduling options for each project.

- Select how New Tasks are created from the drop-down list.
- Indicate how Auto scheduled tasks are scheduled from the drop-down list.
- Select how Duration is entered from the drop-down list.
- Select the time unit for Work from the drop-down list.
- Select a default task type from the drop-down list.
- Check the boxes to indicate how the listed items are handled.

Step 3: If you have more than one project open, select the Project from the drop-down list to set schedule alert options for that project. You can have different schedule alert options for each project.

- Check the boxes to indicate which warnings and suggestions to show.

Step 4: Indicate whether to Calculate the project after each edit.

Step 5: If you have more than one project open, select the Project from the drop-down list to set calculation options for that project. You can have different calculation options for each project.

- Check the boxes to indicate the calculation options.
- Select a Default fixed cost accrual option from the drop-down list.

Setting Proofing, Saving, and Language Options

The Proofing tab in the Project Options dialog box. This tab allows you to control how Project corrects text as you type and how your spelling is corrected.

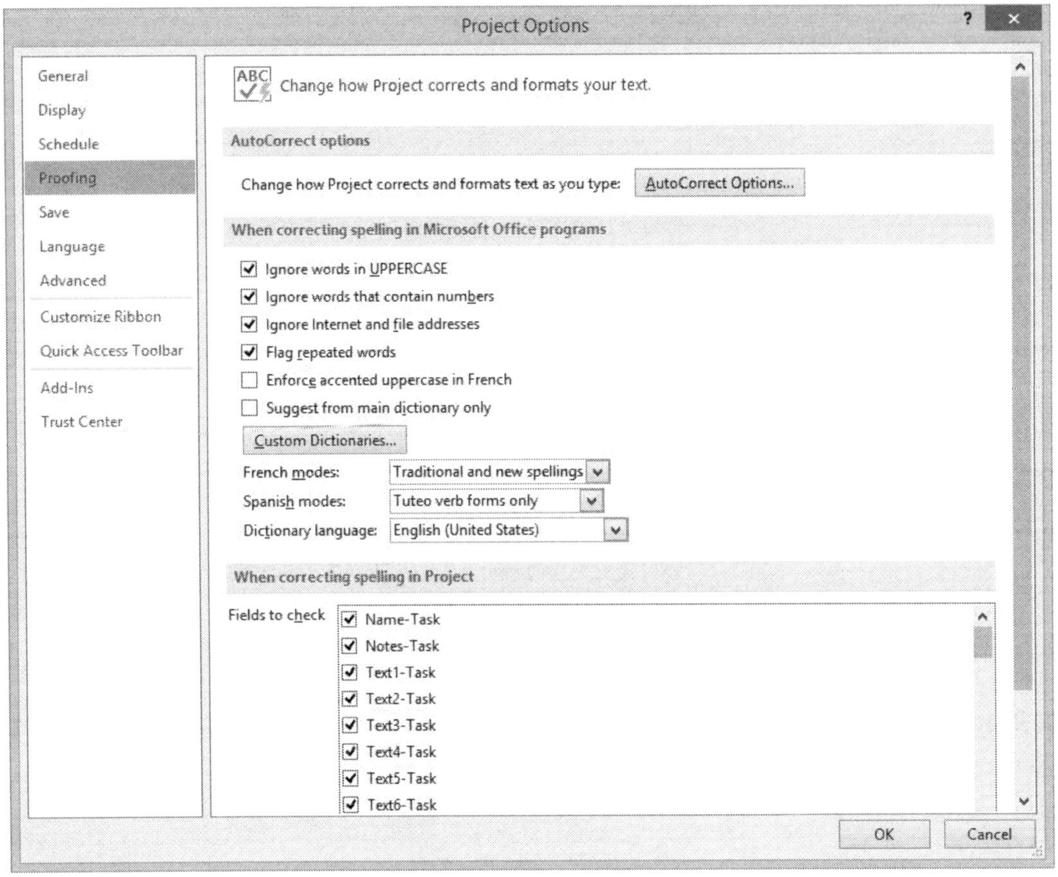

Select the AutoCorrect Options button to open the AutoCorrect dialog box.

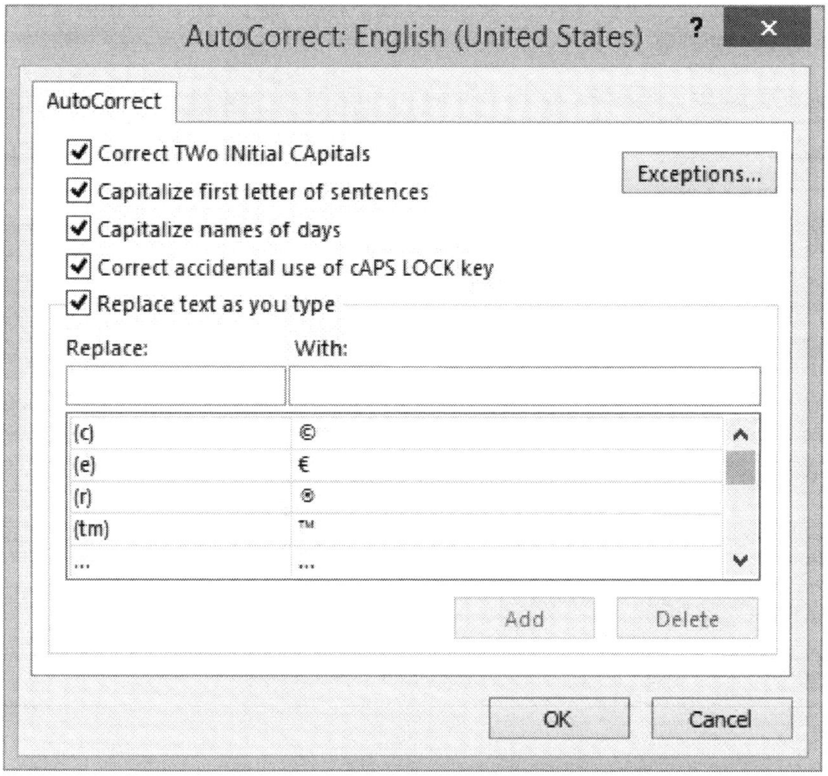

Select the Custom Dictionaries button to open the Custom Dictionaries dialog box.

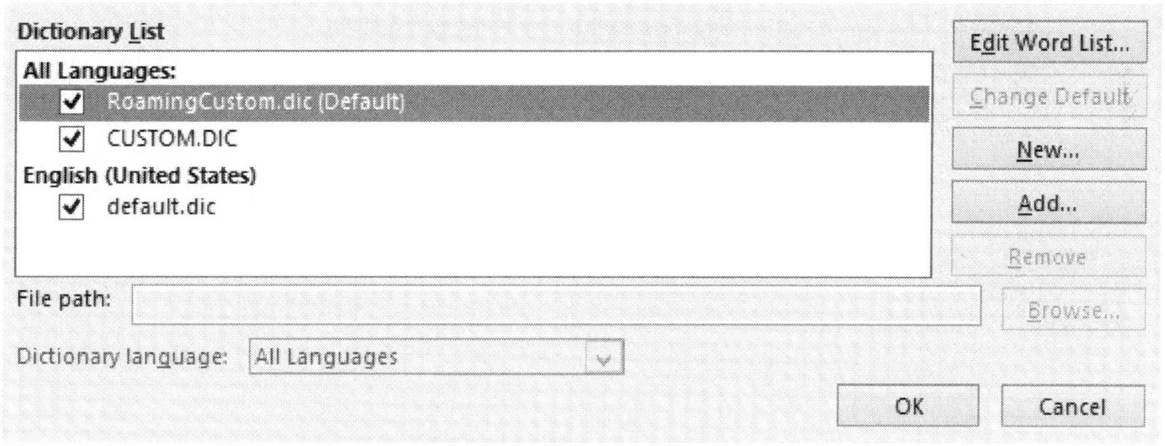

The Save tab on the Project Options dialog box. This tab allows you to control the format and location of your files when they are saved. You can also indicate a frequency for automatic saves. It also controls your template location and cache size and location.

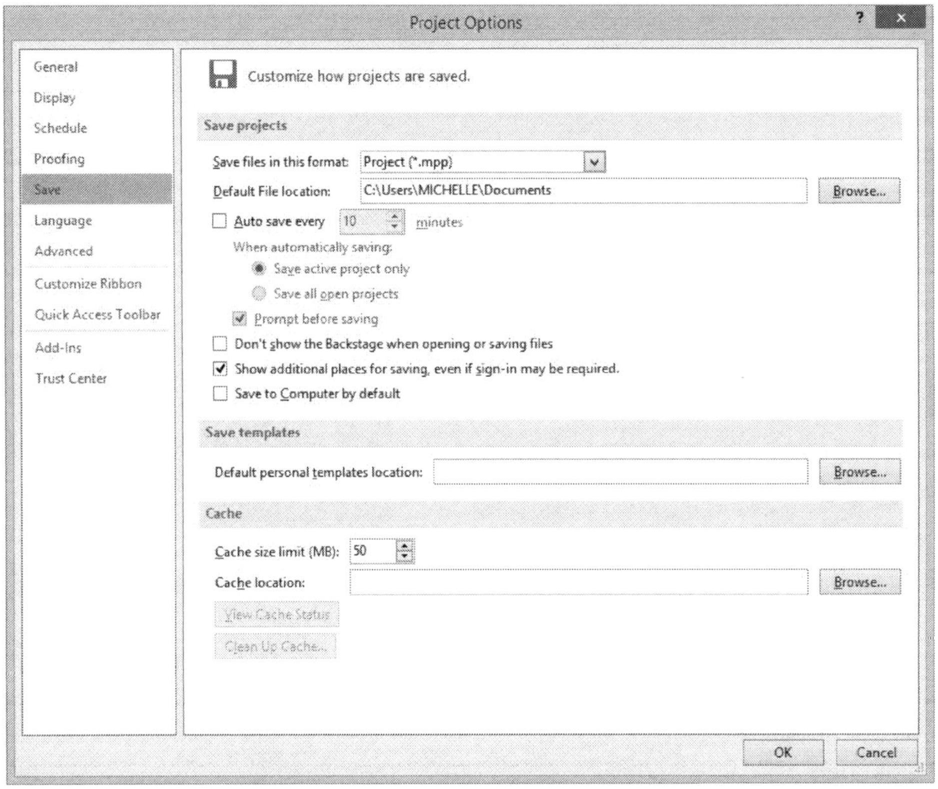

The Language tab in the Project Options dialog box. This tab allows you to control which language is used for proofing and for display and help.

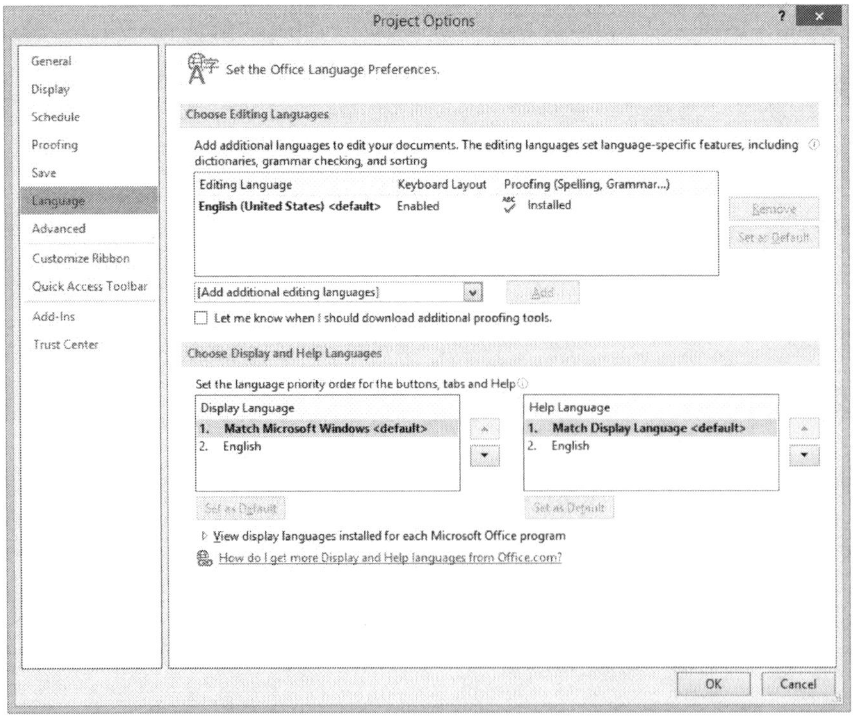

Setting Advanced Options

The Advanced tab on the Project Options dialog box.

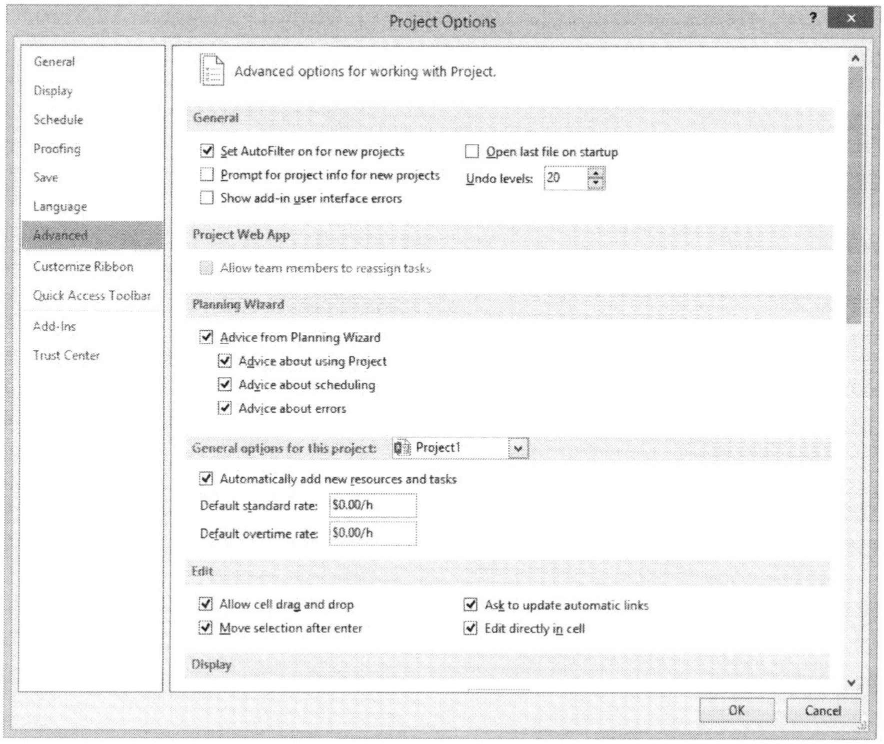

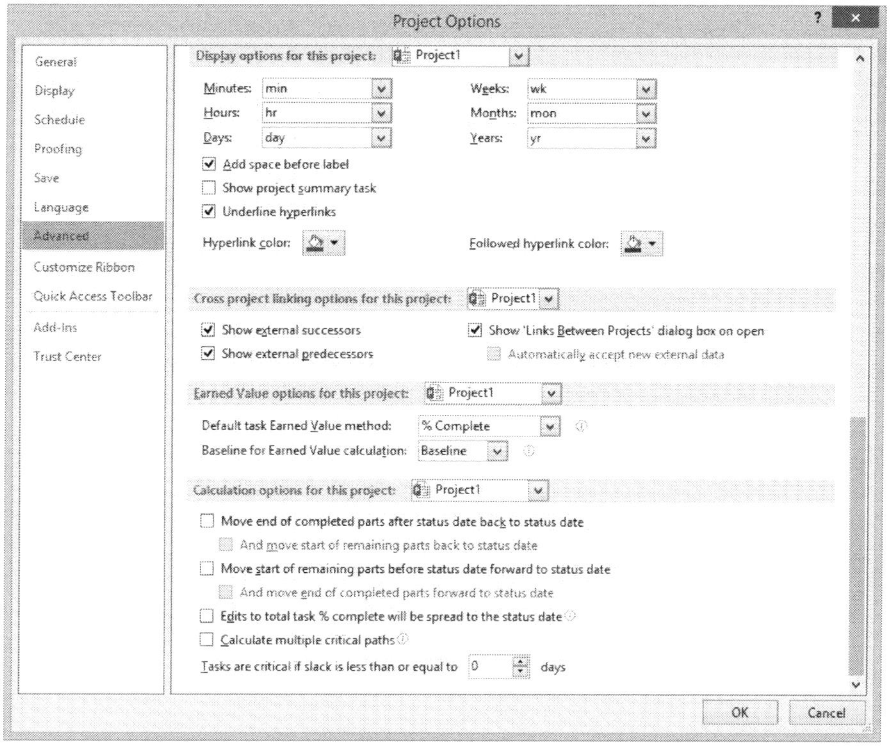

Customizing the Ribbon

To customize the Ribbon, use the following procedure.

Step 1: Select the File tab from the Ribbon to open the Backstage View.

Step 2: Select Options.

Step 3: Select Customize Ribbon from the left side.

In the left column, under Choose Commands From, Project lists the commands available in the application. You can choose a different option from the Choose Commands From drop-down list to change which options are shown or how they are sorted.

The right column shows the available tabs on the Ribbon.

Step 4: To customize the Ribbon, select the command that you want to change on the left column. Select Add. You may need to create a Custom Group before you can add a command.

- Select the Tab where you want the group to appear.
- Select New Group.
- Enter the Group name.

You can also remove commands or rearrange them on the right column.

When you have finished, select OK.

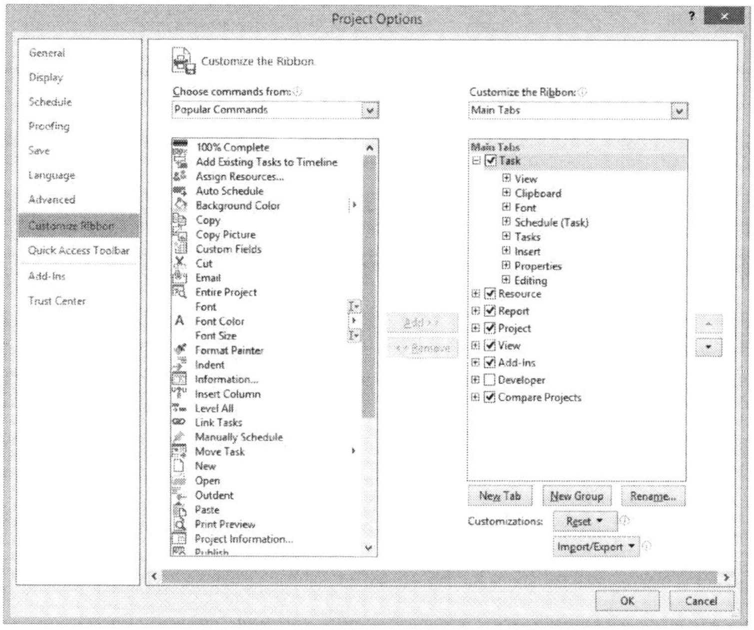

Customizing the Quick Access Toolbar

To customize the Quick Access Toolbar, use the following procedure.

Step 1: Select the File tab from the Ribbon to open the Backstage View.

Step 2: Select Options.

Step 3: Select Quick Access Toolbar from the left side.

In the left column, under Choose Commands From, Project lists the commands available in the application. You can choose a different option from the Choose Commands From drop-down list to change which options are shown or how they are sorted.

The right column shows the available commands on the Quick Access toolbar.

Step 4: To customize the Quick Access toolbar, select the command that you want to change on the left column. Select Add.

You can also remove commands or rearrange them on the right column.

When you have finished, select OK.

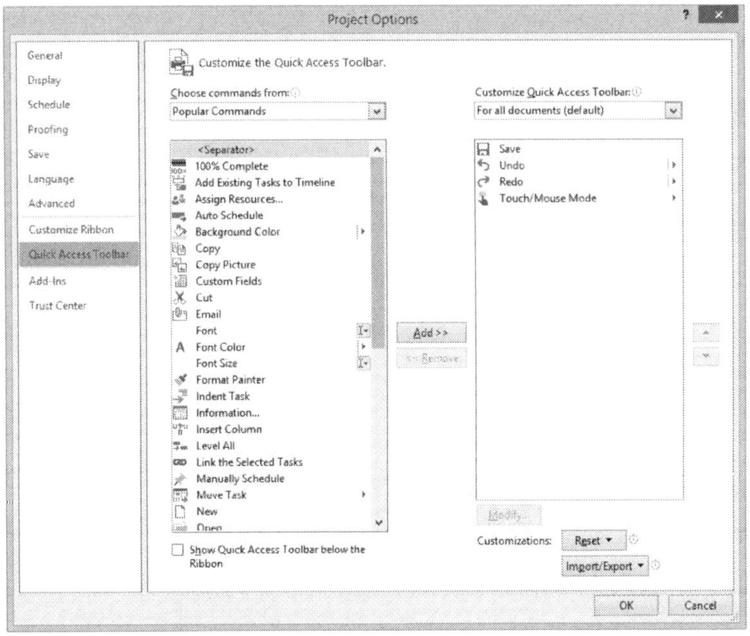

Setting Default Task Types

To set the default task type, use the following procedure.

Step 1: Select the File tab from the Ribbon to open the Backstage View.

Step 2: Select Options.

Step 3: Select the Schedule tab.

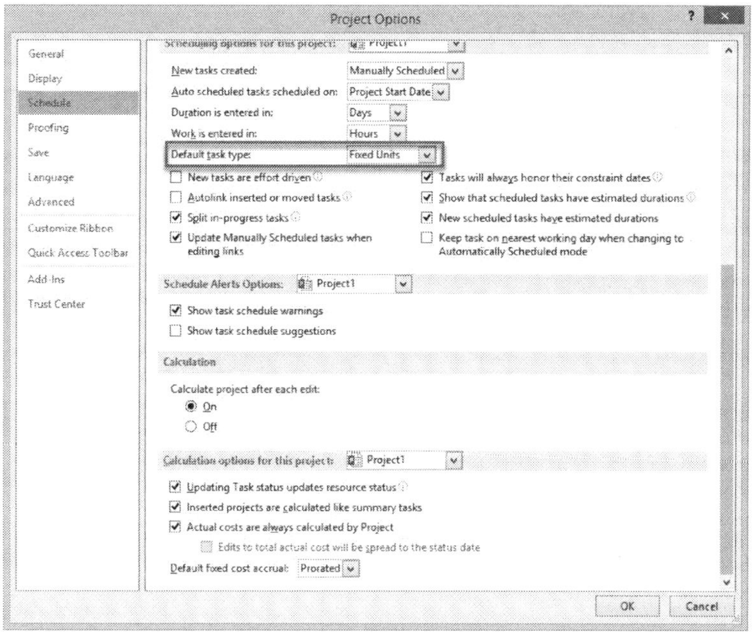

Step 4: Under the Scheduling options header, select the Default Task Type from the drop-down list. You can choose from Fixed Units, Fixed Duration, or Fixed Work.

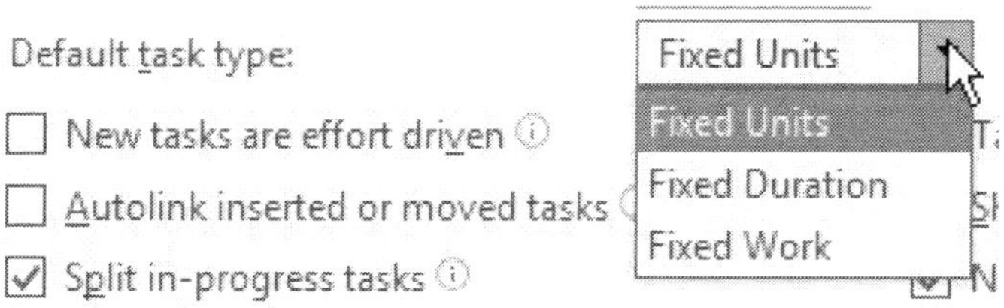

Changing the Default Assignment Unit Format

To set the default assignment unit format, use the following procedure.

Step 1: Select the File tab from the Ribbon to open the Backstage View.

Step 2: Select Options.

Step 3: Select the Schedule tab.

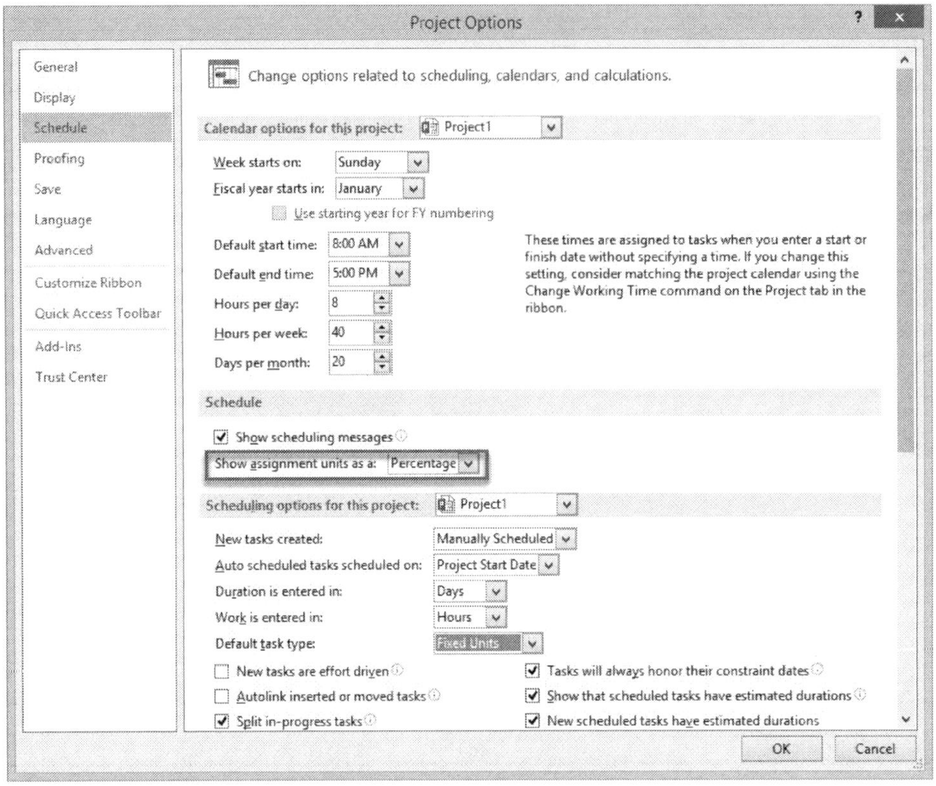

Step 4: Select an assignment unit format from the Show assignment units as a drop-down list under the Schedule heading. You can select Percentage or Decimal.

Chapter 21 – Templates and Other New Project Time Savers

This chapter will help you get your new projects up and running even faster than before. You will learn how to create templates so that you can capture a completed project and use it repeatedly for future projects. You will also learn how to create a project from an existing project. This chapter also explores Project's capabilities of creating project plans from other file formats. First, we will create a project plan from a SharePoint task list. Finally, we will create a project plan from a Microsoft Excel workbook.

Creating a Template from a Completed Project

To save a completed project as a template, use the following procedure.

Step 1: Select the File tab from the Ribbon to open the Backstage view.

Step 2: Select Save As.

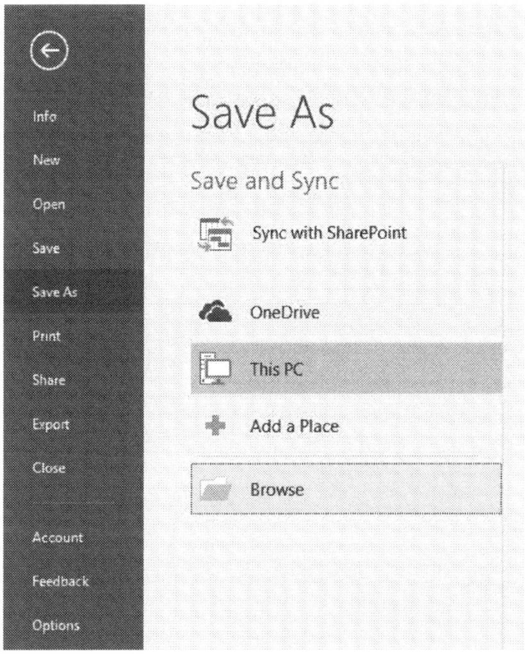

Step 3: In the Save As dialog box, select Project Template from the Save As Type drop-down list.

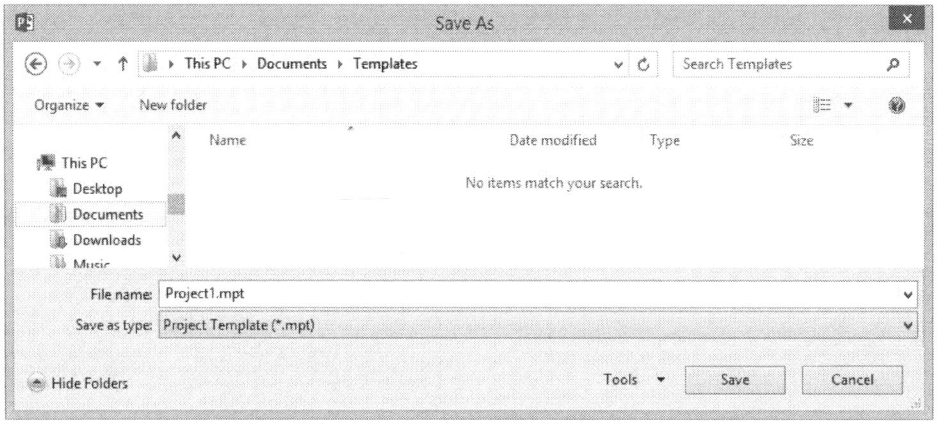

Step 4: To have the template available when creating new Project plans, you will want to keep the template location in the Microsoft/Templates folder that comes up by default.

Step 5: Select Save.

Creating a Project from an Existing Project

To create a new project from an existing project, use the following procedure.

Step 1: Select File from the Ribbon to open the Backstage View.

Step 2: Select New.

Step 3: Select New from Existing Project.

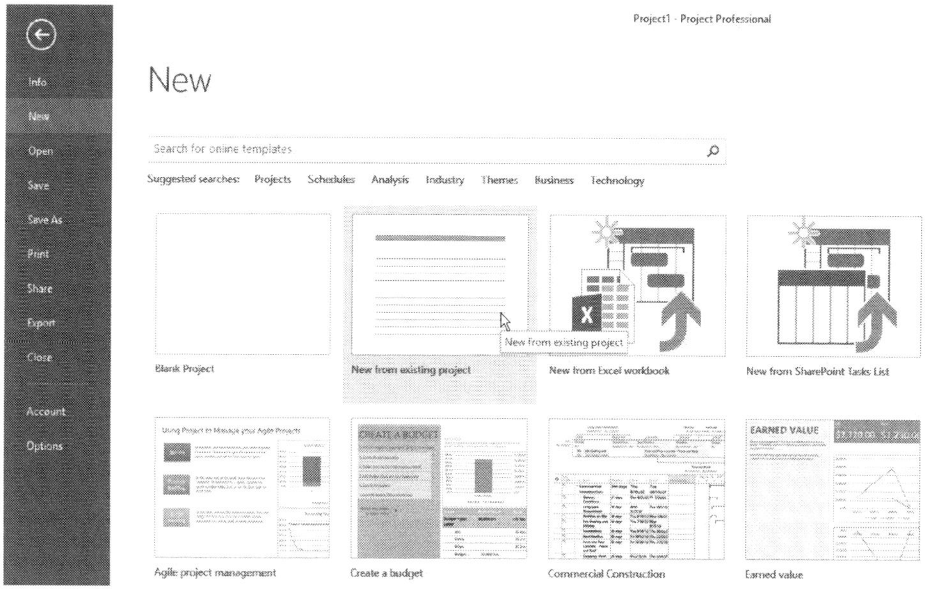

Step 4: In the New From Existing Project dialog box, navigate to the project you want to use to start the new project.

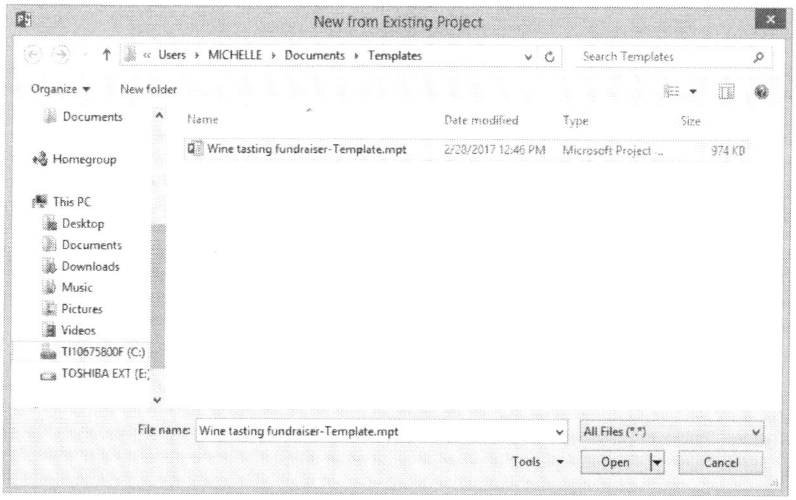

Step 5: Select Create New.

Creating a Project from a Microsoft SharePoint Task List

To create a project from a Microsoft SharePoint task list, use the following procedure.

Step 1: Select File from the Ribbon to open the Backstage View.

Step 2: Select New.

Step 3: Select New from SharePoint task list.

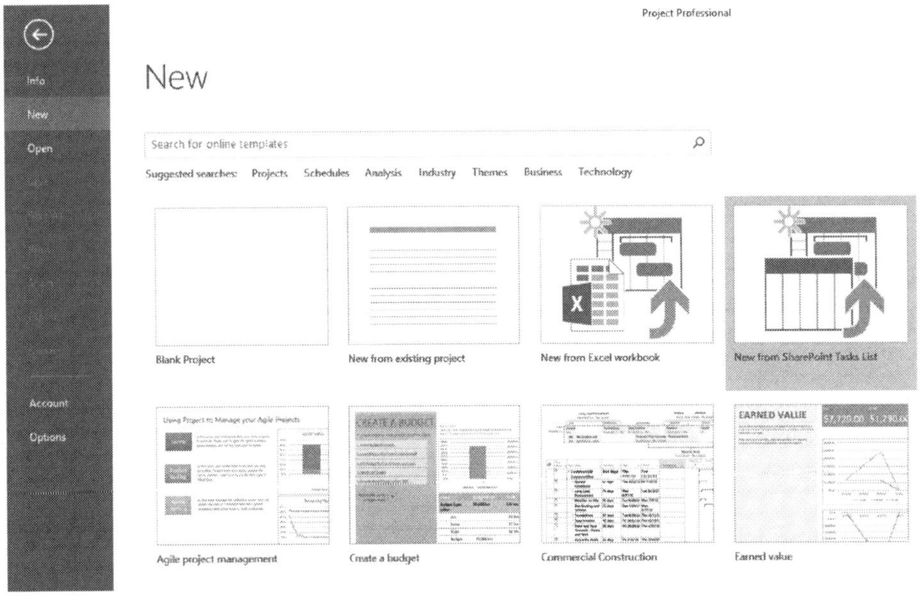

Step 4. In the Import SharePoint Tasks List dialog box, enter (or select from the drop-down list) the URL for your SharePoint site.

Step 5: Select Validate to check the URL.

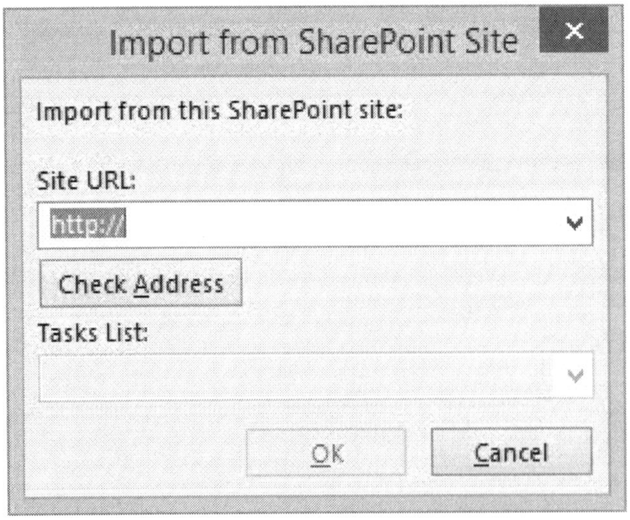

Step 6: Once you have validated the URL, you can choose a Task List from the drop-down list to use your project plan.

Step 7: Select OK.

Creating a Project from a Microsoft Excel Workbook

To create a project from a Microsoft Excel workbook, use the following procedure.

Step 1: Select File from the Ribbon to open the Backstage View.

Step 2: Select New.

Step 3: Select New from Excel Workbook.

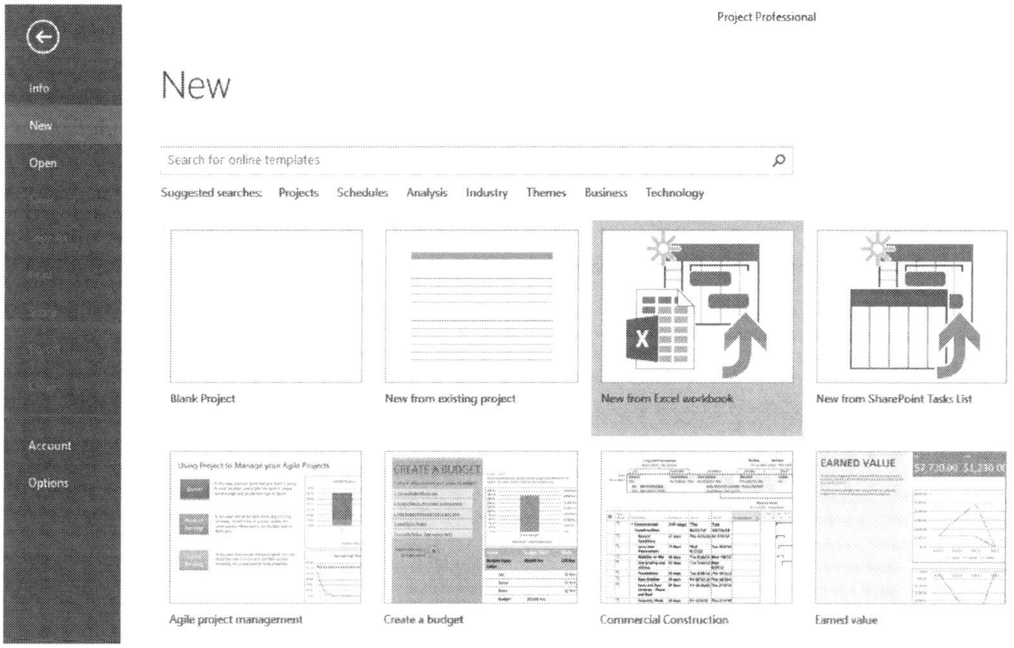

Step 4: In the Open dialog box, navigate to the location of the file you want to use and select Open.

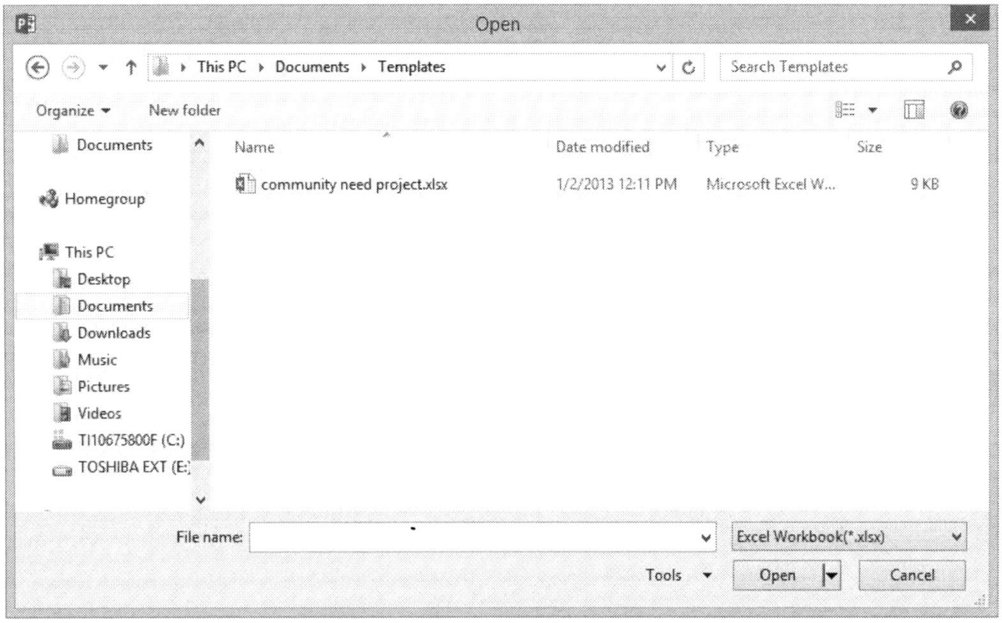

Step 5: In the Import Wizard dialog box, select Next.

Step 6: Select whether to use an existing Map or a New Map. For the purposes of this example, we will select New Map. Select Next.

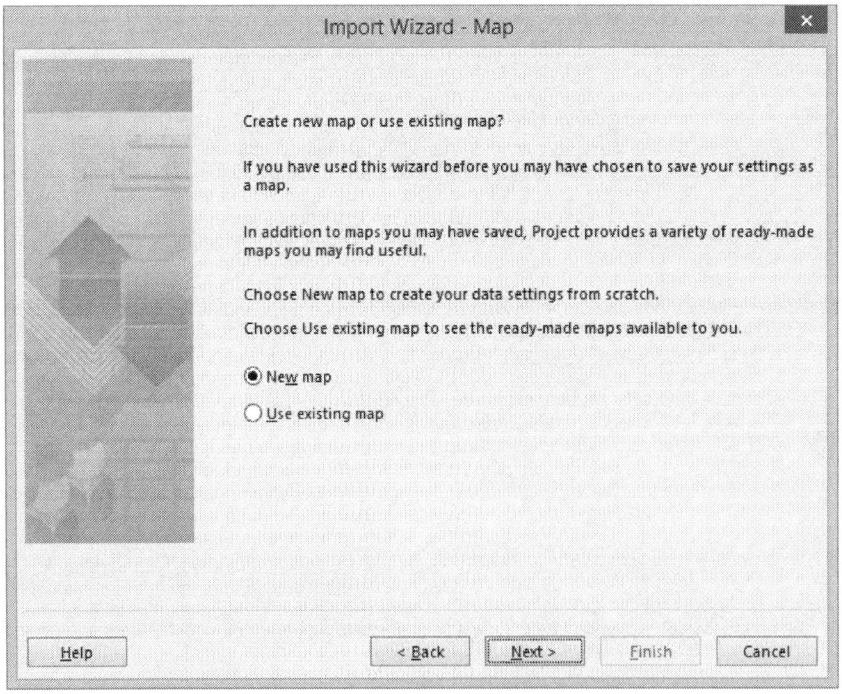

Step 7: Select the option to import the data As a New Project.

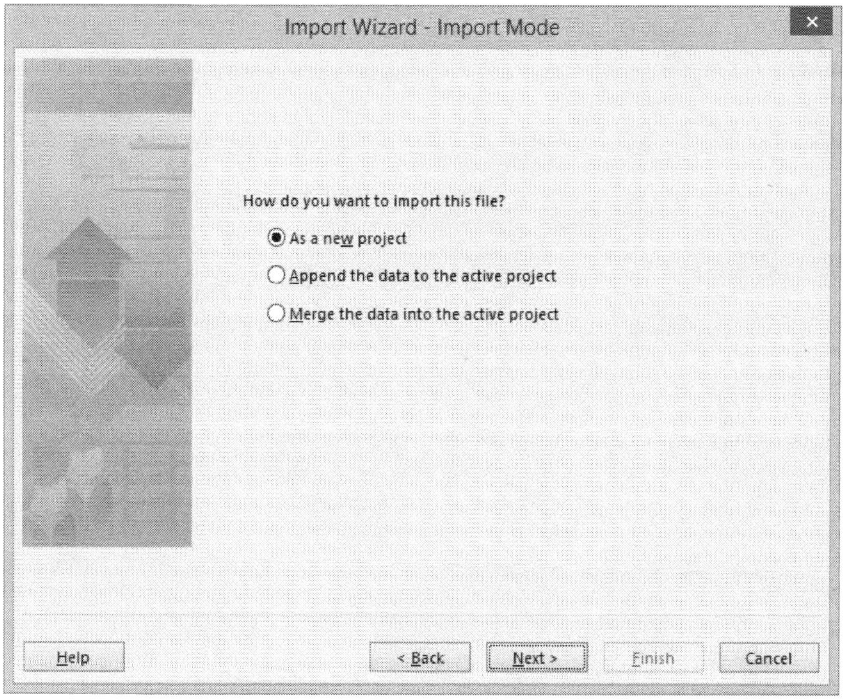

Step 8: The next screen on the Import Wizard allows you to indicate what type of information the Excel file includes. Our example only includes a few tasks and resources, so check Tasks and Resources, as well as the Import includes headers, and uncheck the other boxes. Select Next.

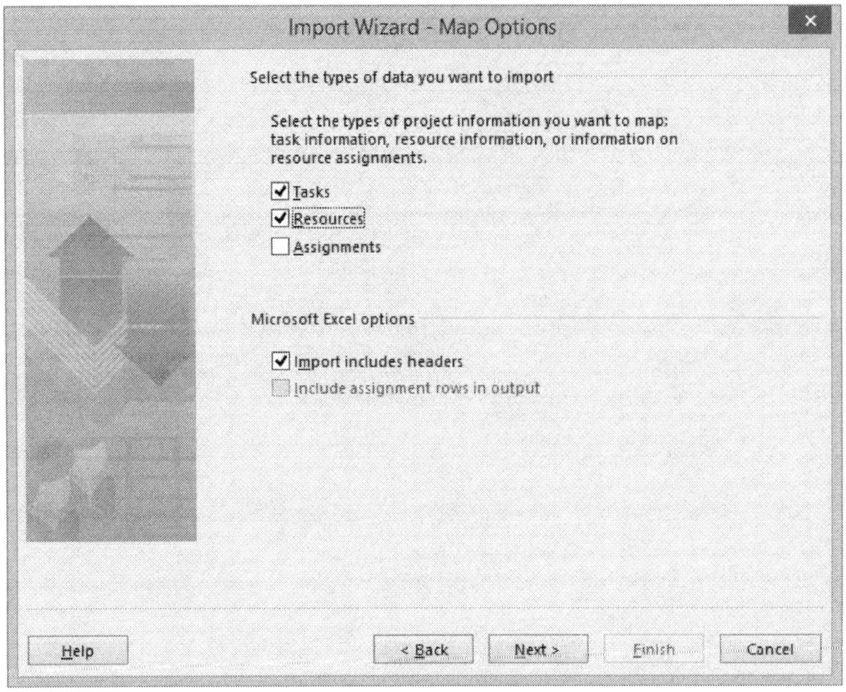

Step 9: The next screen on the Import Wizard maps the actual Excel workbook columns to Project fields for the TASKS only.

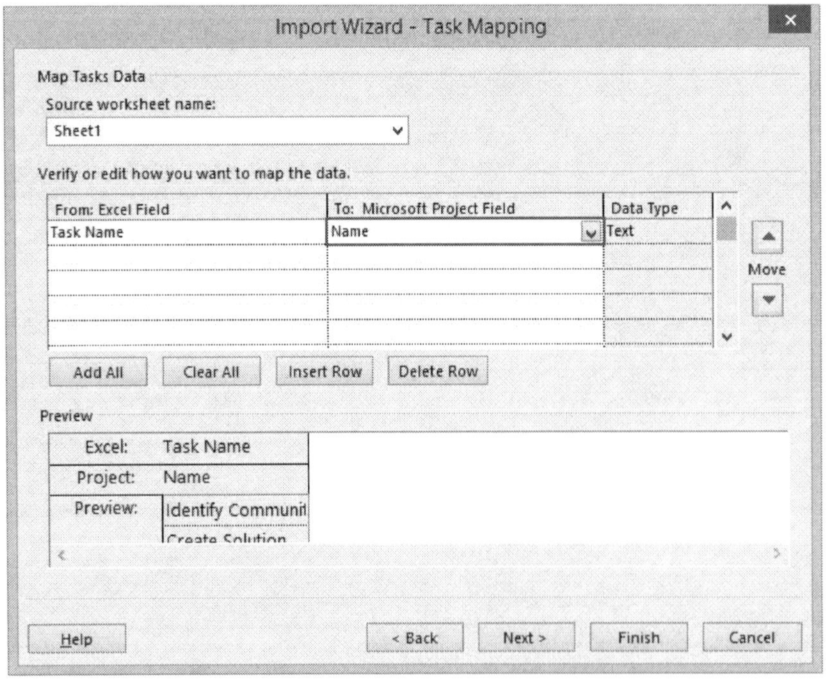

Select the Source Worksheet name from the drop-down list.

- For each column in the Excel worksheet, select a column name from the drop-down list in the first column of the Task mapping dialog box. Then select a Microsoft Project Field from the drop-down list.

- You can use the up and down arrows to rearrange the order of the items.

- You can use Add All, Clear All, Insert Row, and Delete Row to manage the mapping information.

- Continue until you have mapped all the TASK information from the Excel workbook to Microsoft Project fields.

- Select Next.

Step 10: The next screen on the Import Wizard maps the actual Excel workbook columns to Project fields for the RESOURCES only.

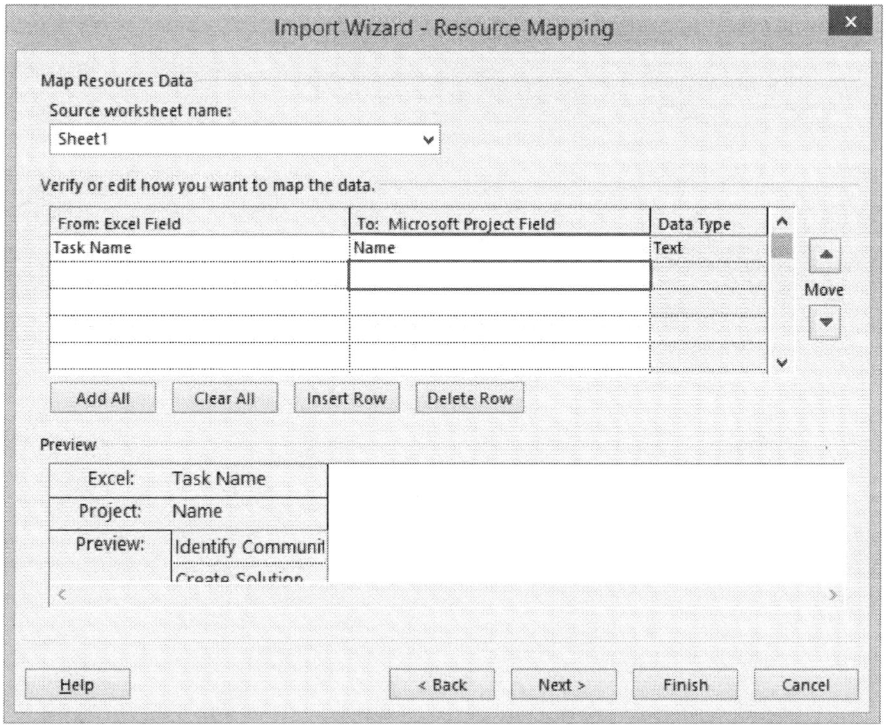

Select the Source Worksheet name from the drop-down list.

- For each column in the Excel worksheet, select a column name from the drop-down list in the first column of the Resource mapping dialog box. Then select a Microsoft Project Field from the drop-down list.
- You can use the up and down arrows to rearrange the order of the items.
- You can use Add All, Clear All, Insert Row, and Delete Row to manage the mapping information.
- Continue until you have mapped all the RESOURCE information from the Excel workbook to Microsoft Project fields.
- Select Next.

Step 11: On the final screen of the Import wizard, you can save your map for future use, or simply select Finish.

Chapter 22 – Working with Custom Fields

This chapter introduces you to the ability to customize a field for a specific use. You will first lean about custom fields. Then you will learn the general procedure for creating a custom field. We will look in depth at using a lookup table and creating formulas, as well as determining graphical indicator criteria. You will learn how to import a custom field from another project to save time. You will also learn how to show that custom field once you have created it.

About Custom Field Types

There are a number of fields available in Project 2016 that you can customize to meet your organization's needs.

You can use formulas, specific value calculations, or graphical indicators, among other customizations to make the fields work for you. Formula fields can include references to other fields. You can also create a list of values for a custom field to ensure fast and accurate data entry. You can use graphical indicators instead of data so that you can quickly see when data in that field meets certain criteria.

Creating a Custom Field

To customize a field, use the following procedure.

Step 1: Select the Project menu from the Ribbon. You can also get to this command from other tabs on the Ribbon.

Step 2: Select Custom Fields.

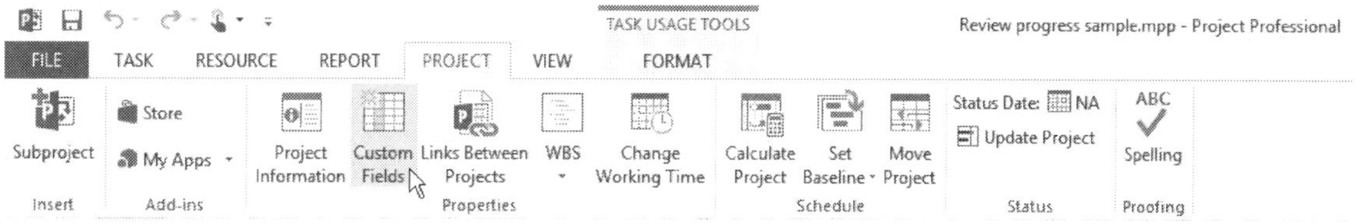

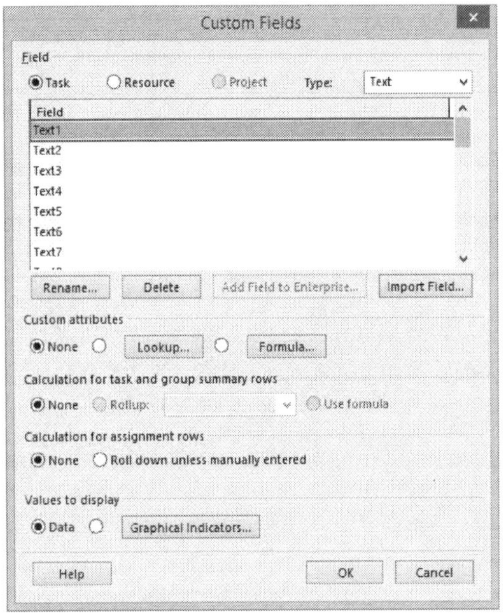

Step 3: Select whether the field will show in Task or Resource views.

Step 4: Select the Type of field from the drop-down list.

Step 5: Highlight the field that you want to customize in the Field list. Each type of field has several customizable fields available.

Step 6: Select Rename to permanently rename the field. Enter the new name and select OK.

Step 7: Indicate the Custom Attributes, if any. We will talk more about lookup tables and formulas in the next two lessons.

Step 8: If you are customizing some types of fields, you can indicate a method of rolling up task and group summary values. This determines how values are summarized at the task and group summary levels. Outline codes and text fields do not roll up. The Use Formula option is available if you defined a formula under Custom Attributes.

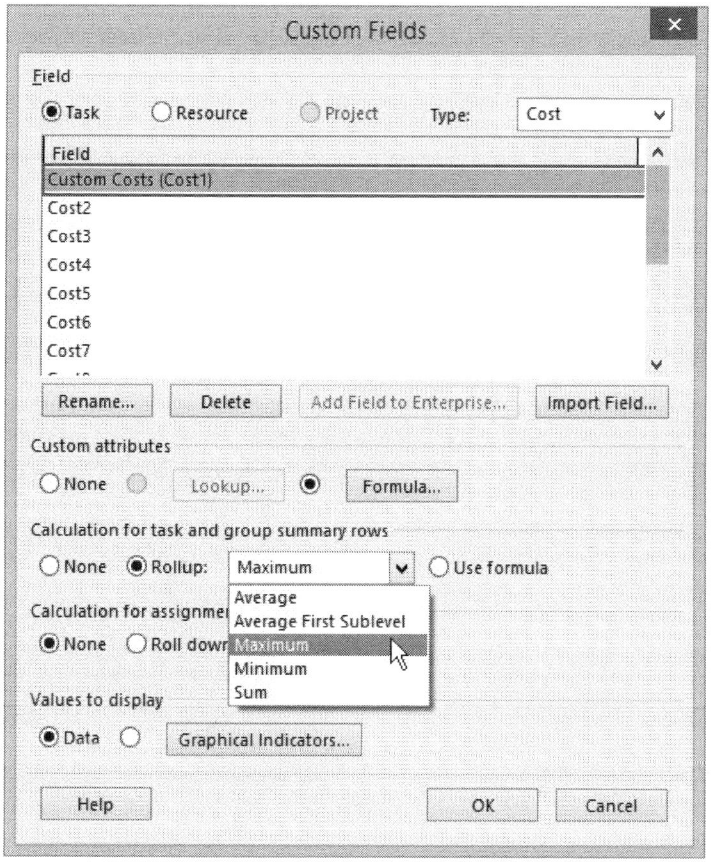

Step 9: Indicate how to calculate assignment rows. None indicates that contents of the custom field are not to be distributed across assignments. Roll down unless manually entered indicates to distribute the contents across assignments.

Step 10: Indicate whether to display values or graphical indicators. We will talk more about graphical indicators later in this chapter.

Step 11: Select OK to save the changes to your field.

Using a Lookup Table

To create a lookup table for a custom field, use the following procedure.

Step 1: Prepare a custom field as defined in the previous lesson, steps 1-6.

Step 2: Select Lookup.

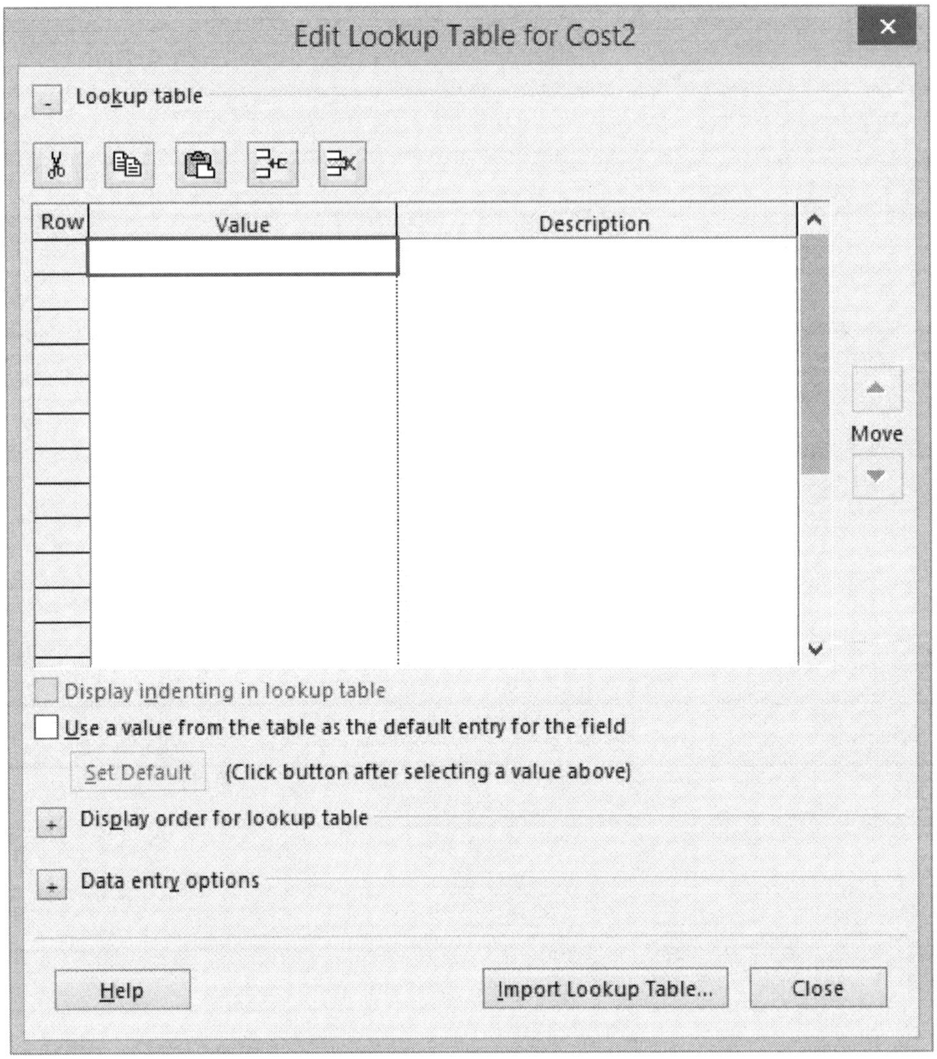

Step3: Enter each list item in the Value column. You can also enter a Description for the value.

The tools at the top of the dialog box allow you to cut, copy, paste, insert, or delete a row. You can also use the Move up or down arrows to rearrange your list.

Step 4: If you want a default value to appear in the field for your list, highlight the value you want to use, check the Use a value box, and select Set Default.

Step 5: You can choose a display order for your values. Select the plus sign next to Display Order and choose By Row Number, Sort Ascending, or Sort Descending.

Step 6: To allow new values to be added during data entry, select the plus sign next to Data Entry Options and check the Allow Additional Items box.

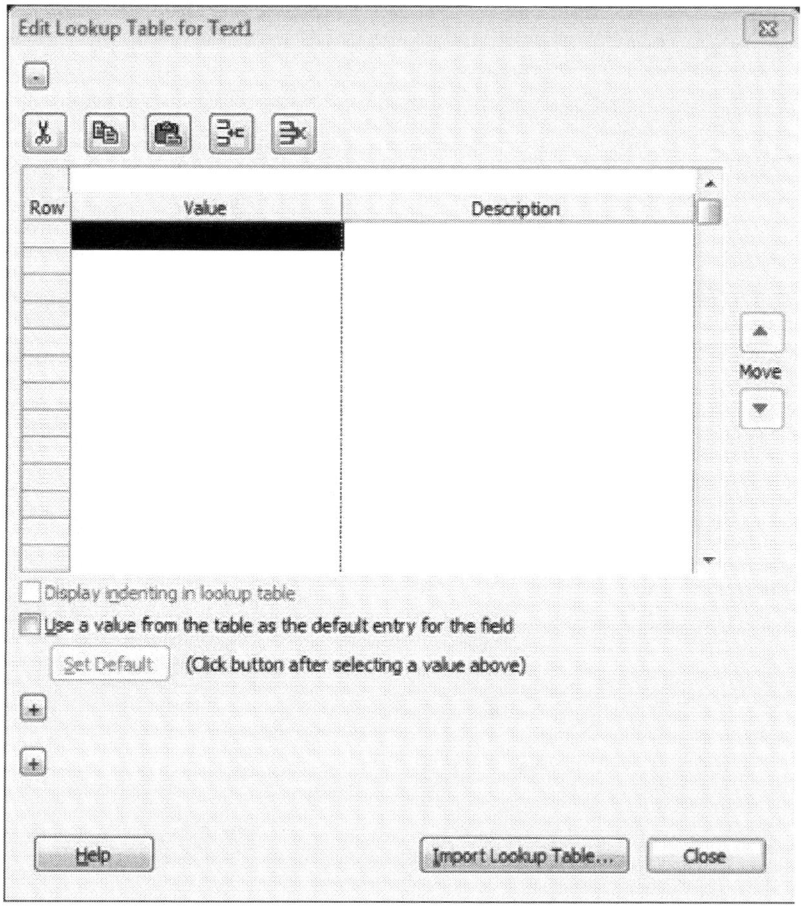

Creating Basic Formulas

To create a basic formula for a custom field, use the following procedure.

Step 1: Prepare a custom field as defined in the custom field lesson, steps 1-6.

Step 2: Select Formula.

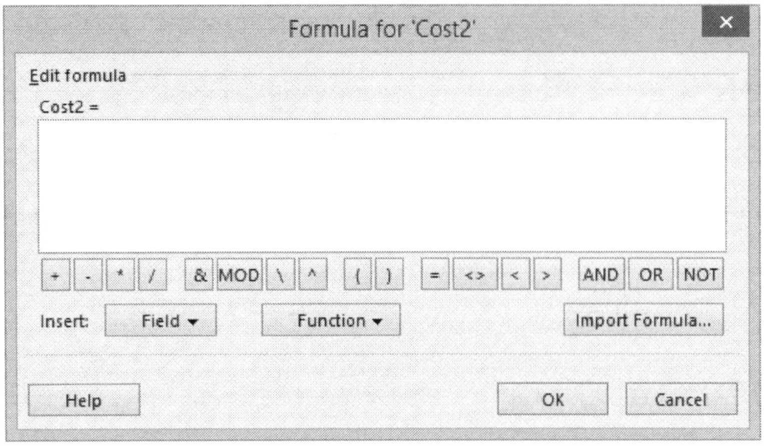

Step 3: You can type the formula in the box, or use the following building blocks to create your formula:

- Field– select a field from the drop-down list to reference another field in the project.
- Function – select a function from the drop-down list to insert a specific function. The functions include placeholder arguments that you can replace with the fields and values you want to use.
- Operators – select an operator button to add that option to your formula.

Step 4: When you have finished building your formula, select OK.

Determining Graphical Indicator Criteria

To create graphical indicator with criteria for a custom field, use the following procedure.

Step 1: Prepare a custom field as defined in the custom field lesson, steps 1-6.

Step 2: Select Graphical Indicators.

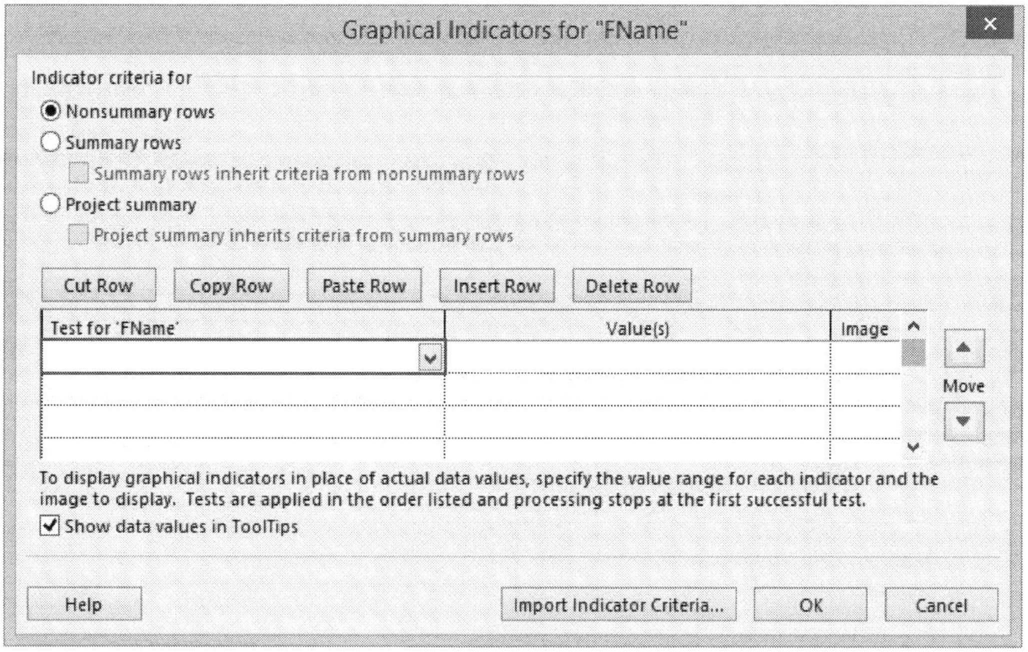

Step 3: Under Indicator criteria for, select either Non-summary rows, Summary rows, or Project summary. If you do not want summary rows or the project summary to inherit test criteria, deselect the corresponding check box.

Step 4: Select a test from the drop-down list (or enter your own) to indicate the criteria for the custom field.

Step 5: Select the value to use for the criteria from the drop-down list to indicate the field where the value is located. You can also enter your own value to use a specific value instead of a value referenced from another field. Project will compare the test to the value in the custom field.

Step 6: Select an Image from the drop-down list to indicate the image you want to display because of the test.

Step 7: Repeat steps 4 – 6 to apply additional tests to the custom field.

Step 8: The tools at the top of the dialog box allow you to cut, copy, paste, insert, or delete a row. You can also use the Move up or down arrows to rearrange your list. The tests are applied in order from top to bottom.

Importing a Custom Field

To import a custom field, use the following procedure.

Step 1: Select the Project menu from the Ribbon. You can also get to this command from other tabs on the Ribbon.

Step 2: Select Custom Fields.

Step 3: Select Import Field.

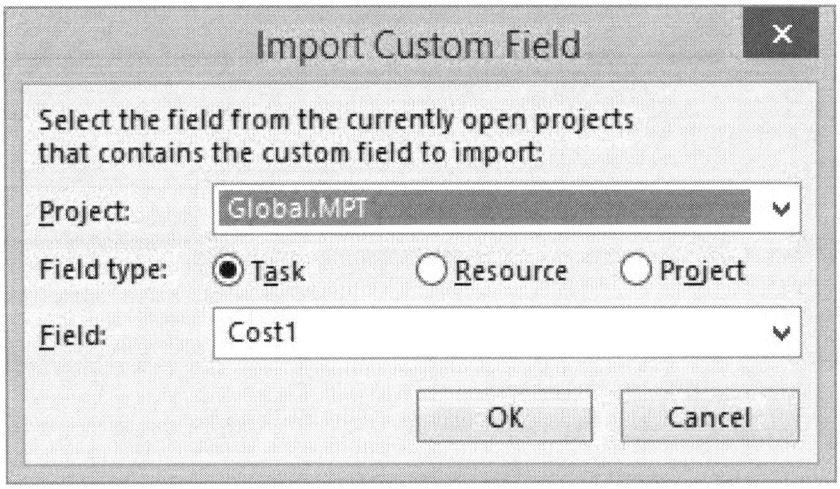

Step 4: Select the project that contains the field you are importing from the Project drop-down list. The other project should be open.

Step 5: Select the type of field as Task, Resource, or Project.

Step 6: Select the name of the field you are importing from the Field drop-down list.

Step 7: Select OK.

Inserting a Custom Field

To insert a custom field into a view, use the following procedure.

Step 1: Your new field will be inserted to the left of the column you select. Select the appropriate column header.

Step 2: Select the Format tab from the Ribbon.

Step 3: Select Insert Column.

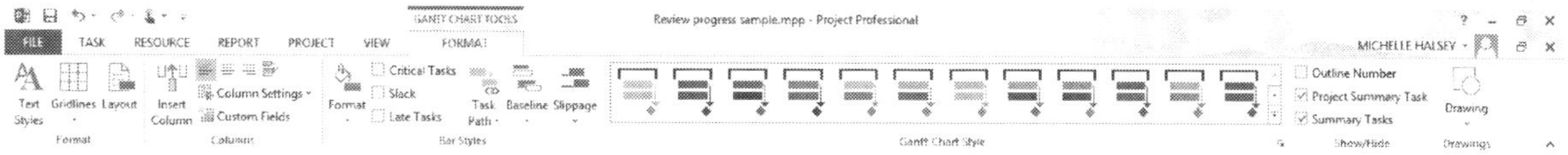

Step 4: Begin typing the name of the field or simply select it from the list.

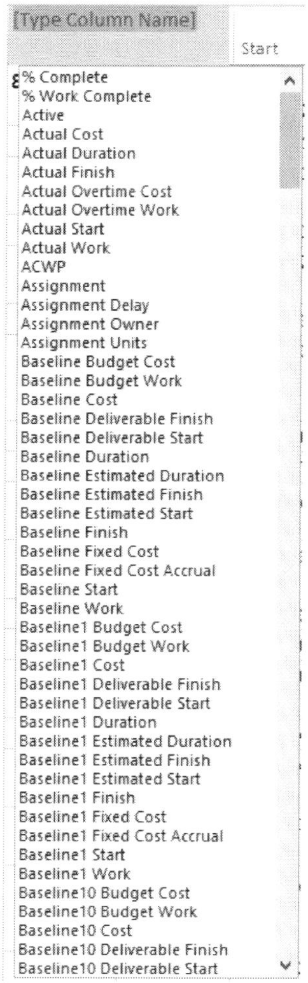

Chapter 23 – Using the Team Planner

This chapter focuses on formatting the Team Planner view so that you can customize the way the information is shown. First, we will look at how to roll up tasks so that you can control the amount of detail shown on the Team Planner. Next, you will learn how to work with the Team Planner gridlines. You will also learn how to change the text styles. You can change the fill and border styles for several different types of tasks, so that you can differentiate each type of task at a glance. This chapter also explains how to prevent over allocations when making changes in the Team Planner view. Finally, we will look at how to show or hide unassigned and unscheduled tasks.

Rolling Up Tasks

To roll up task information, use the following procedure.

Step 1: Switch to the Team Planner view by selecting Team Planner from the View tab on the Ribbon.

Step 2: Select the Team Planner Tools Format tab from the Ribbon.

Step 3: Select Rollup.

Step 4: Select the level of details you would like to view.

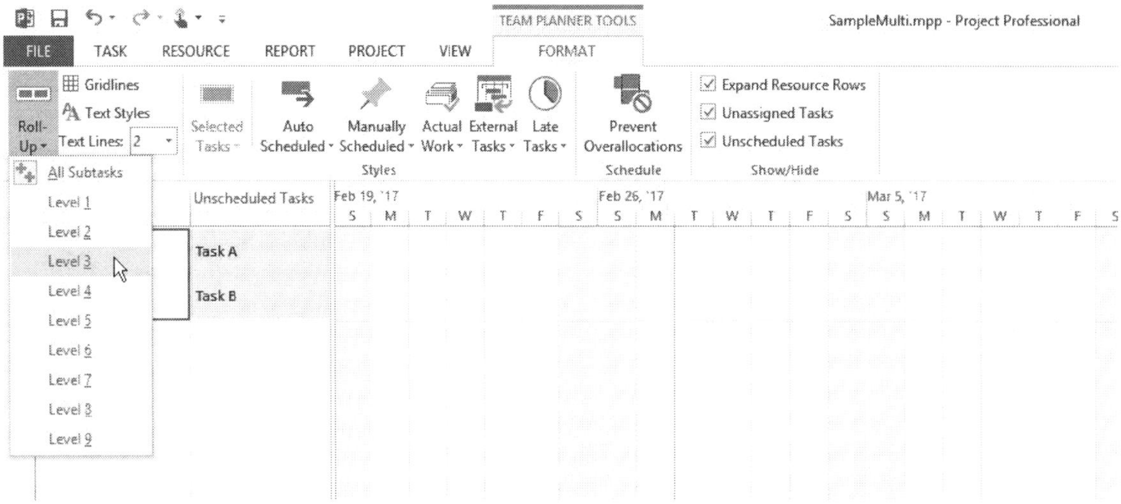

Working with Gridlines

To format gridlines on the Team Planner, use the following procedure.

Step 1: Select the Team Planner Tools Format tab from the Ribbon.

Step 2: Select Gridlines.

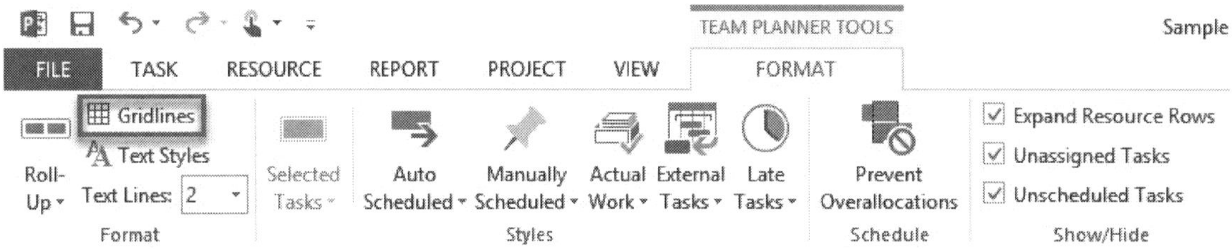

Step 3: In the Gridlines dialog box, select which line you want to change.

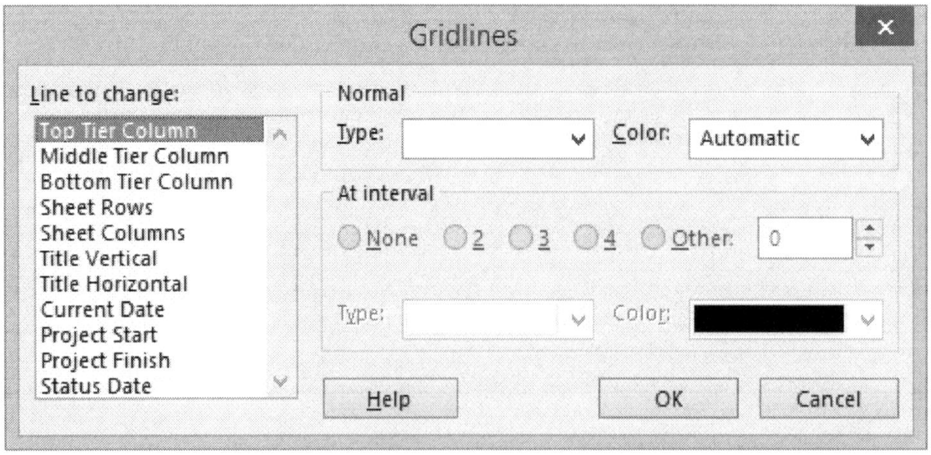

Step 4: Select the Type of gridline you want to use for the selected line from the drop-down list.

Step 5: Select the Color for the selected line from the drop-down list.

Step 6: When you are using Sheet Rows, you can have another type and color for various intervals. Select None, 2, 3, 4, or Other and enter the number of rows. Select the Type of gridline from the drop-down list. Select the Color for the gridline from the drop-down list.

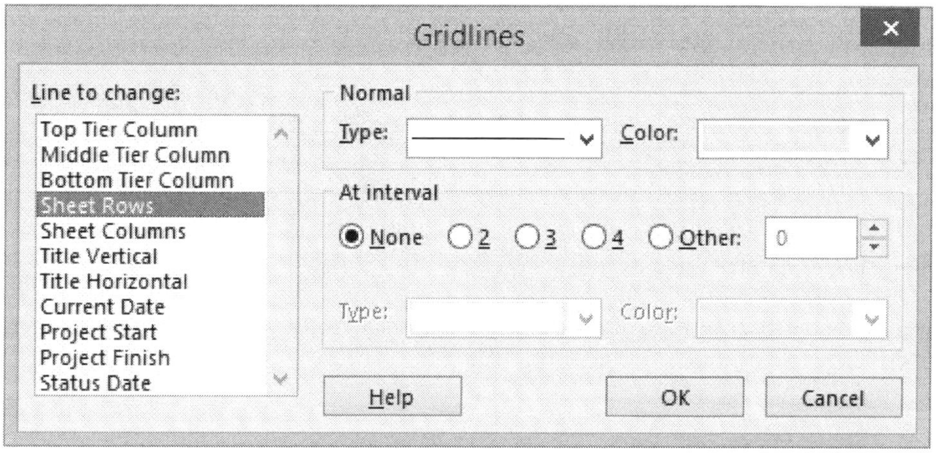

Step 7: Repeat to format additional lines.

Step 8: When you have finished getting all the gridlines the way you want, select OK.

Changing Text Styles

To format the text on the Team Planner, use the following procedure.

Step 1: Select the Team Planner Tools Format tab from the Ribbon.

Step 2: Select Text Styles.

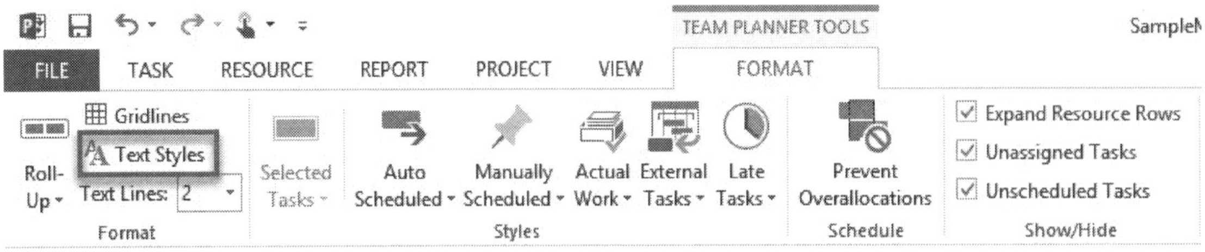

Step 3: In the Text Styles dialog box, select the following:

Step 3a: Choose the Font from the drop-down list.

Step 3b: Select a Font Style from the list box.

Step 3c: Select the Font Size from the drop-down list.

Step 3d: Check the Underline and/or Strikethrough boxes to apply these enhancements.

Step 3e: Select a Color from the drop-down list.

Step 3f: Select a Background Color from the drop-down list.

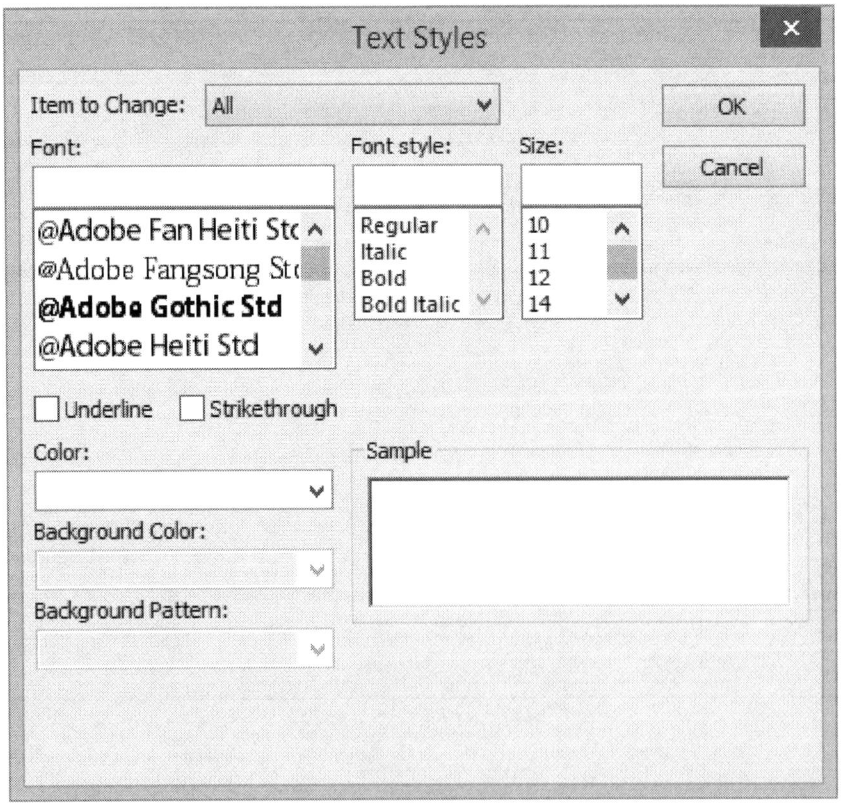

The Sample area shows a preview of your selections.

Step 1: Select OK to apply your Text Styles.

Changing Task Fill and Border Colors

To change fill and border colors, use the following procedure.

Step 1: Select the Team Planner Tools Format tab from the Ribbon.

Step 2: Select one of the following:

- Selected Tasks
- Auto Scheduled Tasks
- Manually Scheduled Tasks
- Actual Work
- External Tasks
- Late Tasks

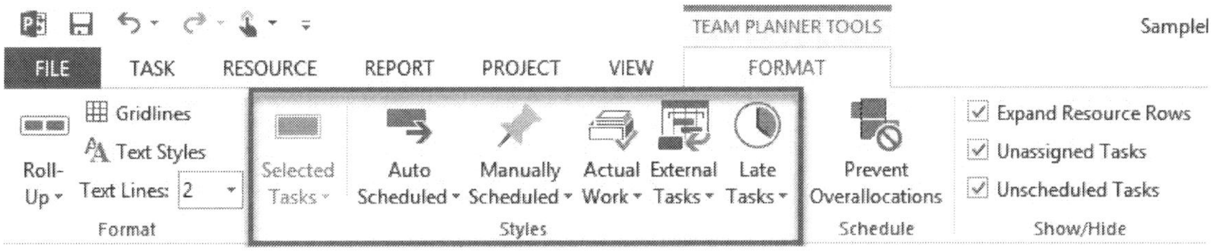

Step 3: Select Fill Color or Border Color.

Step 4: Select the color you want to use from the gallery.

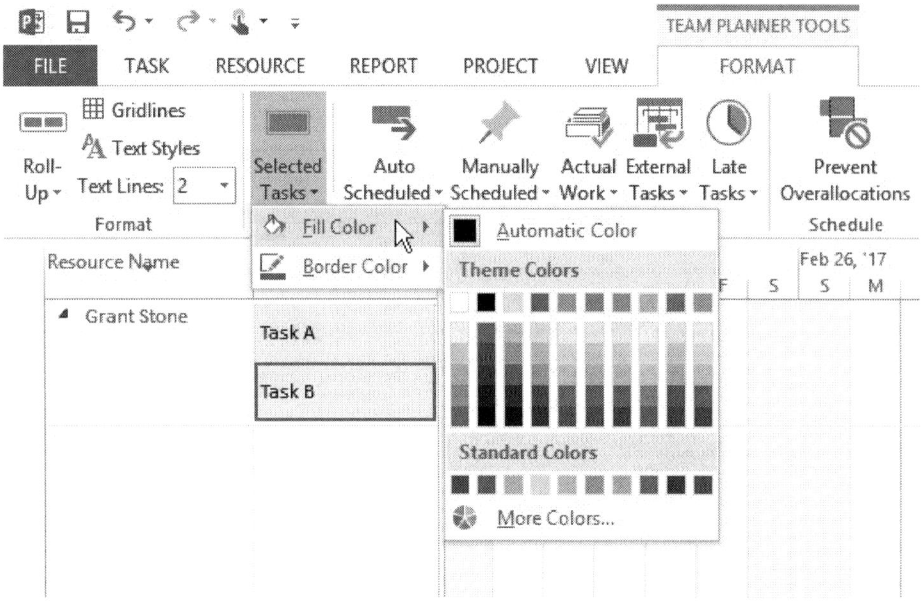

Practice applying different color combinations for different types of tasks and view the results.

Preventing Over Allocations

To prevent over allocations in the Team Planner view, use the following procedure.

Step 1: Select the Team Planner Tools Format tab from the Ribbon.

Step 2: Select Prevent Over allocations.

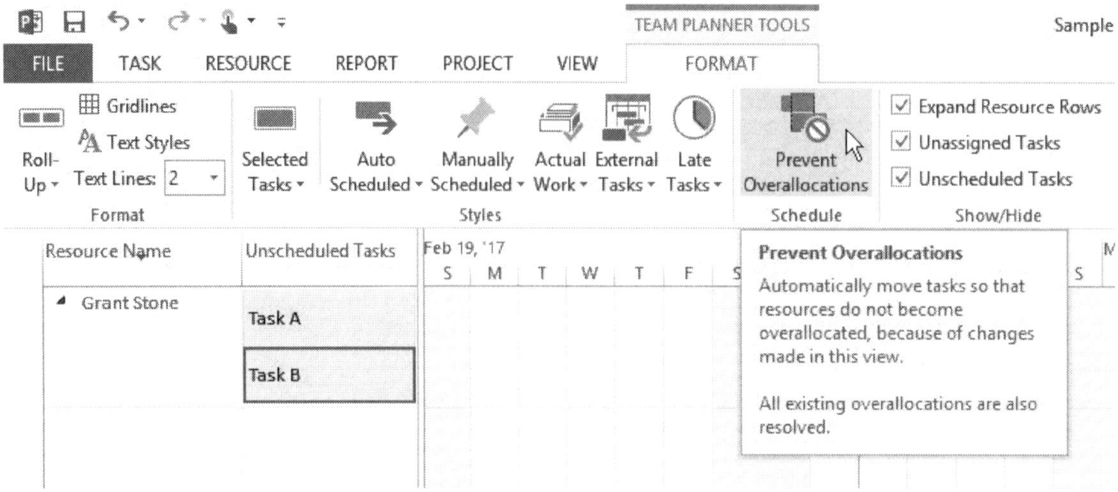

Practice moving tasks to see the results when over allocations might be created.

Showing and Hiding Information

To hide and show unassigned tasks and unscheduled tasks, use the following procedure.

Step 1: Select the Team Planner Tools Format tab from the Ribbon.

Step 2: Clear the Unassigned Tasks box to hide tasks that have not been assigned. Check the box to show them again.

Step 3: Clear the Unscheduled Tasks box to hide tasks that have not yet been scheduled. Check the box to show them again.

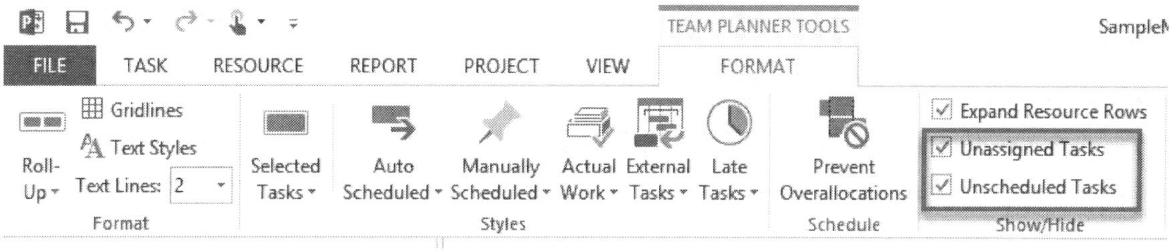

You can also choose whether to expand or condense Resource Rows.

Chapter 24 – Managing Risks and Measuring Performance

This chapter will help you further analyze the information that you have collected with your project schedule. First, you will learn how to review differences between planned, scheduled, and actual work. You will also learn how to find slack in the schedule so that you will know if tasks slip how it will affect the project schedule. This chapter also explains how to compare two versions of a project. Finally, we will look at calculating earned value analysis.

Reviewing Differences Between Planned, Scheduled, and Actual Work

To apply the work table to the Gantt chart, Task Usage, or Resource Usage views, use the following procedure.

Step 1: Select the View tab from the Ribbon.

Step 2: Select Tables.

Step 3: Select Work.

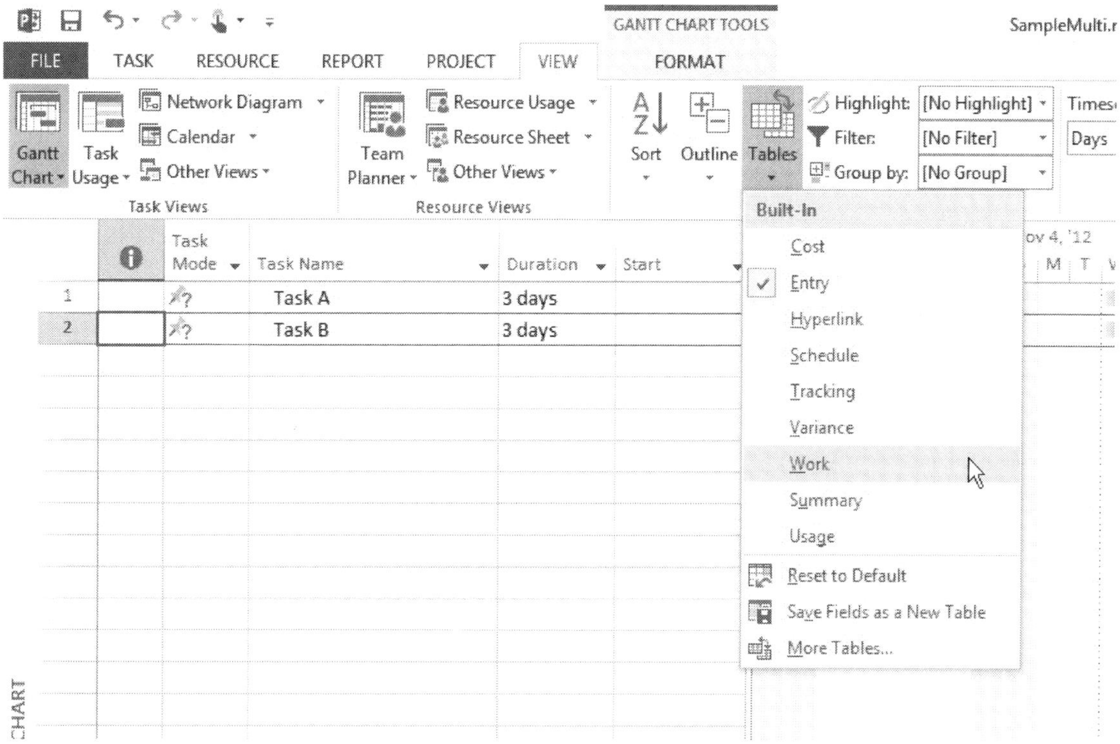

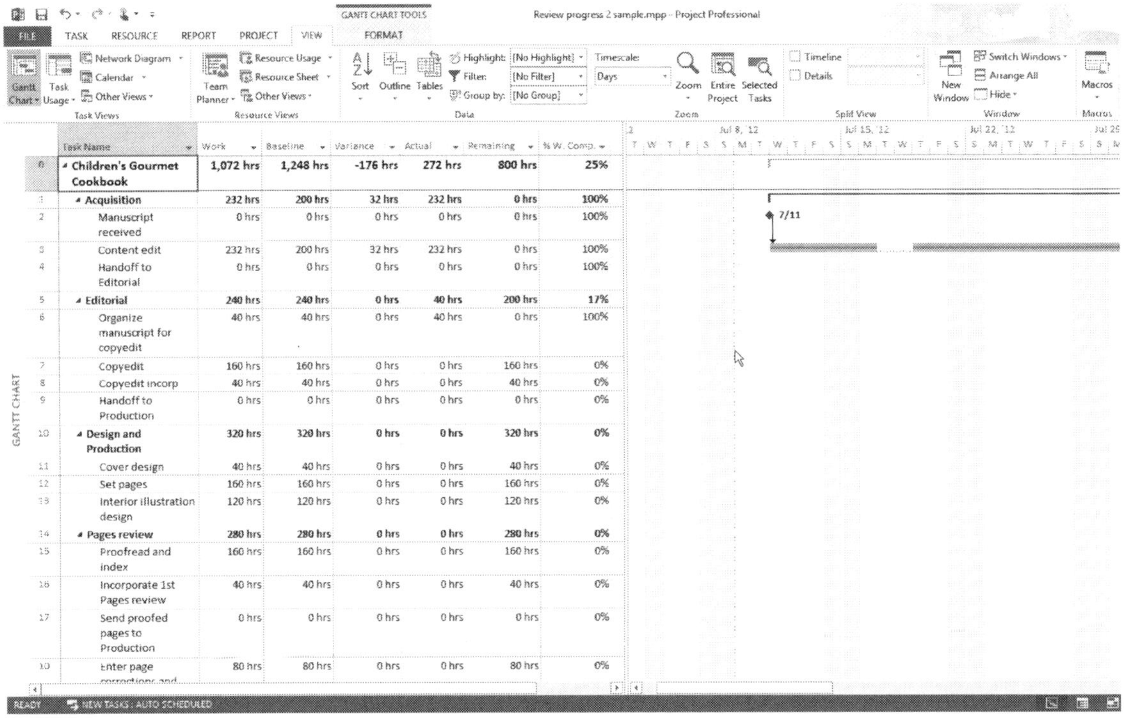

Work and Actual columns have the same amount in completed tasks. Compare the amounts in the Work column (the total of Actual Work plus the Remaining Work) and the Baseline column (a snapshot taken at some point earlier in the project). For finished tasks, the Variance field represents the difference between actual and planned work.

The Task Usage sheet with the work table applied.

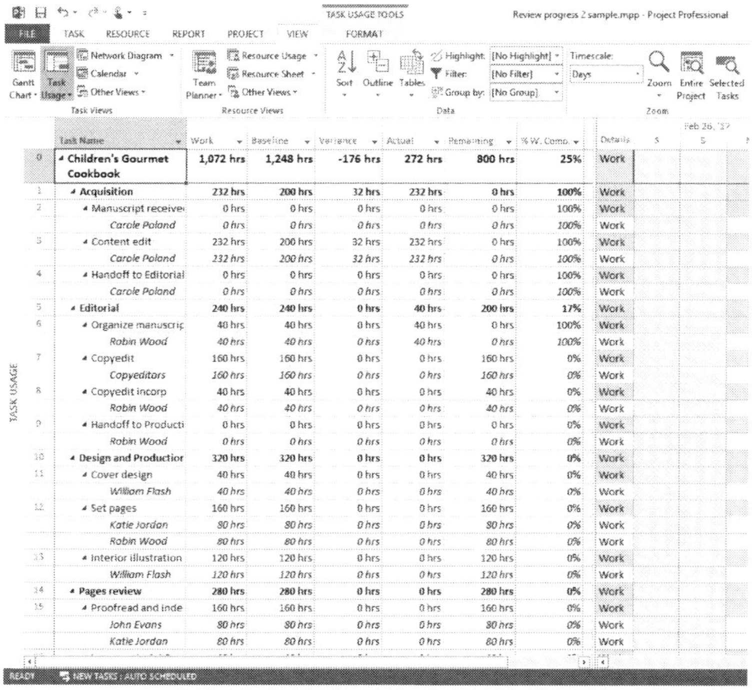

The Resource Usage table with the Work table applied.

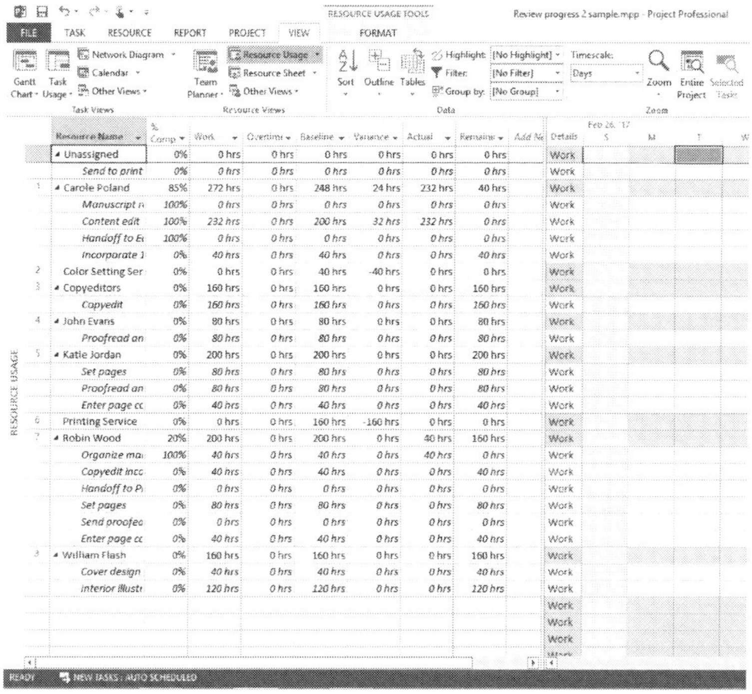

Finding Slack in the Schedule

To find slack in the schedule, use the following procedure.

Step 1: Select the View tab from the Ribbon.

Step 2: Select the arrow under Gantt Chart.

Step 3: Select More Views.

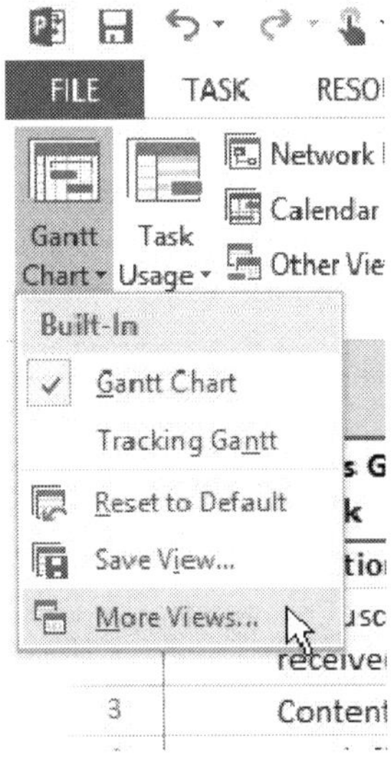

Step 4: In the More Views dialog box, select Detail Gantt. Select Apply.

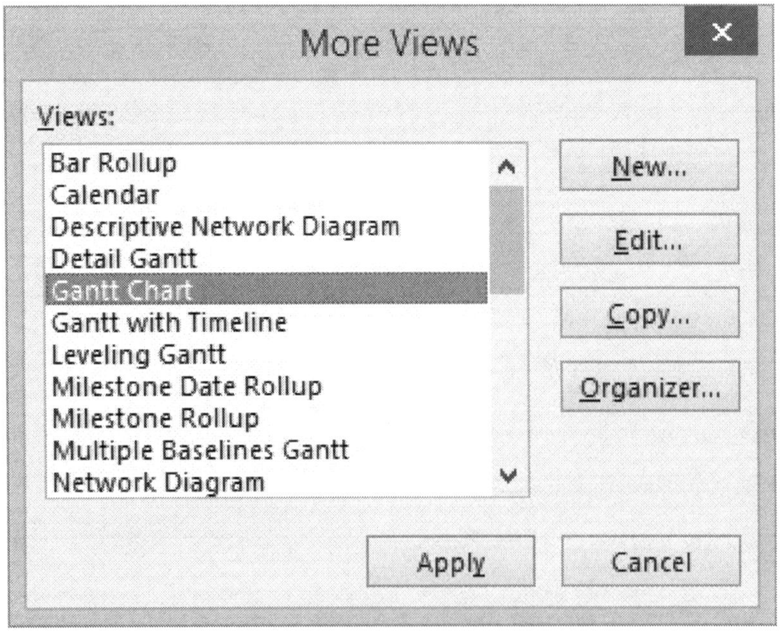

Step 5: Select Tables on the View tab.

Step 6: Select Schedule.

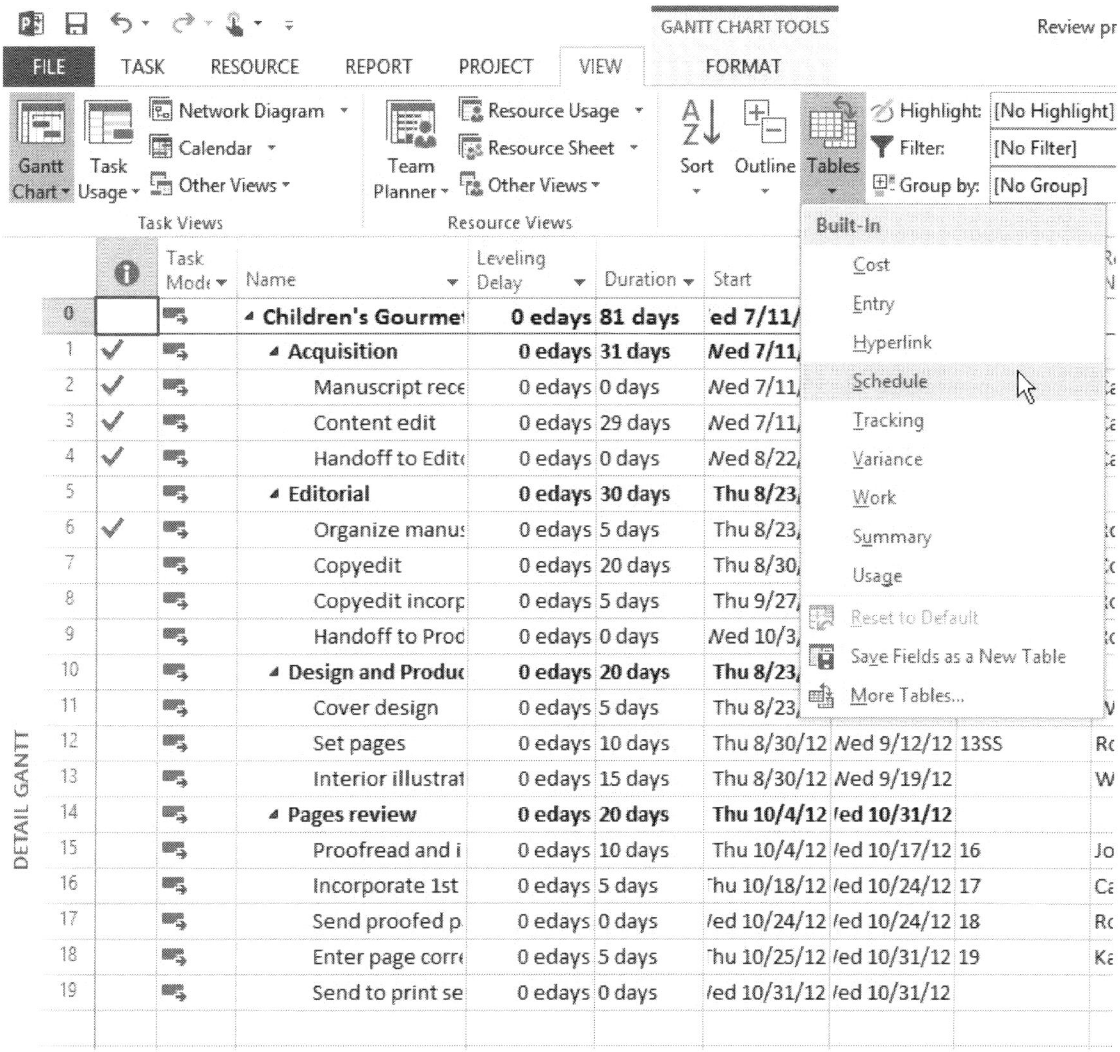

The slack is shown in the Free slack and Total Slack columns, as well as on the chart as thin bars to the right of tasks.

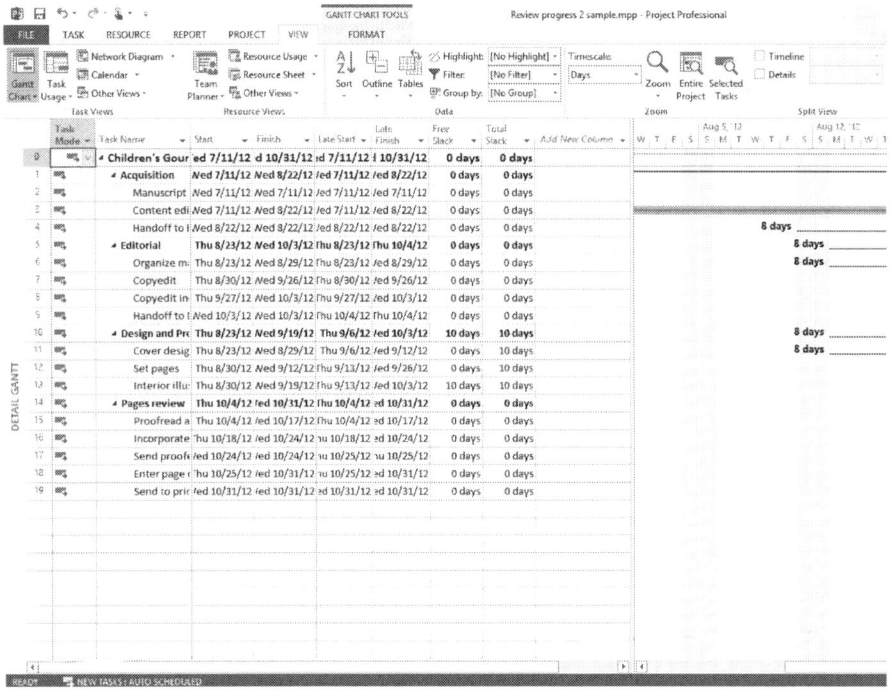

Comparing Two Versions of a Project

To compare projects, use the following procedure.

Step 1: Select the Project tab from the Ribbon.

Step 2: Select Compare Projects.

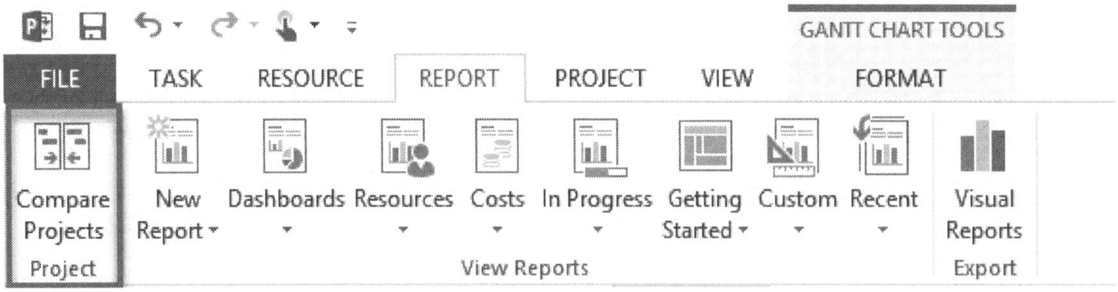

Step 3: In the Compare Project Versions dialog box, select the other project you want to use in the comparison from the drop-down list (if it is open) or select Browse and navigate to the location of the other project.

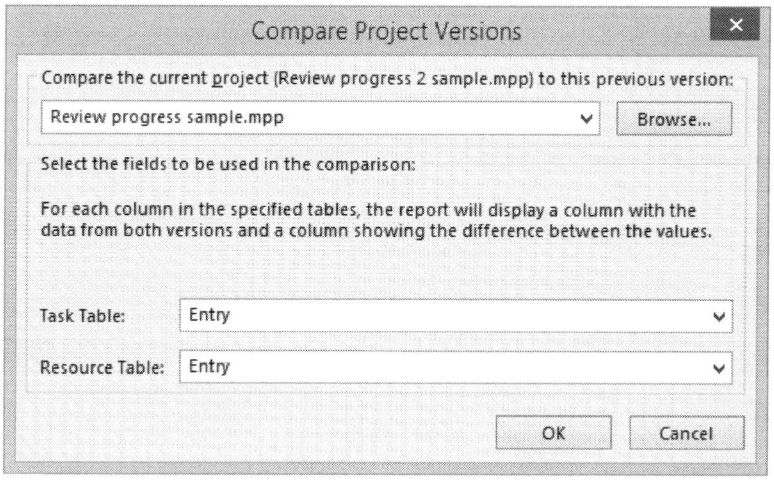

Step 4: Select the Task Table and Resource Table from the drop-down lists to indicate the tables that have the data you want to compare. You can also select None.

Step 5: Select OK.

The Comparison opens. The Legend explains the color coding and indicators for the comparison report on the left side of the screen. The Comparison report contains only the differences between the two versions. It is not a consolidation of the two projects.

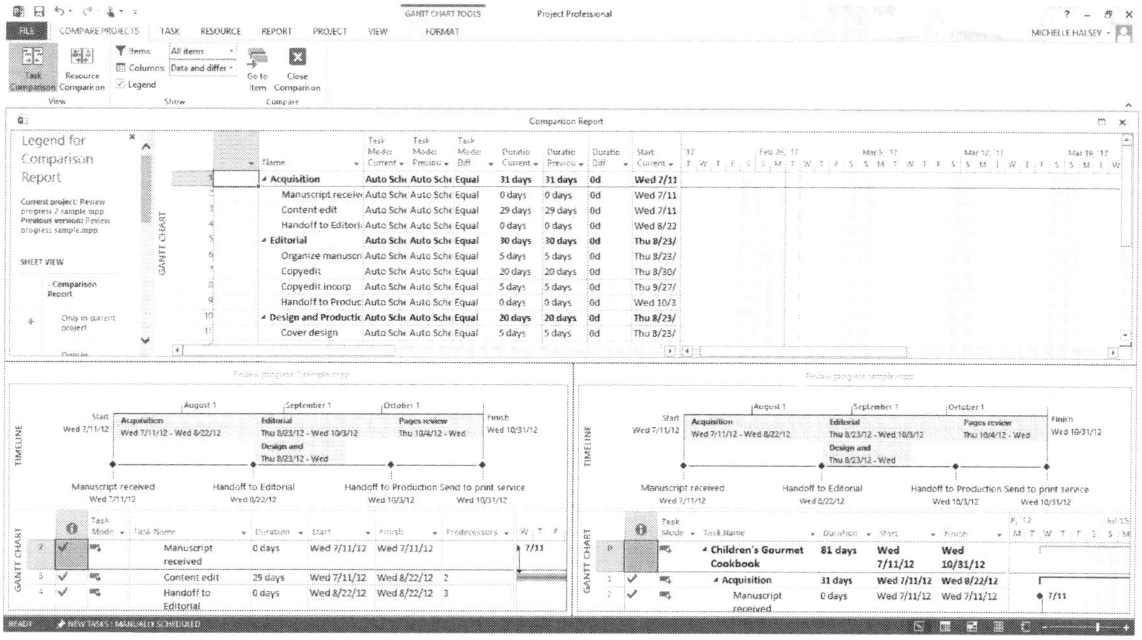

Calculating Earned Value Analysis

To set the type of earned value calculation, use the following procedure.

Step 1: Select the File tab to open the Backstage View.

Step 2: Select Options.

Step 3: Select Advanced.

Step 4: Scroll down to the Earned Value section.

Step 5: Select a calculation method from the Default Task Earned Value Method drop-down list to indicate how Project calculates the budgeted cost of work scheduled (BCWS). BCWS is the baseline cost up to the status date that you choose. Budgeted cost values are stored in the baseline fields, or, if you saved multiple baselines, in the Baseline1 through Baseline10 fields. Note that changing this setting affects tasks added after the setting is changed.

Step 6: Select a Baseline for Earned Value Calculations from the drop-down list.

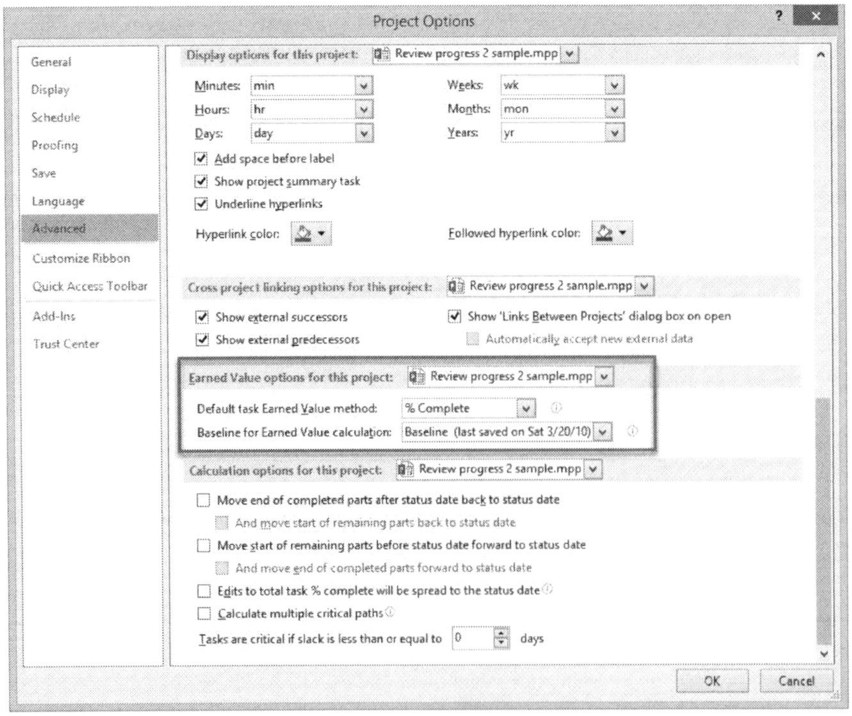

Step 7: Select OK.

Now we will view the calculations. To view the earned value analysis, use the following procedure.

Step 1: Select the View tab from the Ribbon.

Step 2: Select Other Views.

Step 3: In the More Views dialog box, select Task Sheet.

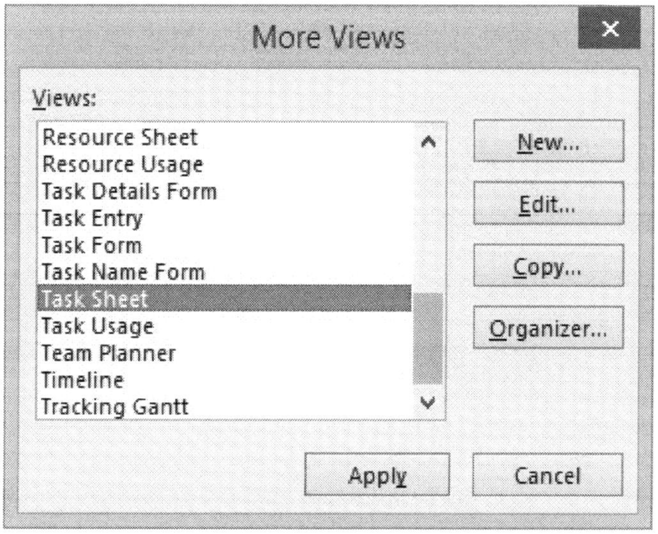

Step 4: Select Apply.

Step 5: Select Tables from the View tab.

Step 6: Select More Tables.

Step 7: From the More Tables dialog box, select Earned Value, Earned Value Cost Indicators, or Earned Value Schedule Indicators.

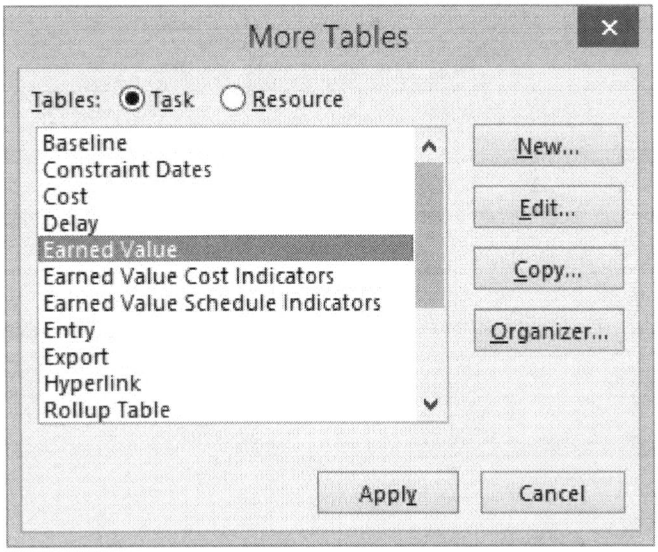

Step 8: Select Apply.

Hover your mouse over the column headings to see a description of each one.

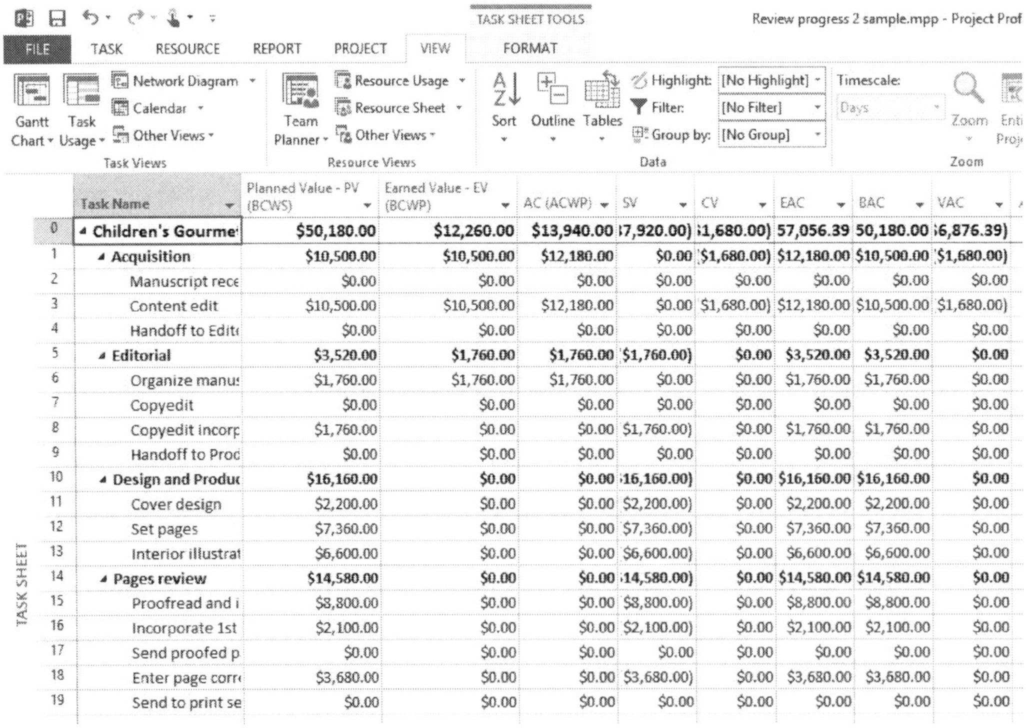

	Task Name	Planned Value - PV (BCWS)	Earned Value - EV (BCWP)	AC (ACWP)	SV	CV	EAC	BAC	VAC
0	**Children's Gourme**	**$50,180.00**	**$12,260.00**	**$13,940.00**	**;7,920.00)**	**;1,680.00)**	**57,056.39**	**50,180.00**	**;6,876.39)**
1	**Acquisition**	**$10,500.00**	**$10,500.00**	**$12,180.00**	**$0.00**	**($1,680.00)**	**$12,180.00**	**$10,500.00**	**($1,680.00)**
2	Manuscript rece	$0.00	$0.00	$0.00	$0.00	$0.00	$0.00	$0.00	$0.00
3	Content edit	$10,500.00	$10,500.00	$12,180.00	$0.00	$1,680.00)	$12,180.00	$10,500.00	$1,680.00)
4	Handoff to Edit	$0.00	$0.00	$0.00	$0.00	$0.00	$0.00	$0.00	$0.00
5	**Editorial**	**$3,520.00**	**$1,760.00**	**$1,760.00**	**($1,760.00)**	**$0.00**	**$3,520.00**	**$3,520.00**	**$0.00**
6	Organize manu:	$1,760.00	$1,760.00	$1,760.00	$0.00	$0.00	$1,760.00	$1,760.00	$0.00
7	Copyedit	$0.00	$0.00	$0.00	$0.00	$0.00	$0.00	$0.00	$0.00
8	Copyedit incorp	$1,760.00	$0.00	$0.00	$1,760.00)	$0.00	$1,760.00	$1,760.00	$0.00
9	Handoff to Proc	$0.00	$0.00	$0.00	$0.00	$0.00	$0.00	$0.00	$0.00
10	**Design and Produc**	**$16,160.00**	**$0.00**	**$0.00**	**;16,160.00)**	**$0.00**	**$16,160.00**	**$16,160.00**	**$0.00**
11	Cover design	$2,200.00	$0.00	$0.00	$2,200.00)	$0.00	$2,200.00	$2,200.00	$0.00
12	Set pages	$7,360.00	$0.00	$0.00	$7,360.00)	$0.00	$7,360.00	$7,360.00	$0.00
13	Interior illustrat	$6,600.00	$0.00	$0.00	$6,600.00)	$0.00	$6,600.00	$6,600.00	$0.00
14	**Pages review**	**$14,580.00**	**$0.00**	**$0.00**	**;14,580.00)**	**$0.00**	**$14,580.00**	**$14,580.00**	**$0.00**
15	Proofread and i	$8,800.00	$0.00	$0.00	$8,800.00)	$0.00	$8,800.00	$8,800.00	$0.00
16	Incorporate 1st	$2,100.00	$0.00	$0.00	$2,100.00)	$0.00	$2,100.00	$2,100.00	$0.00
17	Send proofed p	$0.00	$0.00	$0.00	$0.00	$0.00	$0.00	$0.00	$0.00
18	Enter page corr	$3,680.00	$0.00	$0.00	$3,680.00)	$0.00	$3,680.00	$3,680.00	$0.00
19	Send to print se	$0.00	$0.00	$0.00	$0.00	$0.00	$0.00	$0.00	$0.00

You can also see a breakdown of the earned value of tasks by period, to determine the exact date that your project will run out of money.

Step1: Select Task Usage from the View tab on the Ribbon.

Step 2: Select the Format tab from the Ribbon.

Step 3: Select Add Details.

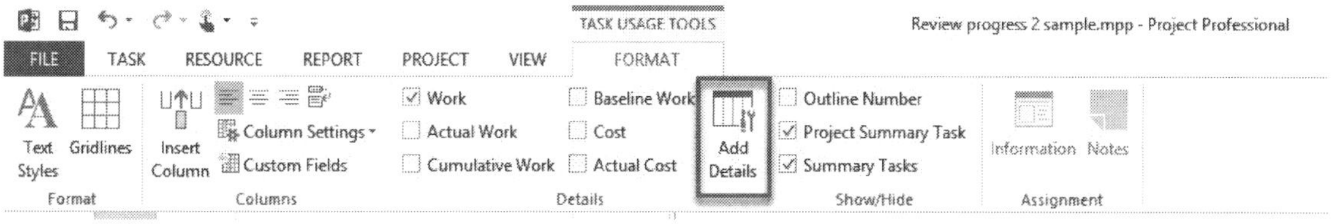

Step 4: In the Detail Styles dialog box, select the Earned Value fields that you want to display. Select Show. Select OK.

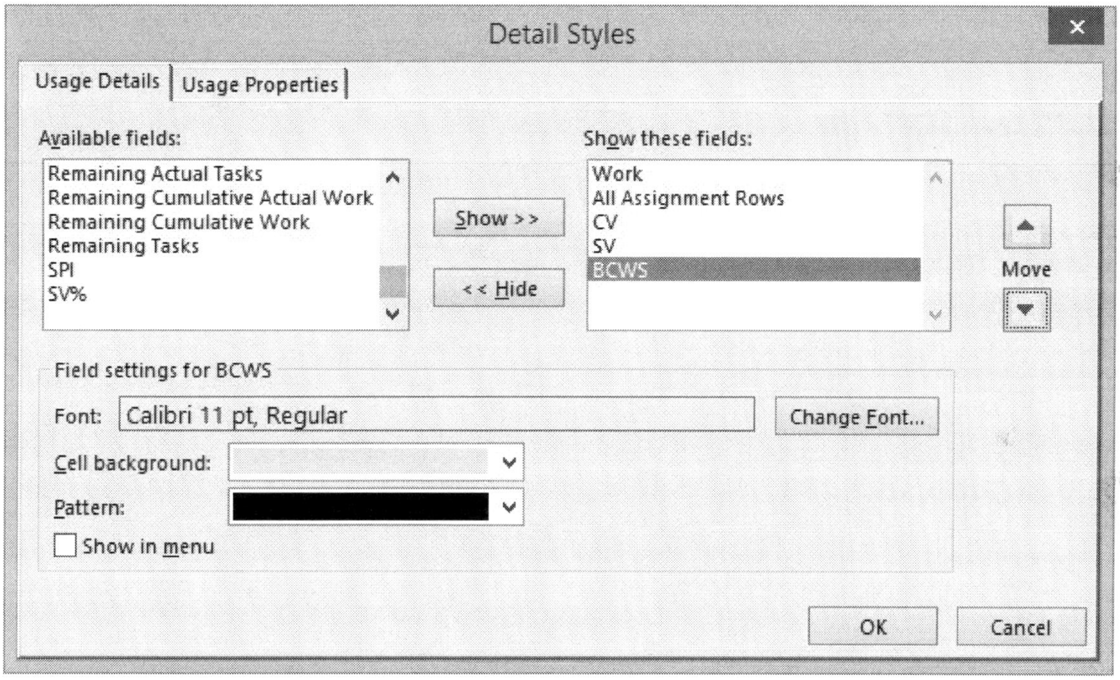

Chapter 25 – Communicating Project Information

Now that you are an expert at managing your project schedule, you will need to know how to communicate that information. This chapter explains some options for communicating Project information with those who do not have Project 2016. You will learn the specific procedures for copying a GIF image of your plan, publishing your project to a SharePoint list, attaching documents and including hyperlinks with your project. You will also learn how to print a view based on a specific date range.

About Sharing information

Project is not always enough, especially when many others who need to see the information do not have Project installed on their computers.

Project Web Access - If your organization is using Microsoft Project Web App, you can ask the project manager to publish the project and then give you permission to view the project. Then you can view the project in Project Web App.

Copy Picture - The person who created the project can use the Copy Picture feature to copy a .GIF image of the plan. The .GIF image can then be sent to you in e-mail, printed out, or added to a Web page.

Copy to another Office application - Project 2016 contains enhanced functionality for basic copying and pasting project information while retaining formatting and column heading information. The person who created the project can copy the information from Project 2016 and paste it into applications like Word, Excel, PowerPoint, or Outlook. You can then view project information in those Office applications.

Visual reports - The person who created the project can create a visual report that is designed for importing into and then viewing in either Excel or Visio.

SharePoint - The person who created the report can save the report to a SharePoint site, without using Project Web App. You can then go to the SharePoint site to view and edit task and resource information associated with the Project. The project information that is edited on the SharePoint site will be copied back the original project automatically.

Copy the Timeline View - The Timeline view in Project 2016 can be copied and pasted in an e-mail message or in any Office application for a quick view of Project progress along a graphic timeline.

Copying a .GIF Image of Your Plan

To copy of .GIF image of the project, use the following procedure.

Step 1: The screen should show the view and timescale that you want to copy. You may instead want to select the rows that you want to copy.

Step 2: Select the Task tab from the Ribbon.

Step 3: Select the arrow next to the Copy icon.

Step 4: Select Copy Picture.

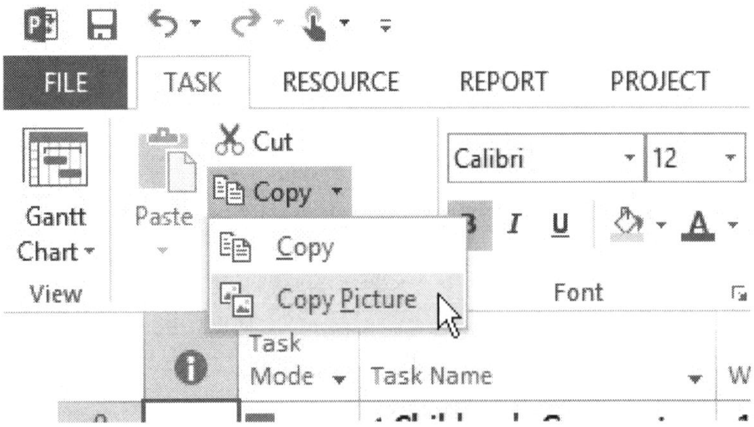

Step 5: In the Copy Picture dialog box, select To GIF image file. Select Browse to indicate the location where the file should be saved. You can change the file name, if desired.

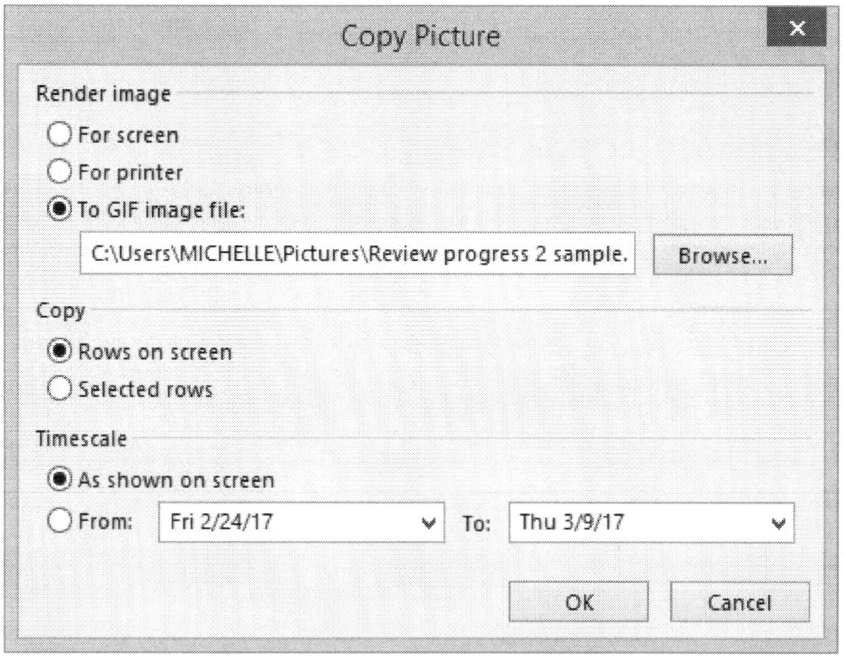

Step 6: Select whether to copy the Rows on screen or the Selected rows.

Step 7: Select whether to copy the timescale as shown on screen or for selected dates by indicating the start date and the end date to copy.

Step 8: Select OK.

With the GIF file, you can copy or insert it into another application like PowerPoint, Word, or an email message.

Publishing a Project to a SharePoint List

To publish a project to a SharePoint list, use the following procedure.

Step 1: Select the File tab to open the Backstage View.

Step 2: Select Save & Send.

Step 3: Select Sync with Tasks Lists.

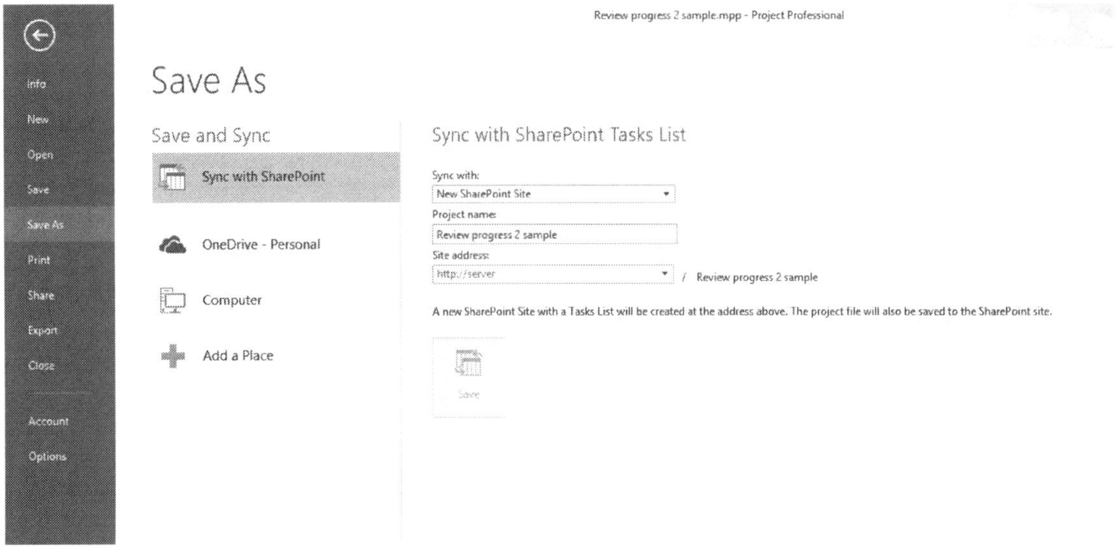

Step 1: In the Site URL drop-down list, select the URL name of the SharePoint site. Do not select a URL with the name of the task list you want to use, just the site at this point.

Step 2: Select Validate URL.

Step 3: In the Select an existing tasks list drop-down list, select the name of the SharePoint task list to which you want to sync. If you want to create a new list, enter the name.

Step 4: Select Sync.

Attaching Documents

To attach a document, use the following procedure.

Step 1: Double-click on the item to which you want to attach a document to open the Information dialog box for that item. In this example, we will use a resource.

Step 2: In the Information dialog box, select the Notes tab.

Step 3: Select the Insert Object icon.

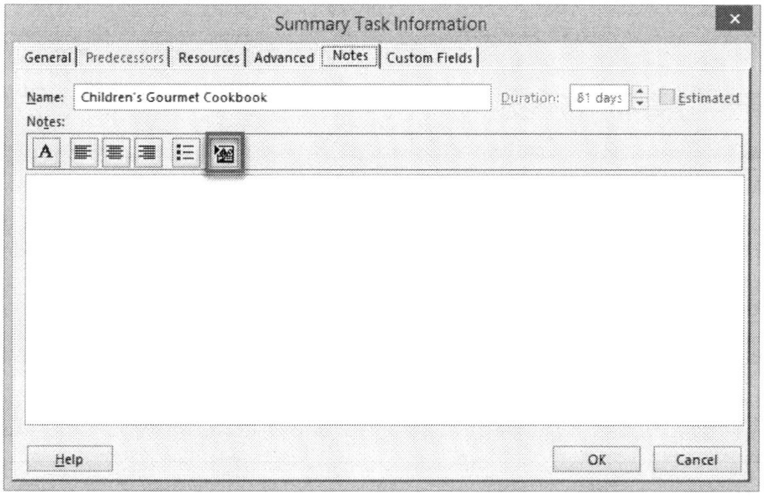

Step 4: In the Insert Object dialog box, select Create from File.

Step 5: Select Browse and navigate to the location of the file you want to use.

Step 6: To create a link so that if the document is updated, the information in Project is also updated, check the Link box.

Step 7: To display an icon for the document rather than the contents, check the Display as Icon box.

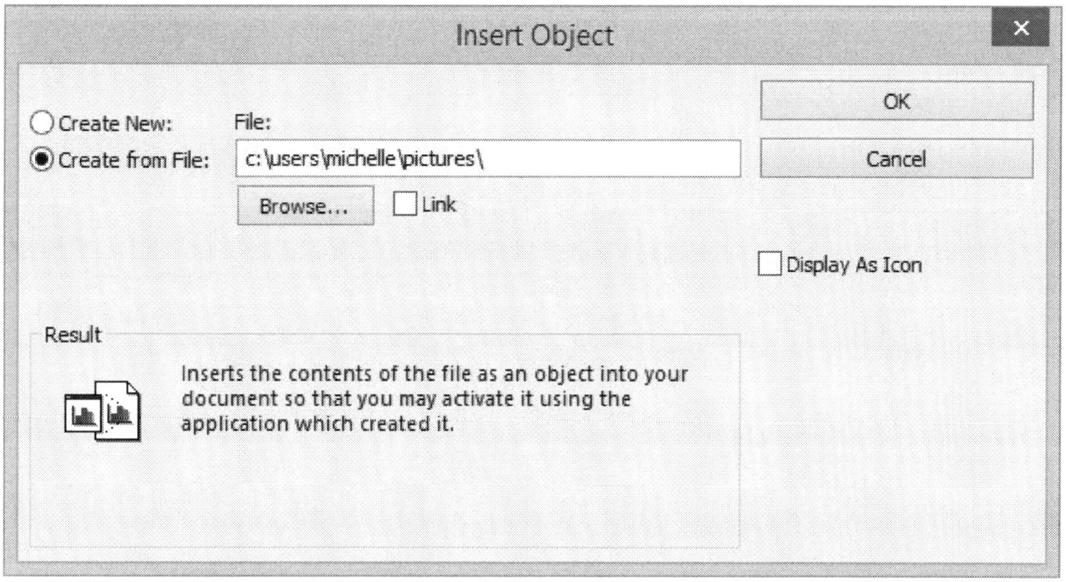

Step 8: Select OK.

Inserting Hyperlinks

To insert a hyperlink, use the following procedure.

Step 1: Select the task or other field in Project where you want to include a hyperlink.

Step 2: Right-click on the item.

Step 3: Select Insert Hyperlink from the context menu.

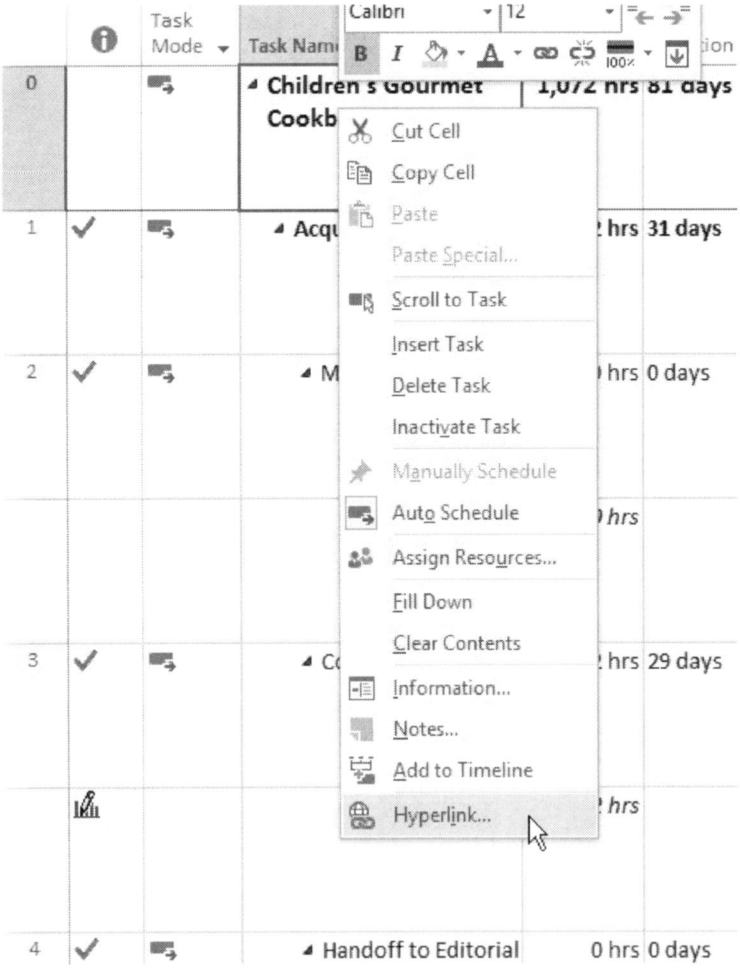

Step 4: In the Insert Hyperlink dialog box, use the Look In tools to find the location of the hyperlink destination. You can also enter or copy in the Address.

Step 5: Enter the Text to display.

Step 6: Select OK.

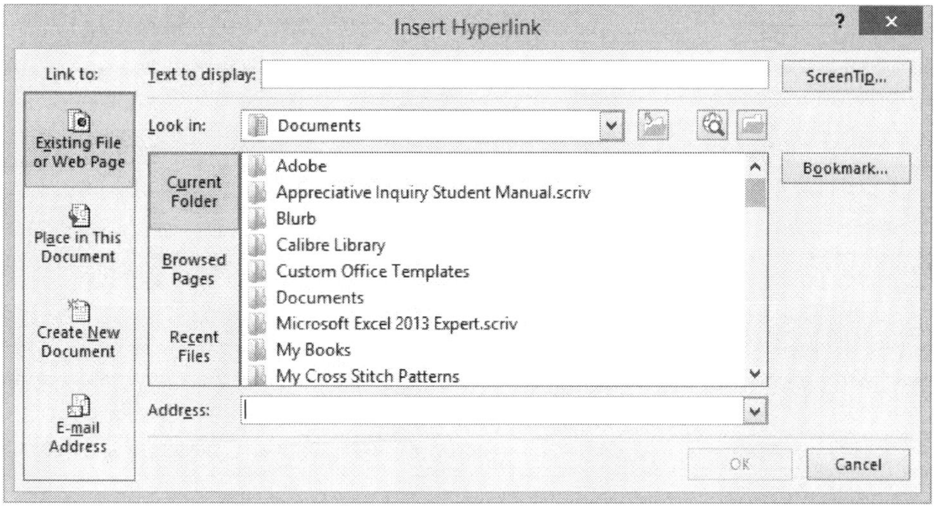

Step 7: Project displays the hyperlink icon in the Indicators column.

Printing Based on a Date Range

To print based on a date range, use the following procedure.

Step 1: Open the view that you want to print.

Step 2: Select the File tab from the Ribbon to open the Backstage view.

Step 3: Select Print.

Step 4: Use the Print Preview area to review your selection.

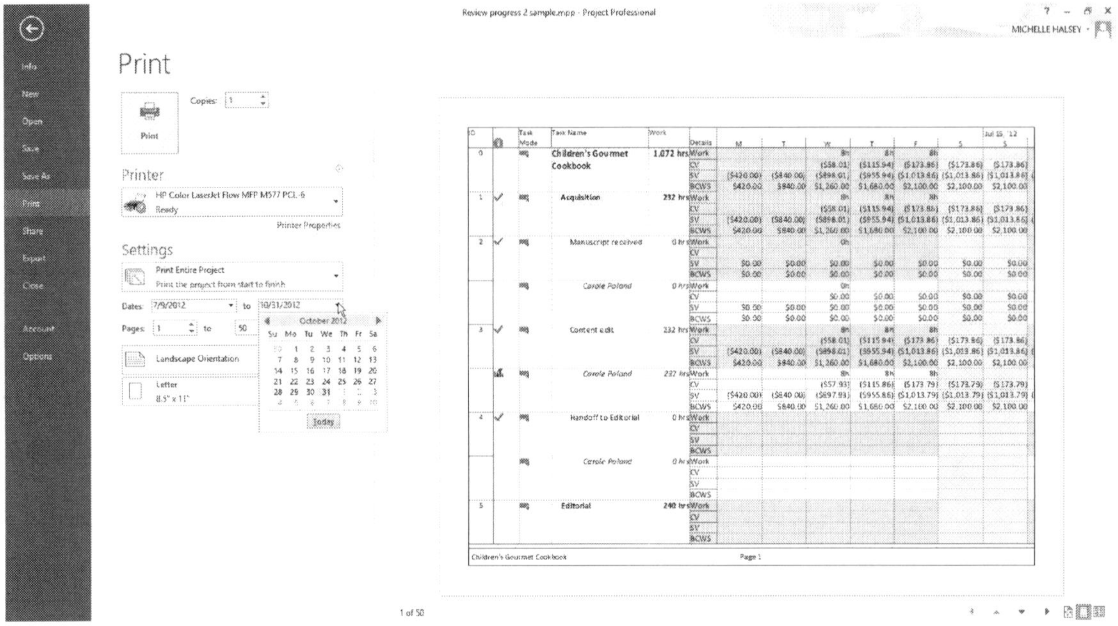

Step 5: In the Dates area, select the Start date and the End date that you want to print.

Step 6: Select Print.

Chapter 26 – New Features in Microsoft Project 2016

Tell Me

Instead of searching the online help or in Project, you can use the Tell Me feature to look for the solution you need.

Add-ins

It is now easier to integrate Microsoft Project Add-ins with your Project. You can download partner applications in the same manner as templates from the Office Store.

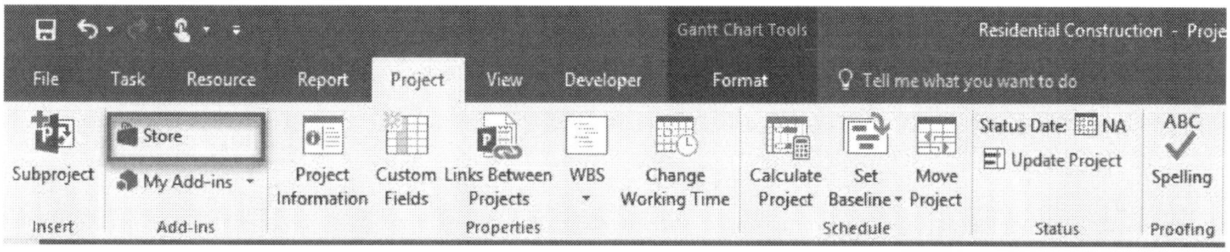

Step 1: Select the Project ribbon.

Step 2: Click Store in the Add-ins menu.

Step 3: Search for the desired add-in in the Store and download and install the add-in.

The MY Add-ins drop-down menu will show the currently installed Add-ins from the online Microsoft Store.

New Themes

Microsoft Project 2016 provides additional themes to your own colors. You will find these themes on the Format ribbon of the Gantt Chart Tools options.

Multiple Timeline View

In Project 2016, you can add a second timeline bar, to show multiple timelines. You can then change the start and end dates for each timeline.

Click the Timeline view, and then click Format > Timeline Bar.

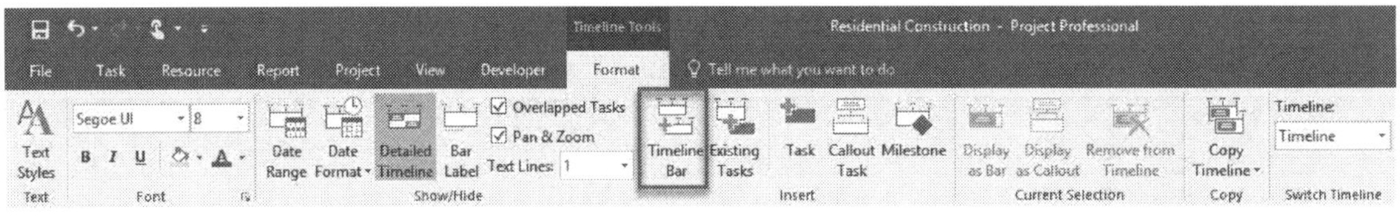

Right-click a timeline and choose Date Range to choose custom start and finish dates.

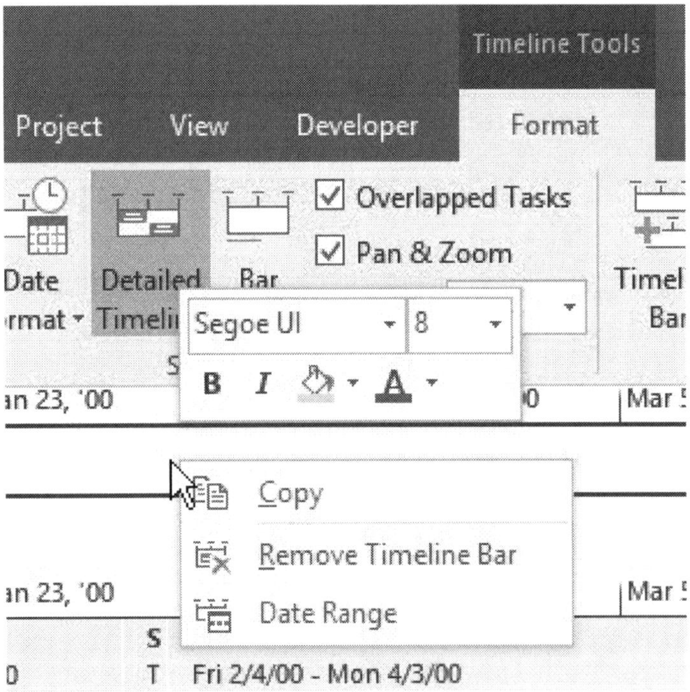

Resource Engagements

You can request a resource in Project and the resource manager can accept or deny the request in the application. If the resource is accepted by the resource manager, then the resource will lock to ensure availability.

Step 1: Go to the Resource Sheet view.

Step 2: Go to the Resource ribbon.

Step 3: Click Add Resource.

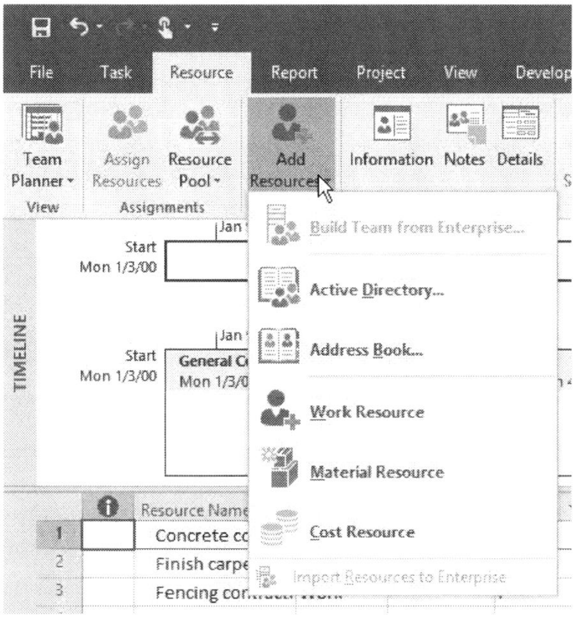

Resource Views

Resource Views is a space designated for resource managers to accept or deny all resource requests. For example, you can view a list of resources who are overallocated

You can see a list of the resources that are overallocated by displaying the Resource Usage view or the Resource Sheet view, and then filter for overallocated resources.

Step 1: Click Resource Sheet or Resource Usage from the View section drop-down.

Step 2: Click the View ribbon.

Step 3: Click the filter drop-down menu in the Data Section and then click Overallocated Resources.

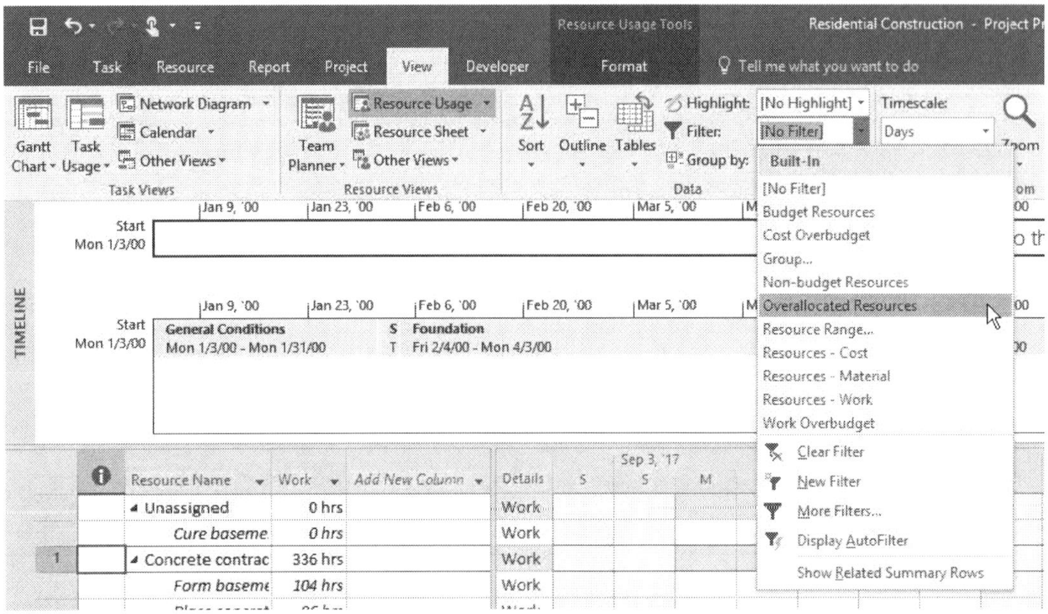

Step 4: Click the Filter dropdown menu and then click No Filter to see the full list of resources again.

Display and Use windows

To do this	Press
Switch to next window.	ALT + Tab
Switch to previous window	ALT + SHIFT + Tab
Close active window	CTRL + W or CTRL + F4
Restore the size of the active window after you maximize the window.	CTRL + F5
Move from one task pane to another pane in the program window (clockwise). F6 may need to be pressed more than once.	F6
Move from one task pane to another pane in the program window (counterclockwise).	SHIFT + F6
Switch to the next window when more than one window is open.	CTRL + F6
Copy a picture of the screen to the Clipboard.	Print Screen
Copy a picture of the selected window to the Clipboard.	ALT + Print Screen

Move Around in Text, Cells, and Tables

To do this	Press
Move 1 character left	Left Arrow
Move 1 character right.	Right Arrow
Move 1 line up	Up Arrow
Move 1 line down	Down Arrow
Move 1 word left.	CTRL + Left Arrow
Move 1 word right	CTRL + Right Arrow
Move to the end of a line	END
Move to the start of a line	HOME
Move up 1 paragraph	CTRL + Up Arrow
Move down 1 paragraph	CTRL + Down Arrow
Move to the end of a text box	CTRL + END
Move to the beginning of a text box	CTRL + HOME
Move to next cell in the table	Tab
Move to preceding cell in the table	SHIFT + Tab
Move to next row in the table	Down Arrow
Move to preceding row in the table	Up Arrow
Insert a tab in a cell of the table.	CTRL + Tab
Start a new paragraph	ENTER
Add a new row to the bottom of the table	Tab at the end of the

To do this	Press
	last row

Use and Access Actions

To do this	Press
Display action menu or message	ALT + SHIFT + F10
Select action next item	Down Arrow
Select action previous item	Up Arrow
Perform an action for the selected item on the action menu	ENTER
Close an action menu or message	ESC

Use Dialog Boxes

To do this	Press
Move to next option or option group	Tab
Move to previous option or option group	SHIFT + Tab
Switch to next tab in a dialog box	CTRL + Tab
Switch to previous tab in a dialog box	CTRL + SHIFT + Tab
Move between options in a drop-down list, or between options in a group of options	Arrow keys
Perform an action assigned to a selected button; select or clear the selected check box	Spacebar
Open a closed list and move to an option in the list	First letter of an option

To do this	Press
	in a drop-down list
Select an option; select or clear a check box	ALT + the letter underlined in an option
Open a selected drop-down list.	ALT + Down Arrow
Close selected drop-down list or Cancel a command or Close a dialog box.	Esc
Perform an action assigned to a default button in a dialog box	ENTER

Use an Edit Box Within a Dialog Box

To do this	Press
Move to beginning of an entry	HOME
Move to end of an entry	END
Move 1 character left or right, respectively	Left Arrow or Right Arrow
Move 1 word left	CTRL + Left Arrow
Move 1 word right	CTRL + Right Arrow

To do this	Press
Select or cancel selection 1 character left	SHIFT + Left Arrow
Select or cancel selection 1 character right	SHIFT + Right Arrow
Select or cancel selection 1 word left	CTRL + SHIFT + Left Arrow
Select or cancel selection 1 word right	CTRL + SHIFT + Right Arrow
Select from an insertion point to the beginning of an entry.	SHIFT + HOME
Select from an insertion point to the end of an entry.	SHIFT + END

File Management

To do this	Press
Display the open dialog box to open a project file	CTRL + F12
Display the open tab in the Backstage view to open a project file	CTRL + O
Save project file.	CTRL + S
Create new project	CTRL + N
Display the Print tab in the Backstage view to print a file	CTRL + P

Use the Open and Save As Dialog Boxes

To do this	Press
Display an open dialog box	CTRL + F12
Display open tab in Backstage view	CTRL + O
Display Save As dialog box	F12
Open selected folder or file	ENTER
Open a folder 1 level above the current open folder	Backspace
Delete selected folder or file	DEL
Display shortcut menu for a selected item	SHIFT + F10
Move forward through options	Tab
Move back through options	SHIFT + Tab
Open the Look in list	F4 or ALT + 1

Use a Network Diagram

To do this	Press
Move to different network diagram box	Arrow keys
Add network diagram boxes to a selection	SHIFT + Arrow keys
Move network diagram box Set manual positioning first by selecting the box to be moved. Select Format, Layout, and then Allow manual box positioning.	CTRL + Arrow keys
Move to top network diagram box in a view or	CTRL + HOME or

To do this	Press
project	SHIFT + CTRL + HOME
Move to lowest network diagram box in a project	CTRL + END or SHIFT + CTRL + END
Move to leftmost Network Diagram box in a project	HOME or SHIFT + HOME
Move to rightmost Network Diagram box in a project	END or SHIFT + END
Move up 1 window height	Page Up or SHIFT + Page Up
Move down 1 window height	Page Down or SHIFT + Page Down
Move left 1 window width	CTRL + Page Up or SHIFT + CTRL + Page Up
Move right 1 window width	CTRL + Page Down or SHIFT + CTRL + Page Down
Select the next field in a network diagram box	ENTER or Tab
Select previous field in a network diagram box.	SHIFT + ENTER

Work with OfficeArt Objects

Move OfficeArt Shapes

To do this	Press
Nudge shape up, down, right, or left	Arrow keys
Increase shape width by 10%	SHIFT + Right Arrow
Decrease shape width by 10%	SHIFT + Left Arrow
Increase shape height by 10%	SHIFT + Up Arrow
Decrease shape height by 10%	SHIFT + Down Arrow
Increase shape width by 1%	CTRL + SHIFT + Right Arrow
Decrease shape width by 1%	CTRL + SHIFT + Left Arrow
Increase shape height by 1%	CTRL + SHIFT + Up Arrow
Decrease shape height by 1%	CTRL + SHIFT + Down Arrow
Rotate shape 15 degrees to right	ALT + Right Arrow
Rotate shape 15 degrees to left	ALT + Left Arrow

Select and Copy OfficeArt Objects and Text

To do this	Press
Select object with text selected inside the object	Esc
Select object with an object selected	Tab or SHIFT + Tab until the object is selected
Select text within an object with an object selected	ENTER
Select multiple shapes	Press and hold Ctrl while you click the shapes
Select multiple shapes with text	Press and hold Shift while you click the shapes
Cut selected object	CTRL + X
Copy selected object	CTRL + C
Paste, cut, or copy object	CTRL + V
Paste special	CTRL + ALT + V
Copy formatting only	CTRL + SHIFT + C
Paste formatting only	CTRL + SHIFT + V
Group shapes, pictures, or WordArt (after selecting the items to group)	CTRL + G
Ungroup shapes, pictures, or WordArt (after select the group to ungroup)	CTRL + SHIFT + G

To do this	Press
Undo last action	CTRL + Z
Redo last action	CTRL + Y
Add next object to a multi-selection	CTRL + Click
Add next object to a multi-selection Or Allow clicking on text in a textbox to add the textbox to the multi-selection.	SHIFT + Click

Edit OfficeArt Text and Textboxes

To do this	Press
Collapse selection	Esc
Select all text	CTRL + A
Delete 1 word left	CTRL + Backspace
Delete 1 word right	CTRL + DEL
Undo	CTRL + Z
Redo	CTRL + Y
Move 1 word left	CTRL + Left Arrow
Move 1 word right	CTRL + Right Arrow
Move to beginning of a line	HOME
Move to end of a line	END
Move up one paragraph	CTRL + Up Arrow

To do this	Press
Move down 1 paragraph	CTRL + Down Arrow
Move to beginning of an object's text	CTRL + HOME
Move to end of an object's text	CTRL + END

Navigate Views and Windows

To do this	Press
Activate an entry bar to edit text in a field	F2
Activate menu bar	F10 or Alt
Activate project control menu	ALT + Hyphen or ALT + Spacebar
Activate a split bar	SHIFT + F6
Close a program window	ALT + F4
Display all filtered tasks or resources	F3
Display Field Settings dialog box	ALT + F3
Open new window	SHIFT + F11
Reduce selection to a single field	SHIFT + Backspace
Reset sort order to ID order and turn off grouping	SHIFT + F3
Select drawing object	F6
Display task information	SHIFT + F2
Display resource information	SHIFT + F2
Display assignment information	SHIFT + F2

To do this	Press
Turn on or off Add to Selection mode	SHIFT + F8
Turn on or off Auto Calculate	CTRL + F9
Turn on or off Extend Selection mode	F8
Move left, right, up, or down to view different pages in Print Preview window	ALT + Arrow keys

Outline a Project

To do this	Press
Hide subtasks	ALT + SHIFT + Hyphen or ALT + SHIFT + Minus Sign (on numeric keypad)
Indent the selected task	ALT + SHIFT + Right Arrow
Show subtasks	ALT + SHIFT + = or ALT + SHIFT + Plus Sign (on numeric keypad)
Show all tasks	ALT + SHIFT + * (on numeric keypad)
Outdent the selected task	ALT + SHIFT + Left Arrow

Select and Edit in a Dialog Box

To do this	Press
Move between fields at the bottom of a form	Arrow keys
Move into tables at the bottom of a form	ALT + 1 (left) or ALT + 2 (right)
Move to next task or resource	ENTER
Move to previous task or resource	SHIFT + ENTER

Select and Edit in Sheet View

To do this	Press
Cancel entry	Esc
Clear or reset selected field	CTRL + DEL
Copy selected data	CTRL + C
Cut selected data	CTRL + X
Delete selected data.	DEL
Delete row with a selected cell	CTRL + Minus Sign (on numeric keypad)
Fill down	CTRL + D
Display the Find dialog box	CTRL + F or SHIFT + F5
In the Find dialog box, continue to next instance of search results	SHIFT + F4

To do this	Press
Use Go To command (Edit menu).	F5
Link tasks	CTRL + F2
Paste copied or cut data	CTRL + V
Reduce selection to 1 field	SHIFT + Backspace
Undo last action.	CTRL + Z
Unlink tasks	CTRL + SHIFT + F2
Set task to manually schedule	CTRL + SHIFT + M
Set task to auto schedule	CTRL + SHIFT + A
Move to beginning of a project (timescale)	ALT + HOME
Move to end of a project (timescale)	ALT + END
Move timescale left	ALT + Left Arrow
Move timescale right	ALT + Right Arrow
Move to first field in a row	HOME or CTRL + Left Arrow
Move to first row	CTRL + Up Arrow
Move to first field of the first row.	CTRL + HOME
Move to last field in a row	END or CTRL + Right Arrow
Move to last field of the last row	CTRL + END
Move to last row	CTRL + Down Arrow

To do this	Press
Move focus between side pane and view on right side	F6
Select different controls in side pane if focus is in side pane	Tab
Select or clear check boxes and option buttons if focus is in side pane	Spacebar
Extend selection down 1 page	SHIFT + Page Down
Extend selection up 1 page	SHIFT + Page Up
Extend selection down 1 row	SHIFT + Down Arrow
Extend selection up 1 row	SHIFT + Up Arrow
Extend selection to first field in a row	SHIFT + HOME
Extend selection to last field in a row	SHIFT + END
Extend selection to start of information	CTRL + SHIFT + HOME
Extend selection to end of information	CTRL + SHIFT + END
Extend selection to first row	CTRL + SHIFT + Up Arrow
Extend selection to last row	CTRL + SHIFT + Down Arrow
Extend selection to first field of first row	CTRL + SHIFT + HOME

To do this	Press
Extend selection to last field of last row	CTRL + SHIFT + END
Select all rows and columns	CTRL + SHIFT + Spacebar
Select a column	CTRL + Spacebar
Select a row	SHIFT + Spacebar
Move within selection down 1 field	ENTER
Move within selection up 1 field	SHIFT + ENTER
Move within selection right 1 field	Tab
Move within selection left 1 field	SHIFT + Tab

Select and Edit in the Entry Bar

To do this	Press
Accept entry	ENTER
Cancel entry	Esc
Delete 1 character to left	Backspace
Delete 1 character to right	DEL
Delete 1 word to right	CTRL + DEL
Extend selection to the end of the text	SHIFT + END
Extend selection to the start of the text	SHIFT + HOME
Turn Overtype mode on or off	Insert

Use a Timescale

To do this	Press
Move timescale left one page	ALT + Page Up
Move timescale right 1 page	ALT + Page Down
Move timescale to beginning of the project	ALT + HOME
Move timescale to end of the project	ALT + END
Scroll timescale left	ALT + Left Arrow
Scroll timescale right	ALT + Right Arrow
Show smaller time units	CTRL + / (on numeric keypad)
Show larger time units	CTRL + * (on numeric keypad)

Made in the USA
Monee, IL
05 November 2026

67816261R00175